Emerging Trends in Mechatronics

Series Editor

Aydin Azizi, Oxford, UK

This book series helps understand how Mechatronics will impact the practice and research of developing advanced techniques to model, control and optimize complex systems. It presents recent state of advances in mechatronics and related technologies including: Automatic Control, Robotics, Agent Based systems, Smart Manufacturing and Industry 4.0. The selected topics give an overview of the state-of-the-art and present new research results and prospects of the future development in the interdisciplinary field of mechatronic systems. The book series will provide up-to-date and useful knowledge for researchers and engineers involved in mechatronics and related fields.

Within the above dimensions, the scope of the book series welcomes high quality original research and review articles that cover a broad range of topics related to theoretical and applied aspects of modern mechatronics. Potential topics include, but are not limited to the following:

Controller design

Many mechatronic systems have challenging control difficulties due to the high nonlinear structures and time-varying dynamic behaviors. In addition to these, there are external disturbances, which cannot be predicted and change according to uncontrolled working environments. Conventional controllers are insufficient to solve the problems and to compensate the environmental disturbances, therefore designing efficient controllers for complex systems has a great importance in the current book series.

Sensor Design & Data Collection Approaches

Essential part of a mechatronics system is the measurement system that senses the variations in the physical parameters, such as temperature, pressure, displacement, and so on, and converts it to voltage or current. In mechatronic systems, accurate measurement of the dynamic variables plays a vital role for the actuators to function properly, therefore on of the aims of this book series, to focus on understanding what should be done to design accurate sensors and efficient data collection methods to improve the performance of mechatronic systems.

Optimization Techniques

Designing a complex mechatronic system involves multiple variables, objectives, constraints, and evaluation criteria as well as their nonlinearly coupled relationships. One of the challenging research areas of this book series is utilizing optimization techniques to design an optimal mechatronic system.

Artificial Intelligence

From the point of view of dynamics, mechatronic systems can be characterized by model uncertainties, high nonlinearities, complicated coupling, and stringent performance requirements, among others. To deal with these facts, this book series would like to focus on investigating and utilizing Artificial Intelligence techniques such as Neural Networks, Fuzzy logic, Neuro-fuzzy, evolutionary techniques etc. to model, design and optimize complex mechatronic systems.

Robotics

In the field of Robotics, there is a need of focusing on kinematic, dynamic, control and optimization of innovative robotic systems. This book series can introduce new and novel robot designs such as hybrid (serial-parallel) redundant manipulators that have a high potential of application in special fields like space robots. State of the art researches on selection of optimum kinematic solutions and path planning based on such an optimization, dynamic analysis and new control methods developments and application of artificial intelligence in this field can be published in this book series.

Smart manufacturing

Application of networked sensors and machinery equipment using internet and intranet in order to monitor and automation of the manufacturing process and improve the performance efficiency and is a point of interest. Advanced robotics, multi input-multi output control approaches combined with the artificial intelligence and optimization techniques, big data processing approaches, multi-scale modelling and simulation of dynamic sub-systems etc. are the hot topics that will be focused in this book series.

AGVs

Automated Guided Vehicle (AGVs) are going to be very popular is industry logistics, surveillance, defense and social applications in order to make processes safer and more cost efficient. Innovative AGVs designs for various missions, Robot Operating System usage on AGV control, Indoor and outdoor navigation and control algorithms, intelligent path planning techniques, sensor fusion, simultaneous application of GPS, INS and offline maps are the topics that are planned to be published in this book series.

Aydin Azizi
Editor

Intelligent Control and Autonomous Systems

Applications in Vehicles, Robotics, and Logistics

Springer

Editor
Aydin Azizi
School of Engineering, Computing
and Mathematics
Oxford Brookes University
Oxford, UK

ISSN 2731-4855 ISSN 2731-4863 (electronic)
Emerging Trends in Mechatronics
ISBN 978-981-95-6401-9 ISBN 978-981-95-6402-6 (eBook)
https://doi.org/10.1007/978-981-95-6402-6

This Springer imprint is published by the registered company Springer Nature Singapore Pte Ltd.
The registered company address is: 152 Beach Road, #21-01/04 Gateway East, Singapore 189721, Singapore

About This Book

This proposed book, titled *Intelligent Control and Autonomous Systems: Applications in Vehicles, Robotics, and Logistics*, is designed to bring together state-of-the-art research and innovative developments in the areas of autonomous vehicles, robotics, intelligent control strategies, and AI-driven logistics systems. It focuses on both theoretical advancements and practical implementations, highlighting applications across Formula Student autonomous racing, robotic systems, and intelligent supply chains.

As industries continue to evolve with growing demands for automation, efficiency, and adaptability, this book offers a timely and comprehensive contribution that will serve as a reference for researchers, practitioners, and students. By covering cutting-edge topics such as model predictive control, dual-mode braking systems, robotic arm control, and AI-driven logistics optimization, the book integrates contributions across multiple domains, providing readers with valuable insights into emerging methodologies and their real-world impact.

Contents

About the Editor

Dr. Aydin Azizi holds Ph.D. in Mechanical Engineering—Mechatronics, M.Sc. in Mechatronics, and B.Sc. in Mechanical Engineering. Certified as a fellow of the Higher Education Academy, official instructor for the Siemens Mechatronic Certification Program (SMSCP), and Editor-in-Chief of the book series *Emerging Trends in Mechatronics* published by Springer Nature Group, he currently serves as Senior Lecturer and Academic Partnership Liaison Manager at Oxford Brookes University. His current research focuses on investigating and developing novel techniques to model, control, and optimize complex systems, with expertise in control and automation, AI, and simulation techniques.

Dr. Azizi is the recipient of the National Research Award of Oman for his AI-based controllers research, DELL EMC's 'Envision the Future' award for the 'Automated Irrigation System,' and 'Exceptional Talent' recognition by the British Royal Academy of Engineering. He has also been recognized for three consecutive years (2023–2025) among the World's Top 2% Scientists by Stanford University and Elsevier for his impactful research contributions. e-mail: aydin.azizi@brookes.ac.uk

Optimal Control of Racing Car Overtaking Manoeuvres Under Aerodynamic Wake Effects

Telmo Prego and Aydin Azizi

Abstract Overtaking in racing is a complex manoeuvre that requires precise driver inputs and is influenced by the vehicle dynamics. In categories with heavy aerodynamic influence in their cars, when following another car, the trailing car experiences significant reductions in downforce and drag, as well as shifts in aero balance, which compromise the performance of the car in all phases of driving; traction, straight-line speed, braking, cornering, and stability. The aim of this study is to develop a proof-of concept simulation framework that applies optimal control methods to investigate the optimal racing line when following another car under wake effects as well as performing overtaking manoeuvres under such conditions. A planar vehicle model representative of a 2024 Formula 3 car was developed by correlating with real telemetry data supplied by Campos Racing. Aerodynamic wake effects were incorporated through a parametric model derived from experimental data from the literature. The optimal control problem was implemented in the Dymos open-source library, with a custom track being generated for this study. The solution strategy consisted of generating a reference trajectory for a single car and then simulating an active car following the leading car and performing an overtake, subjected to wake losses and collision constraints. Results show that the following car exploits drag reduction in straights but avoids the wake in braking, cornering, and traction zones, particularly at corner exit, by optimising its racing line as a compromise between aerodynamic performance and curvature. These findings demonstrate that optimal control can capture realistic overtaking behaviour under aerodynamic wake, providing a foundation for more advanced multi-car interaction studies in motorsport simulation.

Keywords Optimal control · Trajectory optimization · Racing car dynamics · Aerodynamic wake effects · Motorsport simulation

T. Prego · A. Azizi (✉)
School of Engineering, Computing and Mathematics, Oxford Brookes University, Oxford, UK
e-mail: aydin.azizi@brookes.ac.uk

A. Azizi (ed.), *Intelligent Control and Autonomous Systems*, Emerging Trends in Mechatronics, https://doi.org/10.1007/978-981-95-6402-6_1

1 Introduction

The aim of this project is to develop an optimal control simulation framework capable of generating overtaking trajectories while accounting for aerodynamic wake effects on the attacking car. The vehicle model represents a 2024 Formula 3 car and was derived by correlating with real lap telemetry data provided by Campos Racing. Both the vehicle model and the optimal control problem were formulated in Dymos, an open-source optimal control library in Python that leverages direct collocation and nonlinear programming solvers to handle trajectory optimization problems [5].

The approach adopted in this study follows a two-step process. First, an optimal trajectory for a single car around the track was generated. This trajectory was then fixed and used as input to the optimal control problem of the active car, which found its optimal overtaking manoeuvre while experiencing aerodynamic wake effect from the car in front. The aerodynamic wake model used in the work was derived from experimental data available in the literature, enabling a realistic representation of aerodynamic loses during close following.

The results of this study were analysed by comparing the racing line of the active car to that of the single-car simulation, highlighting how aerodynamic wake influences trajectory choices. Particular attention is given to where and why the active car deviates from the ghost trajectory. These adaptations are interpreted in terms of the underlying vehicle physics, including the trade-off between downforce recovery, aero balance stability, and the geometric curvature of the racing line. From this analysis, key trends are identified regarding the strategies that maximise performance while following another car.

This is important because aerodynamic wake is one of the key barriers to overtaking in modern formula racing. The recent study on overtaking fundamentals highlighted that overtaking performance is highly sensitive to wake strength. Stronger wake effects significantly reduce overtaking feasibility, and the pace advantage needed to overtake increases [9]. By investigating how optimal trajectories adapt under wake influence, this project addresses a central challenge in motorsport, identifying how performance can be maximised when following another car and where overtaking opportunities are most likely to arise.

Lap time simulation is a fundamental tool in motorsport, enabling teams and drivers to optimise performance by exploring vehicle dynamics, driving strategies, and circuit characteristics in a virtual environment. Traditional Quasi-Steady-State (QSS) simulations remain widely used due to their computational efficiency. In these methods, the lap is decomposed into a sequence of steady-state manoeuvres, effectively representing snapshots of the vehicle at static equilibrium under given conditions [2]. However, QSS relies on following a predefined racing line along the lap, which makes this kind of simulation impractical for this study.

By contrast, transient-optimal methods employ dynamic models to solve for both the trajectory and control inputs that minimize lap time. These methods treat the problem as a constrained nonlinear optimization, where objectives (e.g., time minimization) and constraints (e.g., dynamics, track boundaries) are nonlinear. In such cases, Nonlinear Programming (NLP) techniques are necessary, since closed-form analytical solutions are not feasible except for the simplest configurations. A critical insight from this approach is that the optimal racing line is not universal but depends on the vehicle's specific dynamics, such as downforce levels, tyre characteristics, and mass distribution [12]. This makes optimal control highly relevant for investigating overtaking and vehicle performance when following another car, given wake is heavily race line-dependent.

Optimal control has become a widely adopted method for lap time simulation because it can co-optimise both trajectory and control inputs while capturing transient vehicle dynamics [4, 10, 12, 13]. This makes it particularly valuable for evaluating scenarios where aero balance or subsystem interactions play a decisive role, but also when the fastest racing line for the given dynamics is not known. Formulated as an NLP problem, the optimal control framework is typically solved using interior-point or sequential quadratic programming methods. Studies have demonstrated applications including active control of devices such as rear wing flaps, optimisation of limited slip differential behaviour, and trajectory planning under physical input constraints. Vehicle models generally employ planar dynamics where it is possible to add complexity to the car model to represent a more realistic behaviour.

The direct method, which discretise the trajectory using approaches such as direct collocation and Dymos' approach, has proven more practical for complex racing problems, while indirect methods offer accuracy but require strong initial guesses and more complex derivations [13]. Overall, optimal control simulations not only provide the theoretical best trajectory but also offer insight into optimal actuation of control systems, informing both vehicle setup and driver guidance.

In this project, a planar vehicle model was adopted to represent the dynamics of the Formula 3 car. This choice strikes a balance between model fidelity and computational efficiency, which is crucial for solving large-scale optimal control problems such as lap time and overtaking simulations.

Previous studies have shown that planar models are not inherently limited in their complexity. For example, they have been extended to include aerodynamic wake effects [10], torque vectoring [4], detailed tyre behaviour including combined slip and nonlinear load sensitivity [12, 13], and aerodynamic loads dependent on ride height or wing angle of attack [3]. These extensions demonstrate that planar models can capture the key dynamics for this study but can also be extended to represent more accurate dynamics.

Aerodynamic wake effects occur when a following car drives through the turbulent airflow shed by a leading vehicle. These effects typically reduce both downforce and drag, alongside destabilising shifts in aerodynamic balance and are strongly dependent on the relative longitudinal and lateral separation between the cars [7, 11]. Existing simulation studies of overtaking often assume that wake effects only reduce drag and downforce without directly affecting aero balance [10]. Incorporating these wake effects within an optimal control framework therefore represents a promising approach for improving the fidelity of overtaking simulations (Fig. 1).

The overtaking and defending problem in motorsport has been studied primarily through Model Predictive Control (MPC) [1, 8] and optimal control frameworks. A

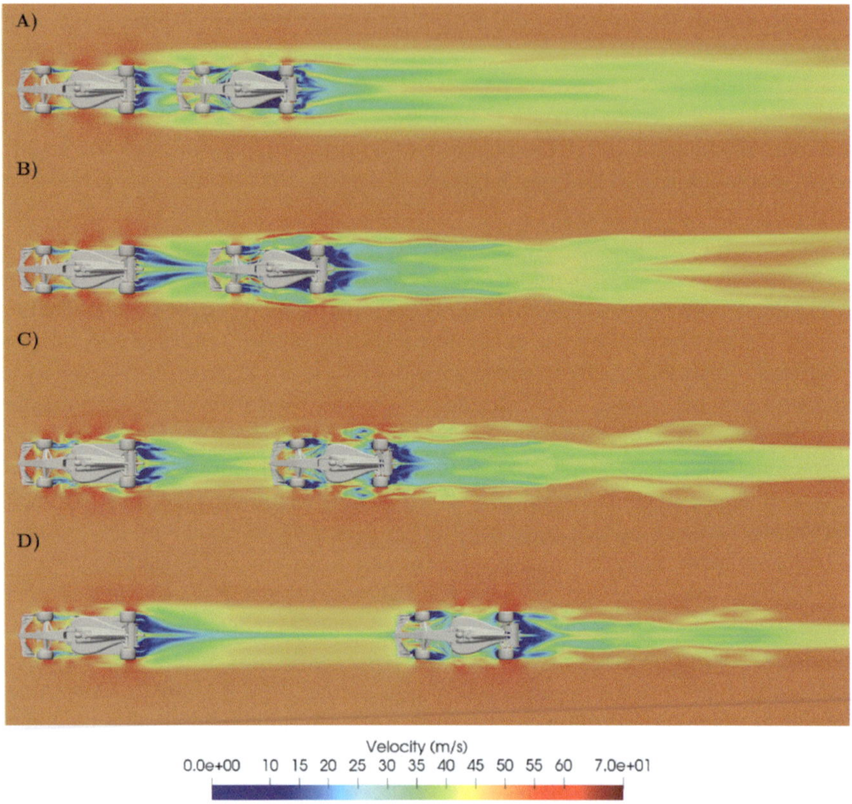

Fig. 1 Flow velocity contours of both cars at distances **a** 0.25 L, **b** 0.5 L, **c** 1 L, **d** 2 L [7]

strategy using optimal control was addressed by Liu [10]. In his study, the trajectory of the lead vehicle is first generated by solving a Minimum Time Manoeuvre (MTM) problem, providing its position, velocity, and timing data along the track. This trajectory is then assumed fixed and forms the upstream car's motion profile. The overtaking car's problem is subsequently formulated as a second MTM, where the lead car's trajectory introduces additional constraints, including a hard collision avoidance zone rather than a penalisation in the cost function. Liu's formulation extends the classical MTM to include the aerodynamic wake of the lead car, allowing the trailing vehicle's trajectory and controls to be optimised under disturbed aerodynamic conditions. The problem was transcribed using Legendre–Gauss-Radau collocation and solved as a large nonlinear programming problem with GPOPS-II and IPOPT. A key assumption in Liu's work is that the leading car does not change its trajectory, without any defensive behaviour. And only shows results of the racing line and overtaking in straight line. While this demonstrates the feasibility of incorporating aerodynamic wake into an optimal control framework, it does not capture the additional complexity introduced by cornering. In this project, the analysis is extended to the full racing line, including cornering phases, where aerodynamic wake not only reduces downforce and drag but also alters the optimal trajectory itself. This broader treatment allows investigation of wake effects on racing lines and overtaking dynamics in more realistic circuit conditions.

Building on the work of Liu [10], this study applies an optimal control framework to the overtaking problem. Optimal control is adopted because of its unique ability to determine not only the control inputs but also the optimal racing line, making it highly suited to analysing vehicle performance when following another car. Extending beyond previous work, the analysis presented here includes cornering scenarios, where aerodynamic wake plays a significant role in altering both the trajectory and vehicle dynamics, thereby influencing overtaking opportunities.

The remainder of this study is organised as follows. Section 2 presents the methodology, including the optimal control problem formulation, vehicle modelling, aerodynamic wake modelling, track generation, and solution strategy. Section 3 reports the results of the simulations, focusing on overtaking scenarios in both low-speed and high-speed contexts, and discusses their implications in the context of existing literature. Section 4 provides the conclusions of the study, outlining the main findings, limitations, and directions for future work.

2 Methodology

2.1 Optimal Control Problem Formulation

The primary objective of this study is to determine the time-optimal trajectory of a race car around a prescribed track using optimal control methods. The problem is posed as an Optimal Control Problem (OCP), where the goal is to minimise the total lap time subject to nonlinear vehicle dynamics and physical constraints on the system.

An OCP seeks to identify the state trajectories and control inputs that optimise a performance criterion while ensuring that the system evolves consistently with its governing dynamics. In this case, the cost function is the minimisation of lap time, the system dynamics are defined by a planar 3-degrees-of-freedom (DOF) vehicle model, and constraints enforce realistic bounds such as maximum power, track limits, and tyre friction circles.

To transcribe and solve the problem, the Dymos library within the OpenMDAO framework was employed. Dymos is a Python-based optimal control tool that implements multiple direct collocation schemes, enabling continuous-time OCPs to be discretised into large-scale NLP problems. It provides flexibility in defining phases of motion, boundary conditions, control parameterisations, and path constraints, making it well-suited for trajectory optimisation in lap time simulation contexts. In this work, the OCP was transcribed using the Gauss–Lobatto pseudospectral collocation scheme [5]. This method discretises the continuous trajectory into collocation points defined by the roots of Legendre polynomials. At these points, the system dynamics must be satisfied exactly, while continuity conditions are imposed between successive segments. This approach ensures dynamic feasibility across the trajectory while reducing the continuous optimal control problem to a finite-dimensional NLP.

The resulting NLP was solved using the Interior Point Optimizer (IPOPT). The problem was discretised into 150 segments, each using third-order Gauss–Lobatto polynomials, resulting in a large but tractable optimisation problem. IPOPT iteratively adjusted the control history and corresponding state trajectories to converge on the time-optimal solution.

A critical element of the problem formulation is the choice of the independent variable. While most Dymos examples adopt time as the independent variable, in this work the state equations were reformulated with arc length along the track centreline, s, as the independent variable.

$$\dot{s} = \frac{\mathrm{d}s}{\mathrm{d}t} = \frac{V\cos\alpha - \lambda}{1 - \eta\kappa} \tag{1}$$

$$n' = \frac{\mathrm{d}n}{\mathrm{d}s} = \frac{\dot{n}}{\dot{s}}$$

where n' is the spatial rate of change of the lateral offset n, and $\dot{s}$ is the rate of change of arc length with respect to time (i.e. longitudinal speed along the track). This transformation is applied to all state equations, reformulating the dynamics in the spatial domain. Using arc length as the independent variable naturally parameterises the vehicle's progression along the track, avoiding issues with variable lap times when time is used directly.

Finally, Table 1 describes the controls, states, constraints, and objectives that compose the OCP.

2.2 *Vehicle Model*

The dynamics of the race car were represented using a planar 3-DOF model adapted from De Buck [4]. The model includes longitudinal, lateral, and yaw motion, with forces computed at each tyre. Aerodynamic loads act as downforce and drag contributions, and tyre forces are defined as functions of slip angle and normal load. The model was modified to incorporate wake-dependent aerodynamic coefficients (described in Sect. 2.3) and traction forces that depend on downforce. Movable aerodynamics were excluded, in line with the F3 car being represented in this study.

The governing equations of motion are expressed as:

$$mrV\lambda + m\dot{V} = X_{\mathrm{fl}} + X_{fr} + X_{\mathrm{rl}} + X_{\mathrm{rr}} + \delta(Y_{\mathrm{fl}} + Y_{\mathrm{fr}}) - D$$

$$m\big(rV - \dot{V}\lambda - V\dot{\lambda}\big) = \delta(X_{\mathrm{fl}} + X_{\mathrm{fr}}) + Y_{\mathrm{fl}} + Y_{\mathrm{fr}} + Y_{\mathrm{rl}} + Y_{\mathrm{rr}}$$

Table 1 Optimisation problem formulation

		Symbol	Lower	Upper	Units
Minimize	Time	t			S
Controls	Throttle	σ	-1	1	
	Steering angle	δ	$-\pi$	π	rad
States	Speed	V			ms^{-1}
	Longitudinal acceleration	a_x			ms^{-2}
	Lateral acceleration	a_y			ms^{-2}
	Normal distance to centreline	n	-4	4	m
	Angle relative to centreline	α			rad
	Slip angle	λ			rad
	Yaw rate	r			rad s^{-1}
	Gap to required lateral offset	g			m
Subject to	Front left grip	c_{fl}	0	1	
	Front right grip	c_{fr}	0	1	
	Rear left grip	c_{rl}	0	1	
	Rear right grip	c_{rr}	0	1	
	Drive shaft power	P		310,000	kW
	System dynamics				
	Track boundaries				
	Collision				

$$I_z\dot{r} = a(Y_{fl} + Y_{fr}) - b(Y_{rl} + Y_{rr}) + t(X_{fl} - X_{fr} + X_{rl} - X_{rr})$$

$$\dot{s} = \frac{V\cos(\alpha - \lambda)}{1 - n\kappa} \tag{2}$$

$$\dot{n} = V\sin(\alpha - \lambda)$$

$$\dot{\alpha} = r - \frac{\kappa V\cos(\alpha - \lambda)}{1 - n\kappa}$$

where V is vehicle speed, r yaw rate, λ vehicle heading, n lateral deviation, and κ track curvature. The forces X and Y denote longitudinal and lateral contributions at each tyre (Fig. 2).

Tyre behaviour was initially represented using a linear lateral stiffness model dependent on normal load:

$$F_{ij}(\alpha, N) = N_{ij}k_{\alpha}\alpha_{ij} \tag{3}$$

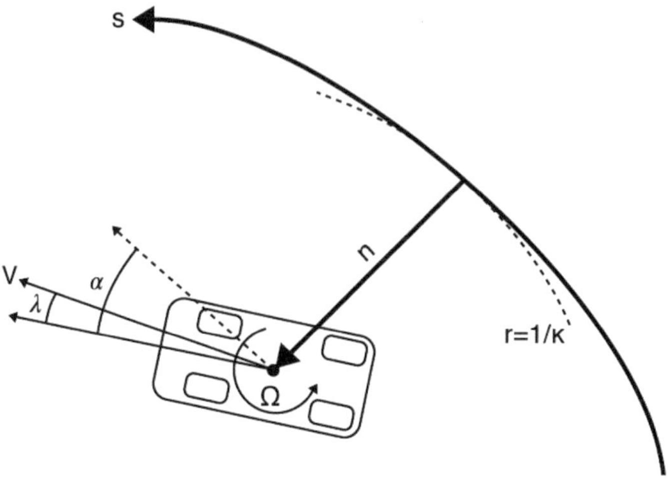

Fig. 2 Vehicle position relative to track centreline and vehicle degrees of freedom [4]

with slip angles calculated from geometric relations:

$$\alpha_{\mathrm{fl}} = \lambda + \delta - \frac{r(a+\lambda t)}{V},\ \alpha_{\mathrm{fr}} = \lambda + \delta - \frac{r(a-\lambda t)}{V},$$
$$\alpha_{\mathrm{rl}} = \lambda + \frac{r(b-\lambda t)}{V},\ \alpha_{\mathrm{rr}} = \lambda + \frac{r(b+\lambda t)}{V}. \tag{4}$$

Normal loads at each tyre were computed as:

$$
\begin{aligned}
N_{\mathrm{fl}} &= \frac{1}{2}mg\frac{b}{a+b} + \frac{1}{2}ma_x\frac{h}{a+b} + \chi m a_y\frac{h}{t} + \frac{1}{2}\frac{1}{2}\rho \mathrm{SC}_z V^2\frac{a+b-\mathrm{CoP}}{a+b}\\
N_{\mathrm{fr}} &= \frac{1}{2}mg\frac{b}{a+b} + \frac{1}{2}ma_x\frac{h}{a+b} - \chi m a_y\frac{h}{t} + \frac{1}{2}\frac{1}{2}\rho \mathrm{SC}_z V^2\frac{a+b-\mathrm{CoP}}{a+b}\\
N_{\mathrm{rl}} &= \frac{1}{2}mg\frac{a}{a+b} - \frac{1}{2}ma_x\frac{h}{a+b} + (1-\chi) m a_y\frac{h}{t} + \frac{1}{2}\frac{1}{2}\rho \mathrm{SC}_z V^2\frac{\mathrm{CoP}}{a+b}\\
N_{\mathrm{rr}} &= \frac{1}{2}mg\frac{a}{a+b} - \frac{1}{2}ma_x\frac{h}{a+b} - (1-\chi) m a_y\frac{h}{t} + \frac{1}{2}\frac{1}{2}\rho \mathrm{SC}_z V^2\frac{\mathrm{CoP}}{a+b}.
\end{aligned} \tag{5}
$$

These terms account for static load distribution, longitudinal and lateral load transfer, and aerodynamic downforce. Suspension dynamics were not modelled explicitly; however, transient effects were partially captured by applying time constants to the load transfer equations.

The aerodynamic model assumes constant coefficients for drag, downforce, and aero balance under free-stream conditions. These coefficients remain fixed except when modified by the wake model of the leading vehicle, covered in Sect. 2.3.

Variations with ride height, yaw, or pitch were omitted to ensure tractability within the optimal control framework.

Engine torque data was derived from an AVL-generated Formula 3 engine model, fitted with third-order polynomials in throttle and RPM (Figs. 3, 4, 5 and 6):

$$\tau(x_1, x_2) = \tau_{\text{rpm}}(x_1) \frac{1}{\tau_{\max}} \tau_{\text{throttle}}(x_2). \tag{6}$$

Fitting with third-order polynomials:

$$\tau = ax^3 + bx^2 + cx + d \tag{7}$$

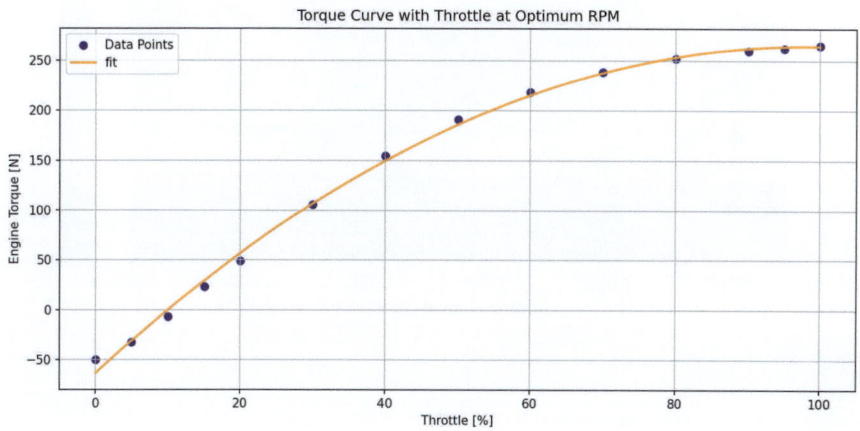

Fig. 3 Engine torque data with throttle input and fitted model

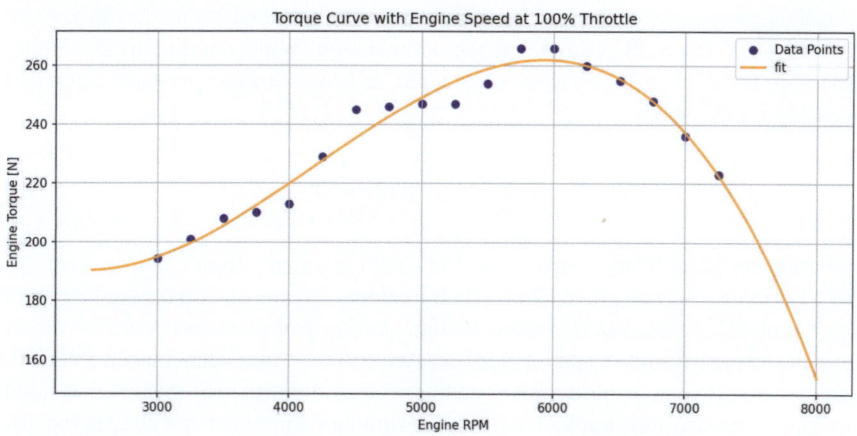

Fig. 4 Engine torque data with rpm and fitted curve

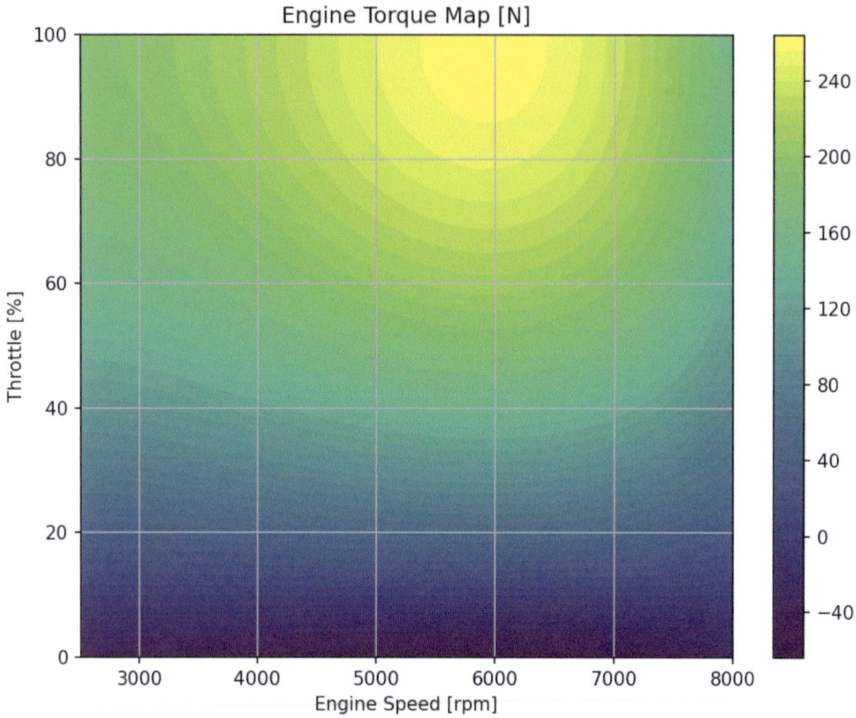

Fig. 5 Map of engine torque

where for rpm $a = -3.4675 * 10^{-9},\ b = 4.369 * 10^{-5},\ c = -0.1524,\ d = 352.4582$.

And for throttle fit $a = 0,\ b = 3.4034 * 10^{-2},\ c = 6.6842,\ d = -63.1975$.

A fixed gear ratio was assumed to avoid the discontinuities associated with discrete gear shifts, which would complicate the optimisation. While continuous smoothing functions for gear changes were considered, a constant ratio provided sufficient accuracy for this study.

$$S_{rl} = S_{rr} = \frac{i_{\text{gear}}}{r_w} \tau_{\text{rpm}}(\text{rpm}) \frac{1}{\tau_{\max}} \tau_{\text{throttle}}(\sigma) \tag{8}$$

Parameter identification was carried out with telemetry from Campos Racing's 2024 Formula 3 season at the Circuit de Barcelona-Catalunya. A baseline F3 model was simulated around the Barcelona layout, starting from known parameter values from the F3 car [6], and key parameters like tyre stiffness, maximum power, and aerodynamic coefficients were iteratively adjusted to minimise error against measured lap data. The Barcelona track model used in simulation predated the 2023 revisions, so only the portion of the lap common to both layouts was used. Nevertheless, this provided sufficient data to identify the vehicle parameters summarised in Table 2.

Table 2 Vehicle parameters

Parameter	Value	Units	Description
m	698	kg	Vehicle mass
a	1.47	m	CoG location to front axle
b	1.47	M	CoG location to rear axle
t_w	1.885	M	Track width
h	0.2	m	CoG height
I_z	450	kg m^2	Yaw inertia
β	0.55	–	Brake balance
χ	0.5	–	Roll stiffness
ρ	1.225	kg m^{-3}	Air density
μ_0^x	1.68	–	Longitudinal base friction coefficient
μ_0^y	1.68	–	Lateral base friction coefficient
k_μ	0	–	Tyre load sensitivity
k_α	10	–	Tyre lateral cornering stiffness
τ_{a_x}	0.2	s	Longitudinal load transfer time constant
τ_{a_y}	0.2	s	Lateral load transfer time constant
r_w	0.33	m	Drive wheel radius
i_{gear}	7	–	Fixed gear ratio
CoP	1.5	m	Centre of pressure location to front axle
C_zA	4.0	m^2	Downforce coefficient
C_dA	1.1	m^2	Drag coefficient

This process produced a calibrated planar model capable of replicating the behaviour of a modern Formula 3 car with sufficient fidelity for optimal control trajectory optimisation, as seen from correlation in Fig. 6.

Figure 6 shows the correlation between the vehicle model and real telemetry data from a Formula 3 car around the Barcelona circuit. The comparison is presented in terms of vehicle speed, yaw rate, and longitudinal and lateral accelerations. Overall, the model captures the main dynamic behaviour of the car, reproducing braking phases, cornering behaviour, top speeds, and accelerations with reasonable accuracy. The speed trace demonstrates close agreement in the straights, major braking zones, and through most corners, although discrepancies appear at Turns 4, 6, and 7, where the model exhibits greater grip than the real car. These deviations likely stem from the simplified tyre and aerodynamic formulations adopted. Nevertheless, the correlation was deemed sufficient to identify and tune the main vehicle parameters, ensuring that the model represents the dynamics of a modern Formula 3 car with adequate fidelity for optimal control simulations.

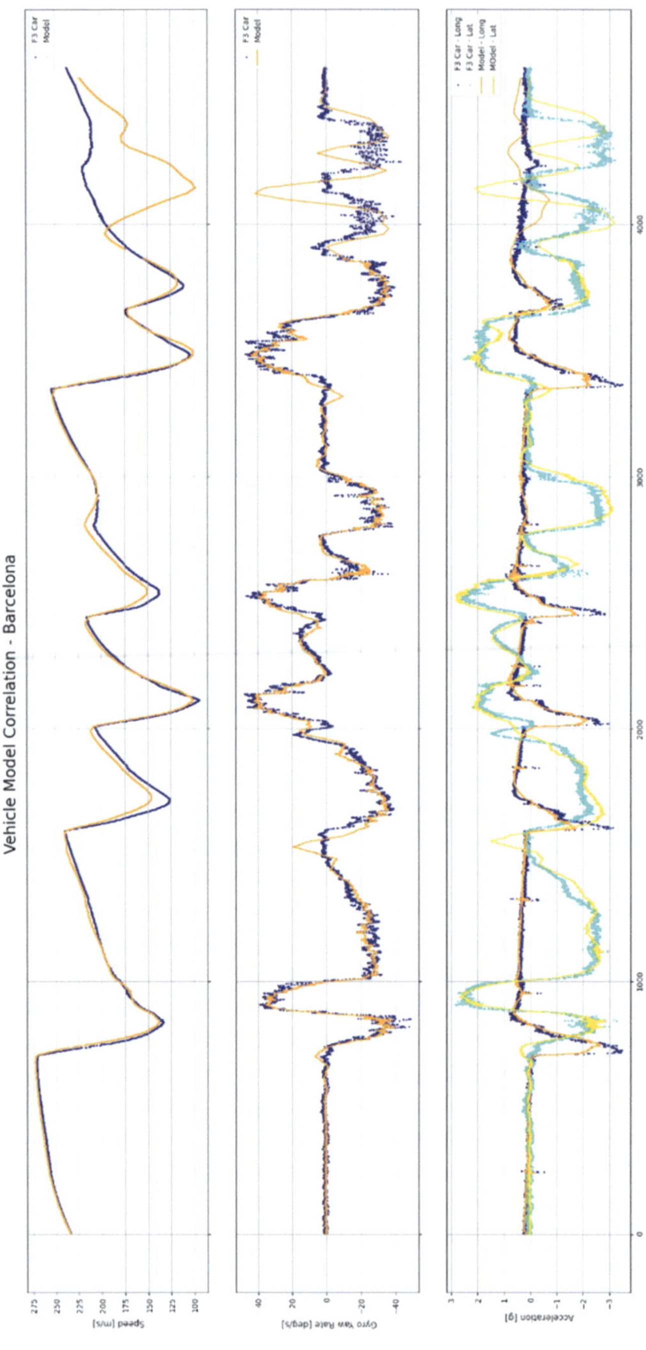

Fig. 6 Vehicle simulation versus real Campos Racing data overlay in Barcelona

2.3 Aerodynamics Wake Model

Because the objective of this project is to investigate overtaking manoeuvres under disturbed flow conditions, it was essential to adopt a model capable of capturing the aerodynamic losses experienced by a trailing car in the wake of a leading vehicle. To incorporate this in an optimal control simulation, a spatially varying wake model was derived based on experimental and simulation data available in the literature [7, 10, 11]. These studies consistently show that the wake of an upstream car results in reductions in downforce, drag, and shifts aerodynamic balance for the following car, with the magnitude of the disturbance depending on the relative longitudinal spacing and lateral offset. Maximum disturbance occurs when the trailing car is directly behind the leader at small longitudinal gaps. The effects decay both as the distance increases and as the following car moves laterally away from the wake centreline.

The model follows the structure of Liu [10], adapted to include explicit effects on aero balance in addition to downforce and drag. The aerodynamic losses are defined by Eq. 9.

$$P_{\text{loss}}(x_1, x_2) = P_0 P_{\text{loss, ini}}(x_1) f(x_2) \tag{9}$$

where x_1 is the lateral offset, therefore $P_{\text{loss, ini}}$ represents the normalised loss when the longitudinal distance is 0 between the cars, x_2 is the longitudinal spacing between the vehicles, and P_0 the maximum possible loss for each aerodynamic parameter. The lateral decay of the wake is represented in Eq. 10.

$$P_{\text{loss, ini}} = \xi u_{\text{US}}^2 e^{\left(\frac{-x_1^2}{\sigma}\right)} \tag{10}$$

with u_{US} the upstream car's velocity, and ξ, σ, where $\xi = 0.0004$, $u_{\text{US}} = 50$, $\sigma = 0.8$, calibration constants chosen to match trends observed in the literature.

The longitudinal decay is captured by separate functions for downforce, drag and aero balance, given by Eqs. 11, and 12 respectively.

$$f_{\text{downforce}} = f_{\text{drag}} = (1 - \tanh(0.4x_2)) \tag{11}$$

$$f_{AB} = \left(1 - \frac{x_2}{-0.3}\right) e^{\frac{-x_2}{1.25}} \tag{12}$$

The maximum magnitudes of the wake losses were set as 70% for downforce, 42% for drag, and 20% for aero balance, consistent with findings from Guerrero [7] and [11]. These values correspond to the extreme case of very close following along the leader's centreline.

Figures 7, 8, 9 illustrate the resulting loss maps for aero balance, drag, and downforce at a reference speed of 50 m/s. The disturbance is most severe in the near wake, where downforce losses can exceed 60% and significant shifts in aero balance occur.

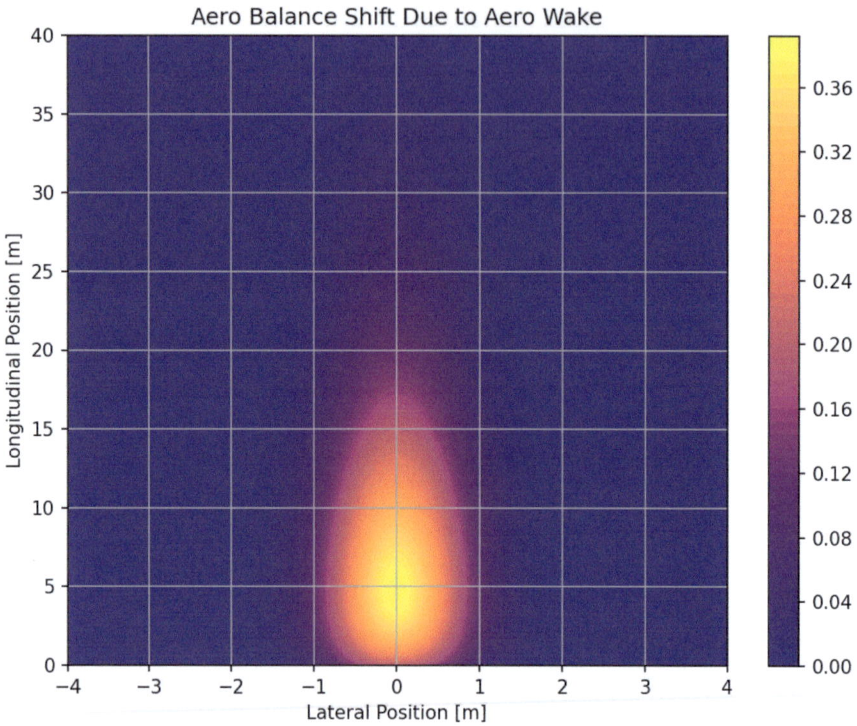

Fig. 7 Map of aero balance shift due to upstream car wake at 50 m/s

As spacing increases beyond 2–3 car lengths, the disturbance diminishes rapidly. Lateral offsets also reduce the effect, with the trailing car recovering nearly full aerodynamic performance once positioned outside the main wake core.

Within the vehicle dynamics model, this wake formulation is applied as a set of correction factors to the baseline aerodynamic coefficients of the trailing car. This is, downforce and drag coefficients are reduced according to their corresponding loss map. Aero balance is also shifter by adding the shift from the map to the baseline free-stream value. Because the functions are continuous and differentiable, they can be directly evaluated within the collocation-based OCP without introducing discontinuities or solver instability. This makes the model computationally efficient while still grounded in experimental evidence.

Several simplifications are inherent to this approach. First, the model is quasi-steady, depending only on instantaneous longitudinal and lateral spacing, and does not account for time-dependent turbulence or vortex shedding. Second, variations in wake structure with yaw angle, ride height, or crosswind conditions are neglected. Despite these limitations, the chosen formulation captures the first-order aerodynamic consequences of wake interaction, providing sufficient fidelity for the purpose of overtaking trajectory optimisation.

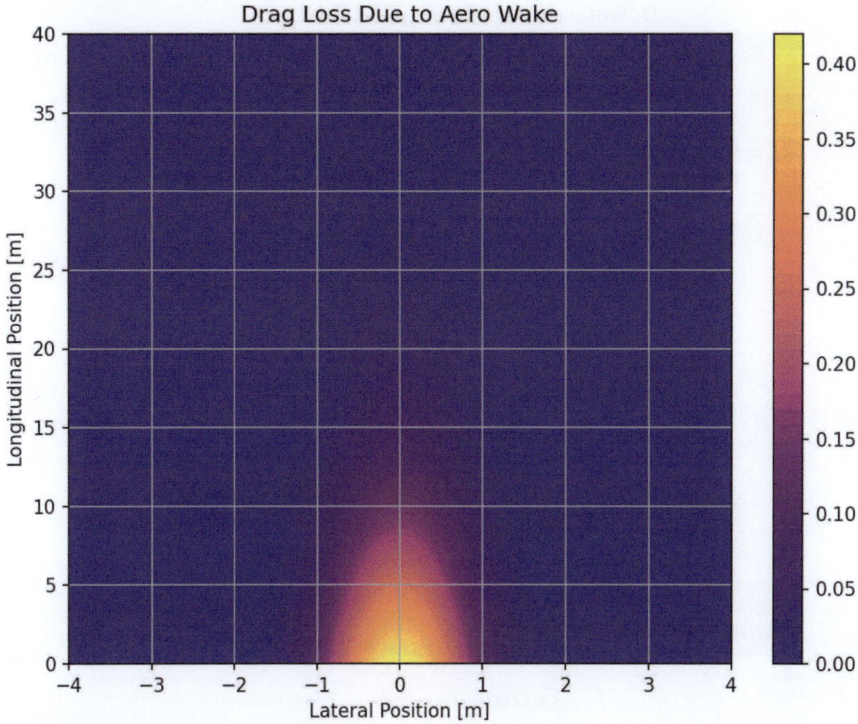

Fig. 8 Map of drag loss due to upstream car wake at 50 m/s

2.4 Track Generation

For this study, a customised track was generated specifically to balance realism with computational efficiency. Since the simulations involve solving large-scale optimal control problems with additional aerodynamic wake and collision constraints, a short open layout was selected to reduce computational load while still capturing the key racing scenarios relevant to overtaking and following a car in the wake. The track configuration was designed to contain both a slow-speed corner and a medium-to-high-speed corner, connecting two straights. This design allows investigation of two different overtaking contexts: (1) the heavy braking zone at the end of the first straight into a tight corner, which represents a classic overtaking opportunity, (2) a faster, more aerodynamically sensitive second corner, were reduced downforce from wake effects plays a critical role in determining grip and exit speed. This track design captures both mechanical and aerodynamic limitations, offering a relevant test case for the overtaking simulations. Figure 10 shows the resulting racing line obtained for the single-car simulation, with velocity distribution mapped along the trajectory.

The layout consists of four segments: (1) a 550 m straight, (2) a 35 m radius corner with a sweep angle of $1.1\pi/2$, representing a low-speed corner after the first

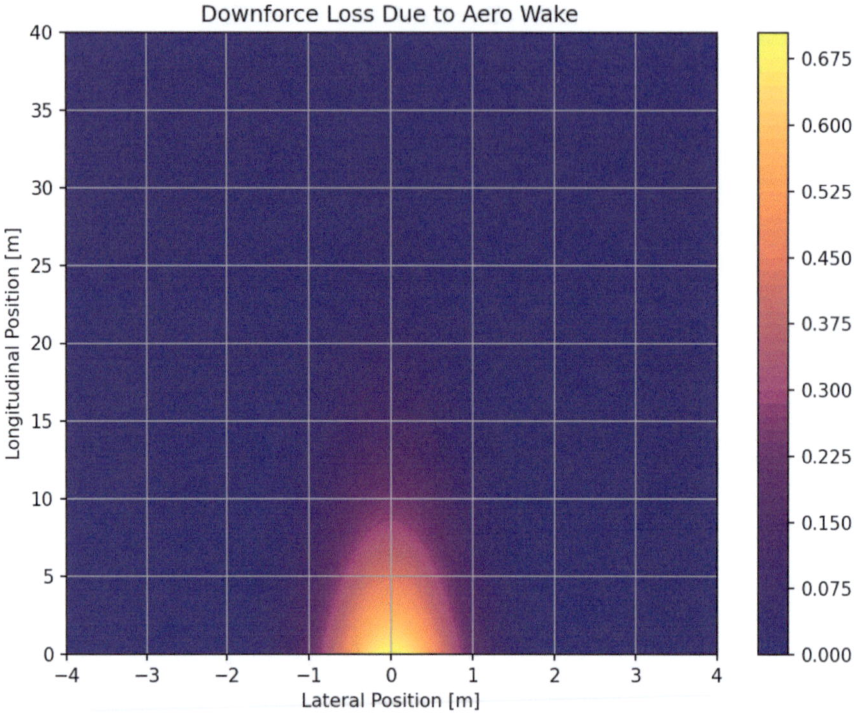

Fig. 9 Map of downforce loss due to upstream car wake at 50 m/s

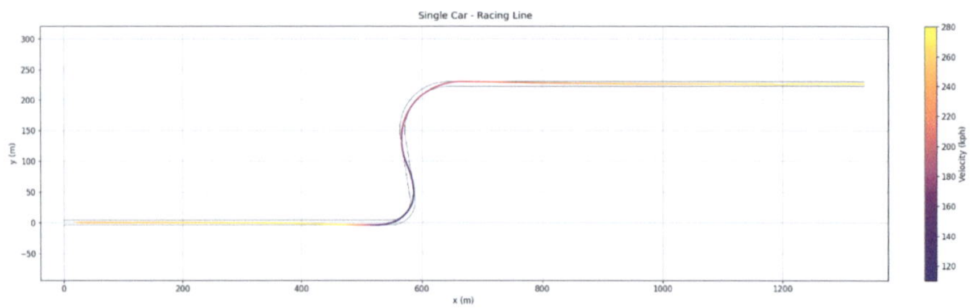

Fig. 10 Track generated and used in study, showcasing racing line of single-car simulation coloured by speed

straight, (3) a 100 m straight connecting to, (4) a 75 m radius corner with a sweep of $1.1\pi/2$, leading into (5) a 700 m straight, where overtaking manoeuvres can be completed. The circuit was defined programmatically using Dymos' track generation utilities. The track is represented as an ordered sequence of parametric segments, each described by four parameters type (straight or corner), length or sweep angle, radius, and direction (left or right).

This compact parametric representation was then passed to a custom track class, which generated the continuous centreline, curvature distribution, and orientation data required for the simulation. These quantities were then used to enforce track boundary constraints in the optimal control problem and to evaluate vehicle dynamics along the circuit.

2.5 Solution Strategy

The solution strategy adopted in this study follows a sequential two-step approach. First, generating the reference trajectory by solving a single-car optimal control simulation to obtain the minimum-time trajectory for a Formula 3 vehicle around the track in free-stream. This trajectory is referred to as the ghost car. Then, using the ghost trajectory as input for the attacking car, perform a dual-car simulation where the active car experiences aerodynamic wake effects and must find an alternative optimal racing line that maximises performance when experiencing wake while avoiding contact with the car in front when executing an overtake [10].

The single-car problem was solved once the vehicle model had been parameterised and correlated with Campos Racing telemetry. The solution provides the theoretical minimum-time lap for the given car–track configuration, including the optimal racing line, velocity profile, and control history. This trajectory serves as the fixed reference for the leading car in all subsequent dual-car simulations. Importantly, the ghost car is assumed to remain on this trajectory without defensive manoeuvres, consistent with the simplifications in Liu [10].

The second step introduces the active car, which begins its lap from behind the ghost car. The active car's dynamics are modified by the wake model described in Sect. 2.3, reducing available downforce and drag as a function of relative spacing. This ensures that the active vehicle experiences the aerodynamic penalties observed in real racing when following closely.

To prevent unrealistic overlap with the ghost trajectory, a collision avoidance constraint was formulated with Eq. 13. The constraint ensures that the active car's trajectory always remains outside the collision zone surrounding the ghost car, in both the longitudinal and lateral directions.

$$
\begin{aligned}
\Delta s &= \left(s_{\text{ghost}} - s_{\text{active}}\right) \\
\text{gap}_{\text{long}} &= \frac{\ln(\cosh(k_{\text{abs}}\Delta s))}{k_{\text{abs}}} \\
\Delta n &= \left(n_{\text{ghost}} - n_{\text{active}}\right) \\
\text{gap}_{\text{lat}} &= \frac{\ln(\cosh(k_{\text{abs}}\Delta n))}{k_{\text{abs}}} \\
w_{\text{close}} &= \frac{1}{1 + e^{-k_{\text{sw}}\left(\Delta s_{\min} - \text{gap}_{\text{long}}\right)}}
\end{aligned}
$$

$$n_{\text{req}} = \Delta n_{\text{close}} w_{\text{close}}$$
$$g = \text{gap}_{\text{lat}} - n_{\text{req}} \geq 0 \tag{13}$$

Here, s represents the longitudinal position along the track centreline and n the lateral offset, while the hyperbolic and sigmoid functions smooth the exclusion zone to ensure differentiability. This formulation is critical, as the OCP requires constraints that are at least C1 continuous to allow gradient and Hessian evaluation. The resulting inequality is used as a path constraint which dynamically varies along the track depending on the longitudinal spacing between the cars, with the required lateral offset increasing to the minimum lateral distance of the cars when overtaking when the active car is closer than the full length of the car and relaxing to 0 when its far.

Once both wake effects and collision constraints were implemented, the active car simulations were run. By construction, these simulations can be interpreted as dual-car optimal control problems, in which the ghost car is fixed and the active car searches for a feasible minimum-time trajectory around it.

The primary goal of this project was to investigate overtakes in two regions of the track. On one hand, at the end of the first straight, getting into a low-speed corner following a heavy braking zone, which is a natural overtaking point in real racing. On the other hand, at the end of the second straight, after exiting a higher-speed corner where aerodynamic effects dominate grip and traction, testing the ability of the wake model to influence overtaking feasibility.

To trigger successful overtakes in these regions, the active car was given a performance advantage relative to the ghost. This was achieved by increasing tyre grip to represent a pace advantage. The initial speed and position of the active car were tuned accordingly to adjust the location where the overtake took place.

The choice of increased grip reflects observations from the article in the Canopy Simulations Blog, a recent study on the fundamentals of overtaking which highlights that overtakes rarely occur between cars of identical pace. Grip therefore serves as a reasonable proxy for this advantage. This work also emphasises the sensitivity of overtaking performance to aerodynamic wake, confirming that stronger wake effects reduce overtaking feasibility [9]. These insights align with the rationale of this project, which aims to explore how the optimal racing line adapts to minimise the negative influence of wake when following another car.

3 Results and Discussion

3.1 Straight 1 Overtake

In this project, a second active car was simulated in which dynamics were affected by the trajectory of the ghost car in front. The active car performed a minimum-time manoeuvre simulation along the circuit, but unlike the single-car simulation, it experienced aerodynamic losses due to the wake of the ghost car. Additionally, when an overtake opportunity arose, the active car adapted its trajectory to avoid collision with the ghost car. This was enforced through the collision constraint described in Sect. 2.5 by Eq. 13.

The first overtaking case was configured to occur in the braking zone before Turn 1 (T1). To enable the manoeuvre at this location, the initial conditions of the active car were adjusted. It started 10 m behind the ghost car, matching its initial speed of 68 m/s. Moreover, it was given a 5% increase in tyre grip, representing the pace advantage. These conditions replicate a realistic overtaking scenario in Formula 3, where overtakes typically require a grip advantage.

The results show behaviour consistent with real-world racing. On the straight preceding T1, the active car follows closely in the wake of the ghost to benefit from the reduction in drag, the slipstream effect. As the braking zone is approached, two critical changes occur. First, the active car moves laterally away from the ghost's wake in order to comply with the collision constraint and initiate the overtaking manoeuvre. Second, this lateral movement allows the active car to exit the wake at the precise moment when maximum braking performance is required.

This timing is important because braking capacity is strongly dependent on available aerodynamic downforce and its distribution between axles. Under wake conditions, both total downforce and aerodynamic balance are degraded, reducing stability and braking performance. By stepping out of the wake just before braking, the active car regains clean airflow, restoring downforce and enabling superior braking compared to the ghost car due to better grip and equal aerodynamic conditions.

The result from Figs. 11 and 12 show a successful overtake at the entry to T1, achieved by combining the slipstream advantage on the straight with restored aerodynamic performance under braking. Note that in Figs. 11 and 13b that the ghost line is shown in green, while the active car's line is coloured by its longitudinal spacing with the reference car, red when it is still behind and blue when it is in front of the ghost car.

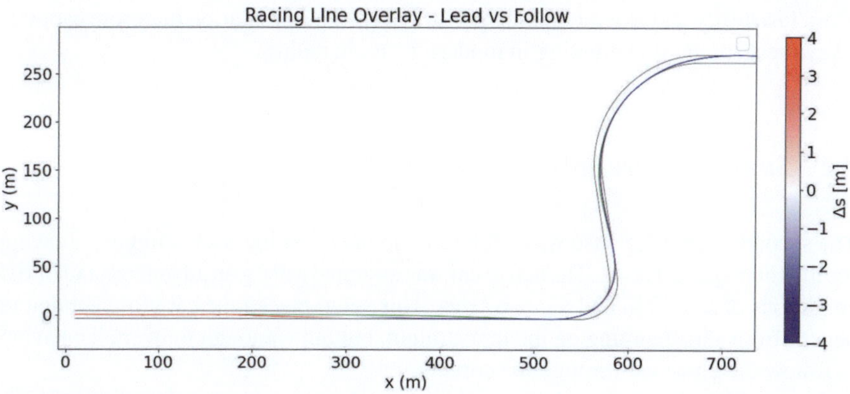

Fig. 11 Comparison of ghost car and active car racing line in overtake 1 scenario, through straight 1, T1 and T2

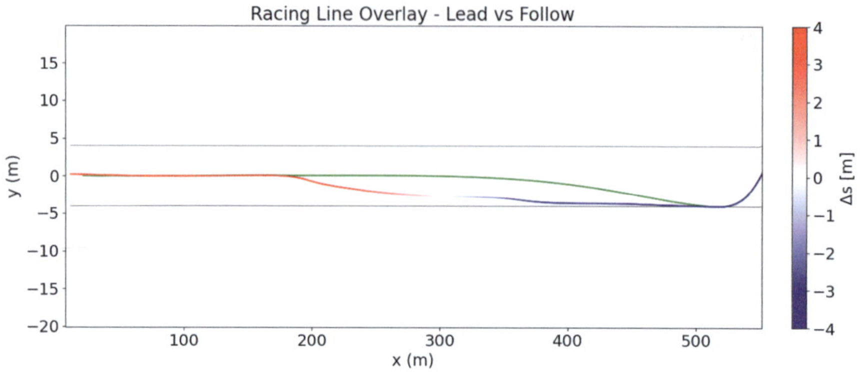

Fig. 12 Comparison of ghost car and active car racing line in overtake 1 scenario, showing overtake manoeuvre in straight 1

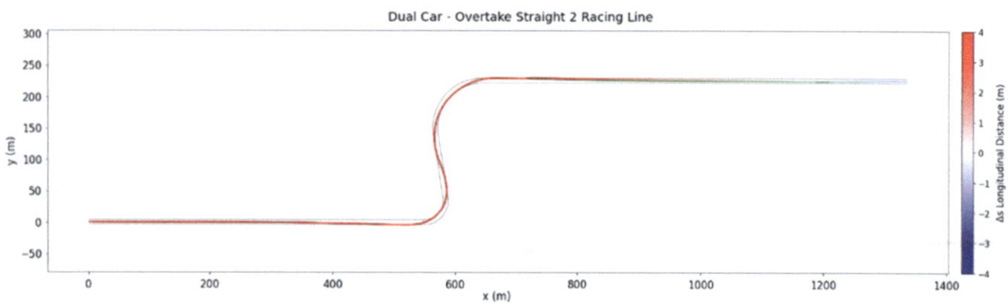

Fig. 13 Comparison of ghost car and active car racing line in overtake scenario 2, through straight 1, T1, T2, and straight 2

This finding reflects overtaking behaviour observed in practice: cars follow closely to maximise drag reduction, but drivers instinctively leave the wake under braking to avoid instability and extend braking performance. The simulation therefore captures a key mechanism of overtaking in modern formula racing.

3.2 *Straight 2 Overtake*

The second overtaking case was configured to occur on the back straight following the medium-speed Turn 2. The active car was assigned a 5% grip advantage and initial conditions of $s_0 = 0$ m and $V_0 = 67$ m/s. This setup placed the car close enough to benefit from slipstreaming on the first straight, but quite enough to overtake, having to follow the ghost car through the corners.

Figure 14 presents the results of this case, showing the speed profiles of both cars, aerodynamic coefficients and balance, and relative gaps in the longitudinal and lateral directions along the track. In green in the second subplot, we see how both the downforce (consistent line) and drag (dashed) decrease exponentially as described by the aero wake model until the point where the active car gets out of the wake to perform the overtake. In red is the aero balance which has a big effect in traction performance as drive is highly dependent on the load acting on the rear axle. In the third subplot, we see the evolution of the relative position of the active car with respect to the ghost car. In red is the longitudinal spacing between the two, starting at 20 m, reducing significantly in T1 braking due to more grip, staying relatively constant through the corners until at the exit and into straight 2 starts decreasing linearly completing the overtake, as can be seen in Fig. 13. In green is the lateral offset of the racing line used by the active car relative to the ghost car and this gives critical insights on how the trajectory changed and where, to maximise grip and traction. The behaviour of the active car demonstrates how the optimal control solution balances ideal racing line with downforce recovery, adjusting its racing line dynamically to maximise performance.

During the first straight, the active car maintains $\Delta n = 0$, staying perfectly in line with the ghost to maximise drag reduction and close the gap. This behaviour reflects the classical slipstreaming strategy observed in open-wheel racing.

As the braking zone into Turn 1 approaches, the active car deviates laterally by approximately 0.5 m, partially exiting the wake, Fig. 14. This lateral movement restores some downforce and improves braking stability, consistent with the discussion of the previous overtake. The limited offset suggests a trade-off: fully leaving the wake would have required moving to the inside line, which reduces corner radius and increases curvature. Instead, the optimisation balances drag reduction with sufficient downforce recovery, resulting in a compromise trajectory.

As expected, the wake-induced downforce loss penalises the active car in Turn 1, producing lower cornering speeds relative to the ghost. However, telemetry shows subtle adaptations of the racing line. At late entry and mid-corner, the trajectory is shifted outwards by ~ 0.12 m, which increases lateral grip; at corner exit, a ~ 0.12 m offset increases total downforce and moves aero balance rearwards, improving traction performance and exit stability. These adjustments are reflected in small gains in downforce and significant reduction in aero balance at Turn 1, Fig. 15.

Through Turn 2, the racing line remains broadly similar to the ghost's, but another lateral deviation occurs at the exit. The active car shifts ~ 0.26 m away from the ghost's wake, again improving downforce and restoring a more favourable aero balance for traction. This strategic move sets up the subsequent overtaking attempt, by generating a speed delta at corner exit, which combined with the slip stream in the straight are favourable conditions for overtaking. Once on the second straight, the active car applies the same principles as in the Turn 1 overtake, as seen in Fig. 16. It aligns closely with the ghost's trajectory to maximise drag reduction, gaining speed relative to the leader. As the collision constraint activates, the active car moves laterally to avoid overlap, conserving its momentum and completing the overtake by the end of the straight, Fig. 17.

Overall, this case demonstrates the dual influence of aerodynamic wake; penalising performance in corners while simultaneously enabling slipstream gains on the straights. The optimised racing line reveals small but deliberate deviations to

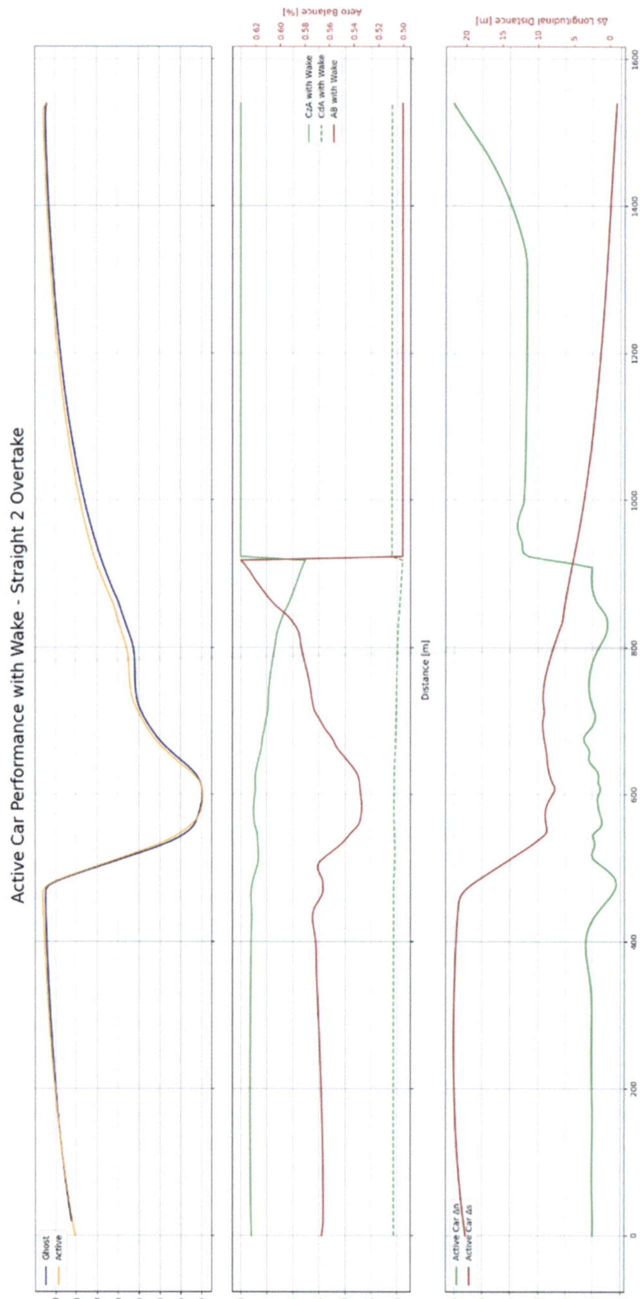

Fig. 14 Simulation data with ghost and active car speed overlay (top), aerodynamic coefficients of active car (middle), and active car relative position (bottom) along the generated track

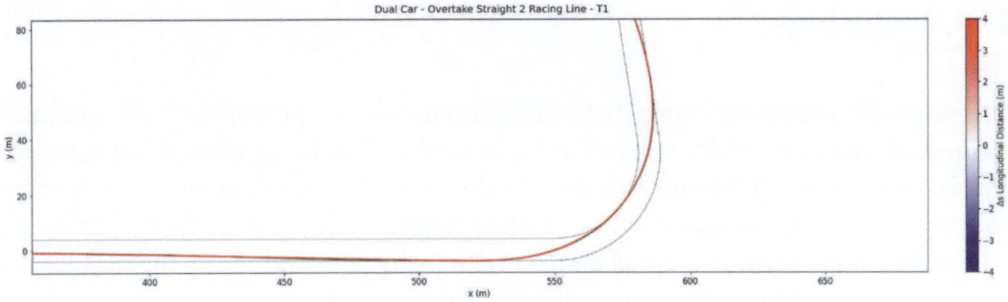

Fig. 15 Comparison of ghost car and active car racing line in overtake scenario 2, into and out of T1

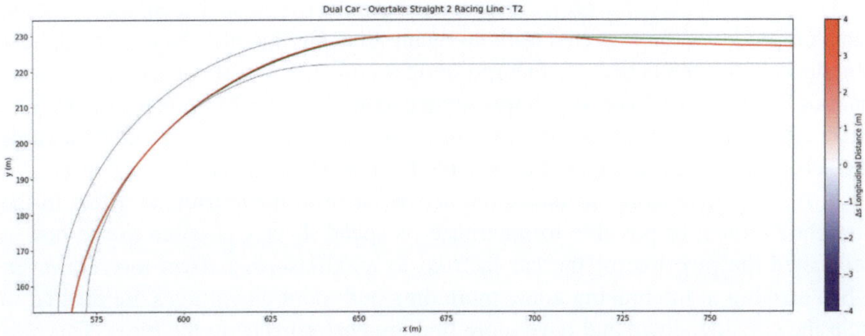

Fig. 16 Comparison of ghost car and active car racing line in overtake scenario 2, out of T2 and at the beginning of straight 2

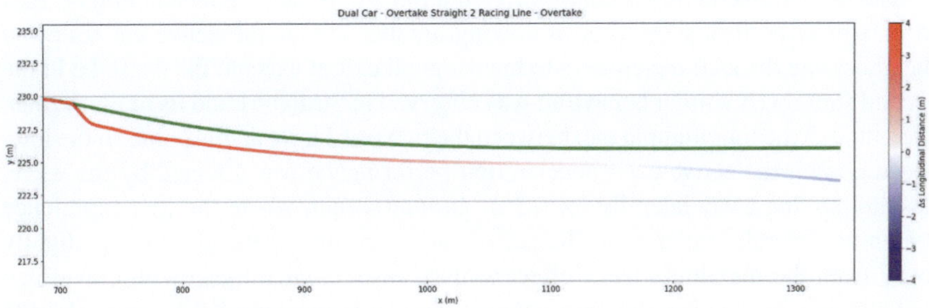

Fig. 17 Comparison of ghost car and active car racing line in overtake scenario 2, showing overtake manoeuvre in straight 2

balance these effects, and the ideal racing line. The results align with observations in F3 racing, where overtakes typically occur after a sequence of wake-induced compromises in the corners, followed by decisive slipstreaming on the straights. These results show an emphasis on the following car trying to maximise traction out of corners when experiencing wake from a car in front.

4 Conclusions

This project has presented a proof of concept for the use of optimal control to optimise racing line trajectory under wake effects, as well as simulating overtaking manoeuvres in motorsport. A telemetry-derived vehicle model representative of a modern Formula 3 car was combined with a wake-sensitive aerodynamic formulation and solved within the Dymos optimal control framework. By structuring the problem in the spatial domain and enforcing collision constraints, the framework successfully produced optimal overtaking trajectories in two distinct scenarios. Overtaking at the end-of-the straight into a heavy braking zone into a slow-speed corner, and in acceleration out of consecutive corners.

The main outputs of the project are on one hand the single-car baseline simulation, which reproduces the expected optimal racing line. On the other hand, the dual-car simulation with the active car incorporating a collision constraint and the aerodynamics losses due to wake, which was setup to complete two different overtakes. The wake effects made the attacking car to suffer from downforce loss and shift forwards the aero balance in the corners but benefited in the straights with reduced drag.

In the first overtaking scenario, the active car used the slipstream effect in the straight as much as possible to maximise its speed. It was possible to see how it replicated the race line of the car in front to maximise this effect too. However, when arriving at the braking zone, more drag and more downforces are desired to help the car slow down and have more braking performance in the tyres. This was captured by the optimal control results, with the active car pulling away from the wake to overtake during the braking phase effectively equalling their aerodynamic conditions and leveraging the given grip advantage.

The second overtaking scenario offered a more complete understanding on how to maximise performance when following another car, as the active car followed the ghost one through the corners to have a good exit at execute the overtake in the second straight. A similar behaviour was observed in Straight 1 and its braking zone, even though the longitudinal gap between the two was larger at this point. In the low-speed corner, the active car's deceleration performance was affected by the wake, despite moving away laterally for 0.5 m. In mid-corner, where the cars experience maximum lateral acceleration, the active car's race line changed, moving slightly away from the maximum wake effect to maximise its lateral grip with downforce and stability by moving the aero balance rearwards outside of the wake. Despite being a low-speed corner and downforce not having such a great effect, the active car found a compromise between maximising traction in corner exit and using a faster line, by moving slightly laterally for better traction. This same behaviour was also seen at the corner exit of Turn 2, with the offset being larger as this was a medium speed corner and downforce on the rear axle at exit has a greater effect, outweighing more the faster line. It is this extra traction outside of Turn 2 which was key to setup the overtake down the second straight. It was seen that it is of primary importance to move away from the wake of the car in front to maximise traction outside of the corners when following another car. And optimal control is shown to be a powerful tool for understanding how race line trajectory adapts to maximise performance when following another car, and how overtaking unfolds under aerodynamic wake.

The effect of the change in grip per se in the racing line was not accounted for, however, it can be assumed that a subtle change in grip will not change the racing line substantially, and therefore that the differences in racing line obtained between the active and the ghost car are due to wake trajectory optimisation.

Several simplifications were made too. The vehicle model was restricted to a planar 3-DOF formulation with simplified tyre dynamics and constant aero coefficients (outside of wake effects). Suspension dynamics, ride height-dependent aerodynamics, and gear shifting were excluded. The wake model was quasi-steady, depending only on longitudinal and lateral spacing, and derived through fitting of published data. The leading car was modelled as a fixed ghost trajectory, with no defensive manoeuvres or reactive behaviour. The track was deliberately simplified to reduce computational cost, rather than replicating a full Grand Prix circuit. These assumptions and simplifications "affect" the specific results in this study but do not undermine the validity of the conceptual framework.

At the same time, it opens other avenues for future work and improvements. First obvious option is improving the vehicle model, incorporating higher-fidelity tyre models such as the Pacejka Magic Formula, suspension dynamics, and ride heights-dependent aerodynamics would allow closer replication of real vehicle behaviour. In modern racing, drivers have access to multiple tools too, including hybrid energy deployment, differential settings, brake balance, and even active aerodynamics like DRS, to modify the dynamics of the car. Future studies could extend the optimal control problem to include such control variables, coordinating their use with the racing line to maximise performance in wake. Another natural step is replacing the ghost trajectory with a defending car that reacts to the threat and attack from the car behind. This could be approached either by precomputing a library of defending trajectories or by formulating a multi-vehicle optimal control problem in which both cars optimise their strategies simultaneously. Such extension would connect the framework with concepts from game theory in racing. The wake model can potentially be improved too, as right now the model in the lateral direction is assumed to have a normal distribution shape, but the wake may have another distribution for instance depending on the downwash of the flow outside the car edges which may have a positive effect in downforce. Finally, techniques such as parallelised solvers, reduced-order modelling, or surrogate modelling could make larger and more complex simulations feasible, including full-lap scenarios with multiple overtaking opportunities.

In summary, this project has demonstrated that overtaking in motorsport can be simulated using optimal control in a way that captures both the benefits and penalties of aerodynamic wake. Although simplified, the framework bridges a gap in the literature by extending wake-aware trajectory optimisation beyond straight-line slipstreaming into cornering scenarios. With further refinement, such approaches could provide valuable tools for motorsport engineers seeking to understand overtaking dynamics, design vehicles, and regulations that promote closer racing, and ultimately enhance performance on track.

References

1. Berntorp K (2017) Path planning and integrated collision avoidance for autonomous vehicles. In: Proceedings of the 2017 American control conference (ACC). USA, pp 4023–4028. https://doi.org/10.23919/ACC.2017.7963560
2. Brayshaw D (2004) The use of numerical optimisation to determine on-limit handling behaviour or race cars. Cranfield University
3. Dal Bianco N et al (2017) Minimum time optimal control simulation of a GP2 race car
4. De Buck P et al (2023) Minimum lap time trajectory optimization of performance vehicles with four-wheel drive and active aerodynamic control. Veh Syst Dyn 61(8):2103–2119. https://doi.org/10.1080/00423114.2022.2101930
5. Dymos OpenMDAO. Optimal control—Dymos documentation. Available at: https://openmdao.github.io/dymos/getting_started/optimal_control.html. Accessed 23 Apr 2025
6. FIA (2024) 2024 FIA F3 technical regulations, Issue 2. Available at: https://api.fia.com/sites/default/files/2024_fia_f3_technical_regulations_-_issue2_-_2024.06.11.pdf. Accessed 10 June 2025
7. Guerrero A et al (2020) Aerodynamic study of the wake effects on a Formula 1 car. Energies 13(19):5183. https://doi.org/10.3390/en13195183
8. Jung C et al (2021) Game-theoretic model predictive control with data-driven identification of vehicle model for head-to-head autonomous racing. IEEE Trans Intell Veh 6(4):681–692. https://doi.org/10.1109/TIV.2021.3079187
9. Jowett R (2022) Overtaking fundamental: action packed 2022 F1 season? Medium. Canopy Simulations Blog. Available at: https://blog.canopysimulations.com/overtaking-fundamentals-action-packed-2022-f1-season-1b9ad121d07f. Accessed 28 Aug 2025
10. Liu X et al (2024) Optimal control of race car with aerodynamic slipstreaming effects. IEEE Trans Control Syst Technol 32(6). https://doi.org/10.1109/TCST.2024.3395722
11. Ravelli U (2019) Aerodynamics of a 2017 Formula 1 car: numerical analysis of a baseline vehicle and design improvements in freestream and wake flows
12. Tremlett AJ et al (2015) Optimal control of motorsport differentials. Veh Syst Dyn 53(12):1772–1794. https://doi.org/10.1080/00423114.2015.1093150
13. Van Koutrik S (2015) Optimal control for race car minimum time maneuvering. Delft University

Development of a Time-Efficient Driver Model Based on Model Predictive Control Approach

Berk Kisa and Aydin Azizi

Abstract Automotive stability at the limits of handling requires controllers that keep yaw rate and sideslip within a safe envelope. A time-efficient driver model was built in MATLAB/Simulink using a bicycle model with a nonlinear Pacejka tyre formulation. Phase-plane analysis (yaw rate vs. front tyre slip angle) was used to define the stability boundaries and to evaluate controller behaviour. The vehicle model was first validated against a high-fidelity ADAMS™ simulation to confirm model fidelity; this validation preceded controller design and did not assess controller performance. The primary controller is a Model Predictive Control (MPC) scheme that anticipates system response and modulates longitudinal acceleration to prevent envelope violation, with steering treated as a measured disturbance. For comparison, PID and fuzzy logic controllers were also implemented: PID offers simplicity and fast response, whereas fuzzy logic provides rule-based robustness to nonlinearities but requires tuning. Simulation results show that all three approaches can constrain yaw and slip under varying conditions, while MPC delivers the most consistent envelope enforcement, albeit with higher computational cost, providing faster and more precise intervention suitable for real-time stability control.

Keywords Stability · Time-efficient · Phase-plane · Vehicle dynamics · Control · MPC · PID · FLC

Abbreviations

ABS	Anti-lock braking system
ADAMS	Automated dynamic analysis of mechanical systems
CoG	Centre of gravity
DoF	Degree of freedom
ESC	Electronic stability control

B. Kisa (✉) · A. Azizi
School of Engineering, Computing and Mathematics, Oxford Brookes University, Oxford, UK
e-mail: 19331410@brookes.ac.uk

A. Azizi (ed.), *Intelligent Control and Autonomous Systems*, Emerging Trends in Mechatronics, https://doi.org/10.1007/978-981-95-6402-6_2

FLC	Fuzzy logic controller
IMU	Inertial measurement unit
MPC	Model predictive control
PID	Proportional–integral–derivative (controller)

List of Symbols

m	Vehicle mass
m_s	Sprung mass
m_u	Unsprung mass
I_z	Yaw moment of inertia
a	Distance from COG to front axle
b	Distance from COG to rear axle
C_f	Front tyre cornering stiffness
C_r	Rear tyre cornering stiffness
V_x	Forward/longitudinal speed
r	Yaw rate
β	Vehicle sideslip angle
α_f	Front tyre slip angle
α_r	Rear tyre slip angle
δ	Steering angle
Y_β, Y_r, Y_δ	Stability derivatives (lateral force due to β, r, δ)
N_β, N_r, N_δ	Stability derivatives (yaw moment due to β, r, δ)
F_x	Longitudinal tyre force
F_y	Lateral tyre force
F_z	Vertical load (tyre normal force)
M_z	Aligning moment
B	Stiffness factor (Pacejka)
C	Shape factor (Pacejka)
D	Peak factor (Pacejka)
E	Curvature factor (Pacejka)
S_H	Horizontal shift (Pacejka)
S_V	Vertical shift (Pacejka)
μ	Road friction coefficient
g	Gravitational acceleration
Δ_t	Sampling time interval
a_x	Longitudinal acceleration (control input)
s_r	Slack variable for yaw rate constraint
s_α	Slack variable for slip angle constraint
N_p	Prediction horizon
N_c	Control horizon
T_s	Sampling time (controller)
K_P, K_I, K_D	PID controller gains

1 Introduction

Vehicle dynamics control has become increasingly important with the growing demand for high-performance and autonomous vehicles [2]. Maintaining stability during extreme manoeuvres is critical for safety yet remains challenging because not all key states are directly measurable. For instance, yaw rate can be measured with onboard gyroscopes and inertial measurement units (IMUs), but the vehicle's front tyre slip angle is difficult to measure directly and is typically estimated using sensor fusion of yaw rate, lateral acceleration, wheel speed, and steering angle sensors [15, 16, 21]. Integration with active front steering or steer-by-wire systems can further enhance slip angle estimation accuracy, as steering rack positions can be measured with high precision [6].

In high-speed manoeuvres or low-adhesion conditions (e.g. snow, ice, gravel), conventional stability systems such as ABS and ESC may be inadequate. The nonlinear tyre–road friction behaviour under such conditions can induce excessive yaw rate and large slip angles beyond the controllable range, potentially leading to loss of control [4, 20]. These limitations motivate the development of more advanced, model-based control strategies to ensure vehicle stability at the limits of handling.

Researchers have proposed a range of approaches to address these challenges. One prominent method is Model Predictive Control (MPC), which predicts future system behaviour over a finite horizon and optimises the control input while respecting state and actuator constraints [9, 18]. In this work, the MPC is designed to modulate longitudinal acceleration (throttle/brake) as the sole manipulated variable, while the driver's steering input is treated as a measured disturbance. The system uses a discrete-time bicycle model with a sampling interval of 10 ms (100 Hz), a prediction horizon of 50 steps (0.5 s look-ahead), and a control horizon of 8 steps (0.08 s). This configuration allows the controller to anticipate near-future behaviour while maintaining real-time feasibility [8]. The controller's objective is not trajectory tracking, but rather enforcing stability envelopes in yaw rate and slip angle, derived through phase-plane analysis.

A known drawback of MPC is its high computational demand, which can limit real-time implementation [7]. While strategies exist to simplify or adapt reference trajectories to reduce this burden [10], MPC generally requires more processing power than classical controllers. To contextualise this trade-off, this dissertation also implements and compares PID and fuzzy logic controllers. PID offers simplicity and fast response [11], whereas fuzzy logic provides robustness to nonlinearities via intuitive rule-based reasoning, though at the cost of more complex tuning and scalability issues [12, 19].

In summary, this research develops a computationally efficient MPC-based driver model that maintains yaw rate and front tyre slip angle within a safe envelope during challenging manoeuvres. The vehicle dynamics model is validated against a high-fidelity ADAMS™ simulation, and comparative evaluation with PID and fuzzy logic controllers is conducted. The key contributions of this work are:

1. Development and validation of a planar bicycle model with nonlinear tyre dynamics.
2. Construction of yaw rate–slip angle phase-plane stability envelopes.
3. Design of an MPC controller that enforces these envelopes by adjusting longitudinal acceleration.
4. Comparative evaluation with PID and fuzzy logic controllers to highlight trade-offs in robustness and responsiveness.

2 Methodology

The methodology followed in this dissertation is summarised in the flowchart below. The study begins with a literature review and background research, providing the theoretical foundation for vehicle dynamics modelling and stability control strategies. Based on this, a bicycle model is developed in MATLAB/Simulink and validated against high-fidelity ADAMS simulations to ensure consistency and accuracy of the simplified model.

Once the model validation is complete, phase-plane analysis (yaw rate vs. front slip angle) is performed to define the stability boundaries. These boundaries serve as the reference framework for designing controllers. The next stage involves the design of three controllers: PID, fuzzy logic, and Model Predictive Control (MPC). Each controller is developed and tuned to maintain the vehicle states within the safe handling envelope.

Finally, the designed controllers undergo optimisation and validation through comparative simulations. Their effectiveness is evaluated by analysing vehicle responses under challenging manoeuvres, allowing a comprehensive comparison of robustness, responsiveness, and computational efficiency (Fig. 1).

2.1 Vehicle Handling Model

The vehicle's dynamic cornering behaviour was represented using a planar bicycle model. This approach focuses on lateral motion and yaw rate, while neglecting pitch and roll effects. The vehicle is idealised as a single-track model. The geometric representation of the bicycle model is shown in Fig. 2.

To simulate and analyse cornering dynamics efficiently, a simplified planar bicycle model was developed in both ADAMS and MATLAB/Simulink. While the ADAMS model provides high-fidelity multibody simulation with detailed kinematics, the Simulink model is built for control development and phase-plane analysis. The two environments complement each other in validating assumptions and developing time-efficient controllers.

To simulate and analyse cornering dynamics efficiently, the planar bicycle formulation was implemented in both ADAMS and MATLAB/Simulink. In ADAMS, the

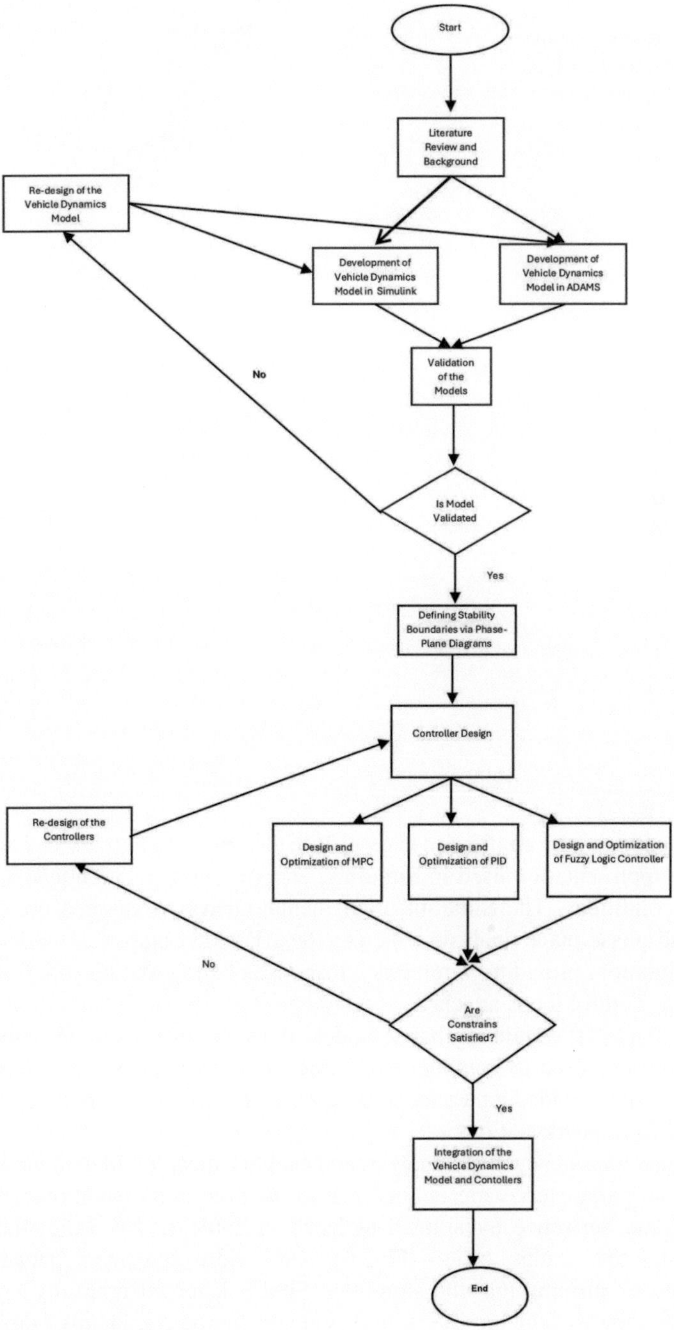

Fig. 1 Research methodology flowchart

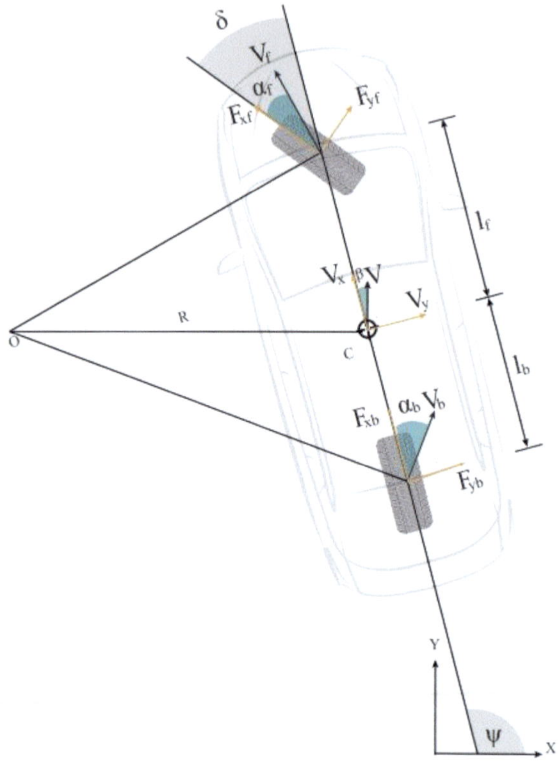

Fig. 2 Geometric representation of a dynamic bicycle model for a car-type vehicle [17]

bicycle equations were configured to reproduce the same single-track kinematics and tyre force representation used in Simulink, ensuring numerical equivalence under identical conditions. The Simulink implementation was developed for controller design and phase-plane analysis, whereas the ADAMS setup was used solely for model validation, providing a reference response of the same bicycle formulation without any control loops attached.

In addition to the simulation-based models, the derivative equations of the bicycle model were also used to compute analytical yaw rate responses for step-steering inputs [3]. This provided a theoretical benchmark to cross-check both the Simulink and ADAMS implementations.

To ensure consistency in the analysis and enable a meaningful comparison, identical vehicle parameters were applied across all three representations (Simulink, ADAMS, and derivative equations), as listed in Table 1. The validation results, presented in the results section (Fig. 6), show close agreement between these approaches, confirming that the simplified Simulink model faithfully reproduces the yaw dynamics of the bicycle formulation and can be confidently employed for phase-plane analysis and controller design.

Table 1 Parameters for the vehicle model

Description	Symbol	Value	Unit	References
Vehicle mass	m	1416	kg	[5]
Sprung mass	m_s	1274	kg	[5]
Unsprung mass	m_u	142	kg	[5]
Yaw moment of inertia	I_Z	1523	Kg m^2	[5]
Distance from front tyre to COG	l_f	1016	mm	[5]
Distance from rear tyre to COG	l_r	1562	mm	[5]
Front tyre cornering stiffness	C_f	66,900	N/rad	Zhang et al. [22]
Rear tyre cornering stiffness	C_r	62,700	N/rad	Zhang et al. [23]

2.1.1 Fundamentals of Planar Vehicle Dynamics

The planar vehicle model is formulated based on Newton's second law and the rotational dynamics about the vertical (yaw) axis. This formulation assumes constant forward speed, motion on a flat surface, and neglects pitch and roll effects.

By applying Newton's second law in the lateral direction, the governing equation for lateral motion is:

$$F_y = m(\dot{v}_y + v_x * r). \tag{1}$$

Here, the yaw rate r couples the lateral velocity with the rotational motion, showing how rotation about the vertical axis influences the lateral acceleration.

For the yaw dynamics, the moment balance about the vertical axis is expressed as:

$$\sum M_z = (I_Z + \dot{r}) \tag{2}$$

and the yaw rate derivative is obtained as:

$$\dot{r} = \frac{l_f F_{yf} - l_r F_{yr}}{I_Z}. \tag{3}$$

Yaw rate, defined as the angular velocity around the vertical axis, is a key stability indicator since it governs how quickly the vehicle rotates during cornering. Excessive yaw rate may indicate oversteer, while insufficient yaw response may indicate understeer [16]. Thus, Eqs. (1)–(3) highlight how yaw rate directly couples with lateral forces and moments to determine cornering stability.

2.1.2 Kinematic Equations

Slip angle, a key term in lateral dynamics, is the angle between wheel heading and motion direction [16]. These angles are essential inputs to the tyre force model, as lateral tyre forces are highly sensitive to slip angles under both linear and nonlinear modelling approaches. The slip angle of the vehicle body and the slip angle at front and rear wheels are expressed using following equations:

$$\alpha_{\mathrm{f}} = \delta - \left(\frac{v_y + l_{\mathrm{f}} * r}{v_x}\right) \tag{4}$$

$$\alpha_{\mathrm{r}} = -\left(\frac{v_y + l_{\mathrm{r}} * r}{v_x}\right) \tag{5}$$

$$\beta = \tan^{-1}\left(\frac{v_y}{v_x}\right). \tag{6}$$

2.1.3 Analytical Derivative Equations of the Bicycle Model

In addition to the simulation models, the bicycle model dynamics were also expressed using derivative (state-space) equations obtained from standard formulations [3]. The equations describe the yaw rate r and sideslip angle β dynamics as coupled first-order differential equations:

$$mV\dot{\beta} + mr = Y_{\beta}\beta + Y_{\mathrm{r}}r + Y_{\delta}\delta \tag{7}$$

$$I_Z\dot{r} = N_{\beta}\beta + N_{\mathrm{r}}r + N_{\delta}\delta \tag{8}$$

where m is the vehicle mass, I_z is the yaw inertia, V is forward speed, a and b are the distances from the centre of gravity to the front and rear axles, and C_{f}, C_{r} are front and rear cornering stiffnesses. The stability derivatives Y_{β}, Y_{r}, Y_{δ}, N_{β}, N_{r}, N_{δ} were computed as functions of these physical parameters.

These equations were selected because they provide an analytical description of yaw dynamics under small-slip assumptions, making them suitable as a theoretical benchmark. Specifically, they allow closed-form step-steer responses to be derived, which can then be compared directly with numerical simulations from Simulink and ADAMS. By including both the force/moment derivatives and the resulting yaw rate ODE, the method captures the essential underdamped and critically damped regimes observed in vehicle responses. This ensures that the simplified bicycle model implemented in Simulink is validated not only against numerical simulations but also against an analytical reference.

2.1.4 Tyre Force Model

Linear Tyre Model

In the case of small slip values, the linear tyre model assumes that the lateral tyre force is directly proportional to the slip angle. This linear formulation is valid under the assumption of small slip conditions and neglects nonlinear effects occurring at larger slip values [14]. The lateral forces can be linearly approximated as:

$$F_{\mathrm{yf}} = (C_F * \alpha_{\mathrm{f}}) \tag{9}$$

$$F_{yr} = (C_r * \alpha_r). \tag{10}$$

Here, C_F and C_r are the cornering stiffnesses of the front and rear tyres, respectively.

Nonlinear Tyre Model

The Pacejka tyre model, commonly referred to as the Magic Formula, is a semi-empirical tyre model specifically designed for use in vehicle simulation environments. Unlike purely physical models, the Magic Formula relies on measured tyre data while incorporating structural elements inspired by physical principles. It provides a mathematical representation of tyre behaviour under steady-state conditions, including lateral force, longitudinal force, and aligning moment responses [14].

$$F_y = D \cdot \sin(C \cdot [B \cdot (\alpha + Sh) - E \cdot (B \cdot (\alpha + Sh) - (B \cdot (\alpha + Sh)))]) + Sv \tag{11}$$

where

- y: output force (either longitudinal F_x, lateral F_y, or aligning moment M_z)
- x: input slip variable (longitudinal slip ratio, slip angle, or a function of them)
- B: stiffness factor
- C: shape factor
- D: peak factor (amplitude, maximum force)
- E: curvature factor
- S_{H}, S_{V}: horizontal and vertical shift parameters.

The Magic Formula can be adapted for different road surfaces by tuning the B, C, D, and E parameters. Table 2 shows example coefficients for four common surface types:

Table 2 Magic formula coefficients for four typical road conditions [13]

Road surface	B	C	D	E
Dry asphalt road surface	10	1.9	1	0.97
Wet asphalt road surface	12	2.3	0.82	1
Snow	5	2	0.3	1
Ice	4	2	0.1	1

2.2 *Phase-Plane Analysis*

Phase-plane analysis provides a powerful graphical framework to assess vehicle stability by plotting yaw rate (r) against front tyre slip angle (α_f) [22]. Each point on the plane represents a unique dynamic state, and system evolution is visualised through state trajectories [4]. Applied to a nonlinear 2-DOF bicycle model with Pacejka "Magic Formula" tyres, this method clearly distinguishes between stable and unstable behaviours.

The key features of the phase plane are:

- Equilibrium points: where trajectories converge, indicating steady-state motion.
- Unstable trajectories: which diverge, signalling oversteer, spin-out, or loss of control.
- Saddle points: acting as separatrices between stable and unstable regions.

From these features, a safe handling envelope is defined. The envelope boundaries are determined by tyre and road friction characteristics:

- Front slip angle limit: taken at the slip angle corresponding to maximum lateral force. For typical conditions, this is $\approx 10°$ (dry), $\approx 6°$ (wet), $\approx 4°$ (snow), and $\approx 2°$ (ice). Beyond this limit, tyres saturate and stability rapidly deteriorates. These tyre force–slip angle curves, from which the limits are derived, are illustrated in the Results section (see Fig. 8).
- Yaw rate limit: defined by the friction-limited lateral acceleration criterion,

$$r_{\max} \approx \frac{\mu g}{v_x} \tag{12}$$

where μ is the road friction coefficient, g gravitational acceleration, and v_x longitudinal speed.

Together, these constraints form a rectangular envelope in the (α_f, r) plane. Inside this region, tyre forces remain effective, and trajectories converge to equilibrium, while outside, tyre saturation leads to instability. Figure 3 illustrates the representative phase-plane envelope, highlighting equilibrium points, saddle regions, and the role of road surface friction in shaping the admissible range.

Since this dissertation focuses on passenger vehicle control, only normal driving conditions are considered. Excessive yaw rate deviations beyond these prescribed

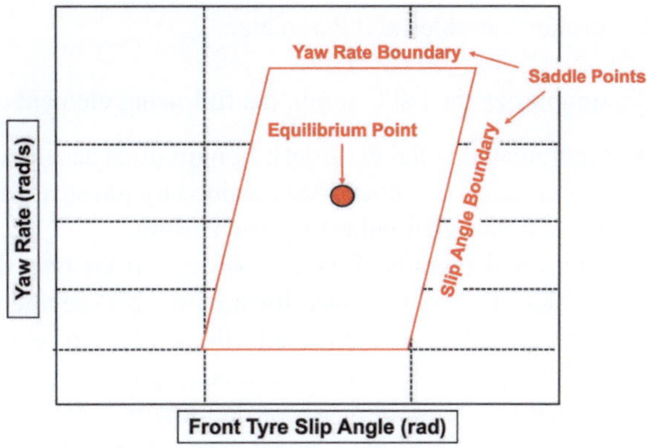

Fig. 3 Representative graph of handling envelope with equilibrium points saddle points

limits are not permitted, ensuring that the proposed controllers (PID, fuzzy logic, MPC) act to confine the system trajectories within the stable envelope.

2.3 *Vehicle Stability Control*

2.3.1 Model Predictive Control

An MPC-based stability controller is designed to maintain vehicle stability during high-speed cornering by modulating the longitudinal acceleration a_x (throttle/brake) as the sole control input. The vehicle is modelled using a discrete-time bicycle model (as described earlier in this document) with a sampling interval of 10 ms (100 Hz). This model captures the lateral–yaw dynamics necessary for stability control, including the vehicle yaw rate (r) and front tyre slip angle (α_f) as key states/outputs.

The prediction horizon is chosen as 50 steps (0.5 s look-ahead), and the control horizon is set to 8 steps (0.08 s), meaning the optimiser computes a sequence of 8 future a_x adjustments at each time step. These horizon lengths are selected to allow the controller to anticipate the vehicle's near-future behaviour (about half a second, comparable to typical yaw response [8], while limiting computational complexity. Past studies have used similar horizon lengths on the order of 0.3–0.5 s for yaw stability [8]. The shorter control horizon (8) constrains the optimiser to only adjust a_x in the immediate future, which is a common MPC strategy to reduce the decision variables and ensure real-time feasibility. The MPC optimisation runs at 100 Hz within Simulink (using MATLAB's MPC toolbox), solving a quadratic program at each step to determine the optimal longitudinal acceleration command.

Controller Variables and Parameters

To summarize the MPC setup, the following elements are defined:

- Manipulated variable (input): Longitudinal acceleration a_x (vehicle throttle/brake command). This input is constrained by physical acceleration limits representing the vehicle's engine/braking capabilities.
- Controlled outputs: Yaw rate (r) and front tyre slip angle (α_f). These are the variables to be constrained for stability. No specific trajectory tracking is imposed on (r) or (α_f); instead, the controller's role is to keep them within safe limits rather than force them to follow reference signals.
- State/update variables: The MPC internal state uses the bicycle model (lateral velocity, yaw rate, etc.), updated based on the vehicle's measured steering input and the chosen a_x. The driver's steering angle can be treated as a measured disturbance or known preview input to the model.
- Prediction horizon: $Np = 50$ steps (0.5 s).
- Control horizon: $Nc = 8$ steps (0.08 s).
- Sampling time: $\Delta t = 0.01$ s (10 ms).

Control Objectives and Constraints

The cost function was formulated to minimise control effort while strongly penalising violations of yaw rate and slip angle constraints. A moderate weight was assigned to changes in longitudinal acceleration (Δa_x) in order to ensure smooth actuation and avoid oscillatory behaviour, thereby preventing unnecessary throttle/brake switching while still allowing sufficiently rapid intervention when stability is at risk.

The quadratic cost function can be expressed in simplified form as:

$$J = \sum_{i=0}^{Np-1} \left(W_{\Delta a}(\Delta a_{x.}(k+i))^2 + W_{\mathrm{slack},r} s_r(k+i)^2 + W_{\mathrm{slack},\alpha} s_\alpha(k+i)^2 \right) \qquad (13)$$

where s_{r} and s_α denote the slack variables for yaw rate and slip angle constraints.

Thresholds for yaw rate and slip angle were derived from phase-plane analysis and validated in simulation scenarios, including double lane-change and high-speed cornering. The chosen horizons ($N_{\mathrm{p}} = 50$, $N_{\mathrm{c}} = 8$) balanced prediction capability with computational efficiency, allowing real-time operation at 100 Hz. Warm-started quadratic programming ensured that optimisation was solved quickly at each step, supporting closed-loop execution without delays.

Simulation results confirmed that the MPC intervention strategy successfully reduced acceleration when limits were approached, allowing the vehicle to carry maximum feasible speed while remaining within the predefined stability envelope.

2.3.2 PID Control

A Proportional–Integral–Derivative (PID) controller was implemented as a baseline strategy due to its simplicity, robustness, and ease of implementation in real-time systems [11]. The control objective is to maintain the vehicle within a predefined safe handling envelope, bounded by yaw rate and front tyre slip angle limits derived from phase-plane analysis.

The PID controller continuously adjusts the longitudinal acceleration command a_x (throttle/brake) to keep both yaw rate (r) and front tyre slip angle (α_f) within their allowable bounds. Steering angle $\delta(t)$ is treated as a driver input and is not actuated by the controller. A representative PID–plant structure to be used in the dissertation is shown in Fig. 4.

The PID compensator is characterised by the transfer function

$$G_\mathrm{C} = \left(K_\mathrm{P} + \frac{K_i}{s} + K_\mathrm{d}s\right). \tag{14}$$

PID Controller Tuning

Controller tuning was carried out in MATLAB/Simulink using the nonlinear bicycle model with Pacejka tyres. Step-steer manoeuvres at a forward speed of 10 m/s were used to identify suitable gains. The yaw rate loop was tuned first by adjusting proportional and derivative gains to achieve a fast but stable transient response, and a small integral term was added to remove steady-state bias. The tuned values used in the study are:

$$K_\mathrm{P,r} = 7, \quad K_\mathrm{D,r} = 1.2, \quad K_\mathrm{I,r} = 0.06\ \mathrm{s}^{-1}$$
$$N = 15, \quad \mathrm{T_s} = 0.05\ \mathrm{s}$$

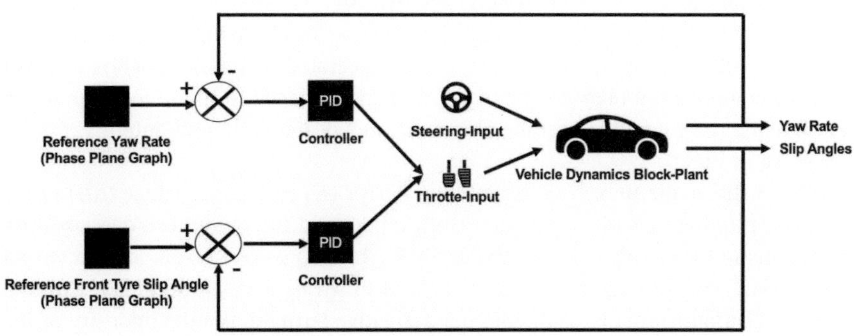

Fig. 4 Representative image of PID controller with plant

A secondary slip angle safeguard loop was activated only when $|\alpha_f|$ approached its nonlinear limit, defined by the threshold.

$$th_{\mathrm{ON}} = \alpha_{\mathrm{f}} - \Delta\alpha \tag{15}$$

This tuning ensured that the PID controller could stabilise the vehicle response across different manoeuvres by reducing speed whenever yaw rate or slip angle limits were approached, preventing envelope violations. While the PID design provided a clear and low-complexity baseline, comparative simulations showed that MPC achieved faster and more precise envelope enforcement (see Figs. 13, 14 and 15: Comparative Time Responses of PID/FLC/MPC).

2.3.3 Fuzzy Logic Controller

In addition to MPC and PID, a fuzzy logic controller (FLC) was implemented as a comparative strategy. Fuzzy logic is attractive for vehicle dynamics applications because it can effectively handle nonlinearities and modelling uncertainties, while allowing intuitive rule-based design [19]. The main advantages of FLC include robustness to parameter variations, relative simplicity of implementation in Simulink, and its ability to embed heuristic or driver-like decision-making [19]. However, there are notable disadvantages: the rule base can become large and complex, making analysis difficult, and tuning often requires time-consuming trial and error, with limited formal guarantees of stability compared to optimisation-based methods such as MPC [19]. Simulation studies confirm that while fuzzy control improves stability in many cases, its performance may degrade outside the trained operating range, whereas MPC maintains stronger consistency under varying conditions [12].

The fuzzy controller was designed and implemented in MATLAB/Simulink using the Fuzzy Logic Toolbox, adopting the Mamdani inference system. The controller takes as input variables the yaw rate error (rad/s) and the front tyre slip angle (rad) error, both normalised in the range – 1 to 1. The output variable is the corrective longitudinal acceleration command. Each input and the output were represented by five triangular membership functions (Negative Big, Negative Small, Zero, Positive Small, Positive Big), creating a total of 25 rules (5×5). These rules, expressed in the intuitive "if–then" format, were automatically generated within the toolbox to cover all possible input combinations. The overall fuzzy inference structure is illustrated in Fig. 5, while the triangular membership functions of the input and output variables are shown in Fig. 6.

Table 3 shows the fuzzy rule base linking error (E) and its derivative (dE) to the output longitudinal acceleration (a_x). Both inputs and the output are described by five linguistic terms (NB, NS, Z, PS, and PB). Large positive errors lead to strong braking (NB), while large negative errors lead to strong acceleration (PB). Around zero error, outputs remain small, and the derivative term adjusts the intensity of the response.

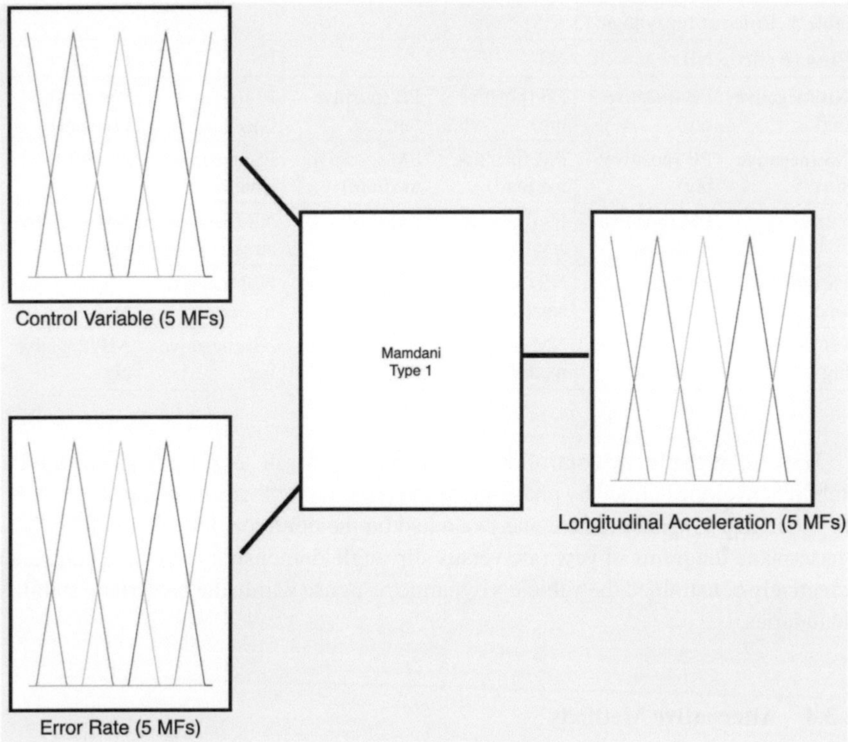

Fig. 5 Structure of the Mamdani type fuzzy inference system with control variable and error rate and one output (longitudinal acceleration)

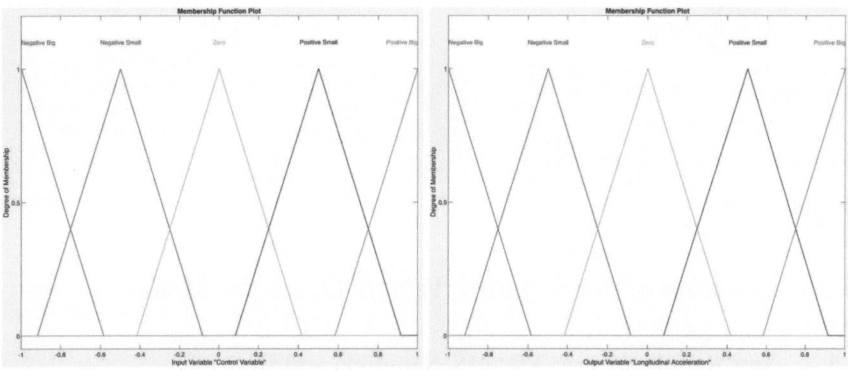

Fig. 6 Membership functions of the input and output variables (defines as negative big, negative small, zero, positive small, positive big)

Table 3 Rules of fuzzy logic (5 × 5)

Error (E)\d$E$	NB	NS	Z	PS	PB
NB (negative big)	PB (positive big)	PB (positive big)	PB (positive big)	PB (positive big)	PM (positive medium)
NS (negative small)	PB (positive big)	PM (positive medium)	PM (positive medium)	PS (positive small)	Z (Zero)
Z (zero)	PM (positive medium)	PS (positive small)	Z (Zero)	NS (negative small)	NM (negative medium)
PS (positive small)	Z (Zero)	NS (negative small)	NM (negative medium)	NM (negative medium)	NB (negative big)
PB (positive big)	NM (negative medium)	NM (negative medium)	NB (negative big)	NB (negative big)	NB (negative big)

The role of the fuzzy controller is to maintain yaw rate and slip angle within the stability envelope defined by phase-plane analysis, rather than enforcing strict trajectory tracking. Its performance was evaluated on the nonlinear bicycle model, where phase-plane diagrams of yaw rate versus slip angle demonstrated that the controller effectively constrained the vehicle's dynamic response within the prescribed stability boundaries.

2.3.4 Alternative Methods

See Table 4.

2.3.5 Assumptions

See Table 5.

3 Results

3.1 Validation of the Simulink Vehicle Dynamics Model

3.1.1 Yaw Rate Response: ADAMS, Simulink, and Derivative Equations

To validate the simplified bicycle model, the yaw rate responses obtained from ADAMS multibody simulations and the MATLAB/Simulink implementation were compared with the analytical solution derived from the derivative equations presented by Balkwill [3]. The vehicle was subjected to a 10° steering step input (0.1745 rad) at

Table 4 Alternatives and the choice of approach [10, 13, 14, 16, 18, 21, 23]

Area	Current choice	Strong alternatives	Justification for the choice
Modelling tool	MATLAB/Simulink	CarSim, IPG CarMaker, Simcenter Amesim, Modelica/CasADi	Validated full-vehicle models, HIL/SIL, faster scenario setup
Validation	ADAMS	CarSim/ASM; physical test	Different fidelity/cost balance, hardware integration
Vehicle model	Bicycle (2-DoF)	Double-track (4-wheel), 7-DoF (roll/pitch + wheel DOFs), multibody	Good balance between simplicity and accuracy for initial control design. It reduces computational load and is widely used for cornering dynamics
Tyre model	Pacejka MF	Brush, Dugoff; Linear (early stage)	Physically grounded; highly accurate to capture nonlinear tyre behaviour at high slip angles. Linear models are only accurate for small slip conditions, and brush models require more physical parameters and are computationally heavier
Stability analysis	Phase plane (r–α_f)	β–r/β–$\dot{\beta}$ portraits; eigenvalue; Lyapunov	Visual boundaries versus analytic guarantees
Control	MPC (MV = a_x) + PID + FLC	LQR, SMC, $H\infty$/μ-synthesis; Hybrid (e.g. LQR + SMC)	Constraints (MPC), robustness (SMC/$H\infty$), low cost (LQR)
Driver model	"Time-efficient" automatic throttle	Preview (MacAdam), PID + delay, RL-based driver	More realistic human delay, minimum-time racing strategy

a forward speed of 10 m/s, and Fig. 7 illustrates the comparison of yaw rate responses predicted by the three approaches.

The results demonstrate a close agreement between the analytical derivative formulation and the high-fidelity ADAMS model, both of which align well with the Simulink response. All three methods indicate that the vehicle exhibits a critically damped yaw rate response (no oscillations and smooth convergence to steady state). This agreement confirms that the Simulink model accurately represents the vehicle's yaw dynamics and can therefore be confidently employed for further analysis and control development.

Table 5 List of assumptions and their impact on the project outcome [16, 18]

Assumption	Impact of the assumption on outcomes
Planar (2D) motion only	Neglects roll dynamics, pitch effects, vertical load transfer, and rollover risks. Assumes a flat road and low lateral load transfer
Bicycle model approximation	Reduces model complexity but ignores left–right asymmetry and individual wheel dynamics. May miss wheel lift or inside/outside wheel slip differences
Rigid body chassis and tyres	Ignores suspension compliance and chassis flex, which influence ride quality and transient responses
Linear tyre model (initially used for validation)	Accurate only for small slip angles and low-speed manoeuvres; fails to model tyre saturation and peak grip accurately
Constant speed during manoeuvres	No throttle or braking dynamics are included. Vehicle's natural response to loss of grip cannot be captured
Symmetric cornering behaviour	Ignores asymmetry in tyre forces due to load transfer during left versus right turns, affecting fidelity in emergency manoeuvres
Neglecting actuator dynamics (e.g. steering servo, throttle response delay)	Controller assumes instantaneous response. Real systems may exhibit lag, leading to reduced tracking performance and possible instability in aggressive manoeuvres
No driver variability or human-in-the-loop behaviour	Assumes perfect control input. Human drivers introduce noise, delays, and adaptive behaviour, which may alter system performance or cause instability near the limit
Fixed tyre parameters	Thermal and wear-dependent effect are neglected

3.1.2 Front Versus Rear Tyre Slip Angle Comparison

Under a 10° step-steering input (0.1745 rad) at a forward speed of 15 m/s, the front tyre slip angle shows an initial transient peak due to the instantaneous steer application and then decays smoothly to a steady value of $\approx$ 0.145 rad (as shown in Fig. 8). The rear tyre slip angle increases monotonically and settles at $\approx$ 0.100 rad. The steady-state relation $\alpha_f > \alpha_r$ confirms an understeering vehicle. Both responses are well-damped with no observable oscillations.

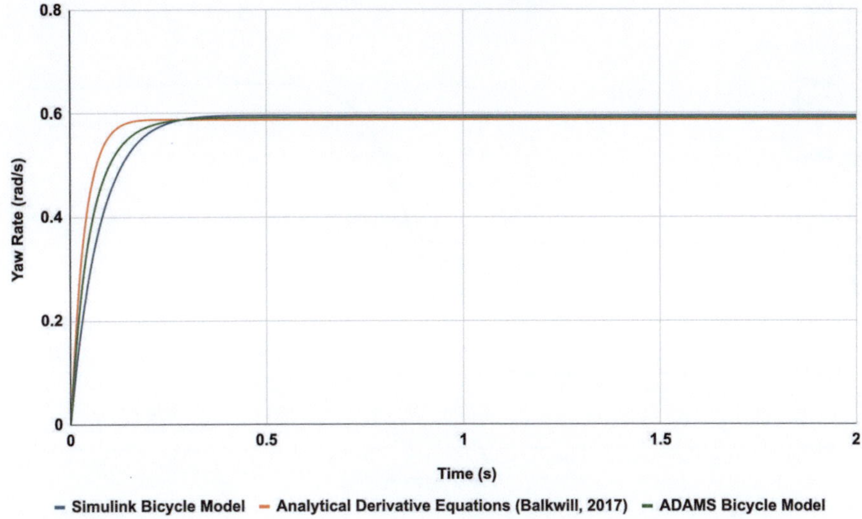

Fig. 7 Comparison of yaw rate responses from Simulink bicycle model, ADAMS multibody model, and analytical derivative equations [3] under 10° step steer input (0.1745 rad) at 10 m/s

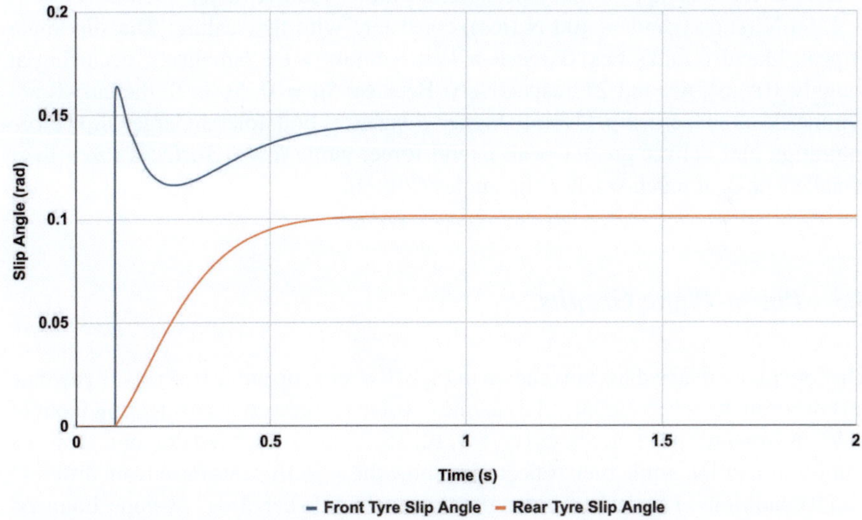

Fig. 8 Slip angle response of front and rear tyres to a 10° (0.1745 rad) step input at 10 m/s

3.2 *Nonlinear Tyre Model*

Under the adopted Magic Formula calibration (table B–C–D–E) and a vertical load of $F_z = 4000$ N, the maximum lateral force follows the expected order Dry > Wet

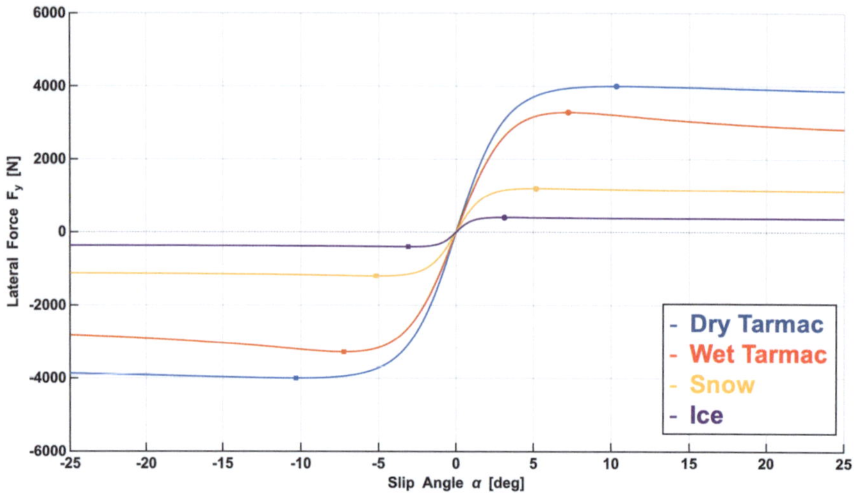

Fig. 9 Slip angle versus lateral force using the Pacejka model for different road conditions ($F_z =$ 4000 N, $S_H = 0$, $S_V = 0$)

> Snow > Ice. The approximate peak values are ~ 4000 N (dry), $\sim$ 3300 N (wet), $\sim$ 1200 N (snow), and $\sim$ 400 N (ice), consistent with the scaling. The slip angle at peak likewise ranks Dry (largest) > Wet > Snow > Ice (smallest), occurring at roughly 10°, 6°, 4°, and 2°, respectively. Because $S_H = 0$, $S_V = 0$, the curves are symmetric about zero slip. Overall, higher-μ surfaces both tolerate larger slip before saturation and deliver greater peak lateral force, while low-μ surfaces reach their (smaller) peak at much smaller slip angles (Fig. 9).

3.3 Phase-Plane Graphs

The four phase-plane diagrams shown in Fig. 10 (front slip angle (α_f) versus yaw rate (r)) illustrate how vehicle lateral dynamics evolve under a constant steering input (δ = 5°) as forward speed increases (v_x = 5, 10, 15, 20 m/s). Each red dot represents the equilibrium point, while blue trajectories show the system's response from different initial conditions. The dashed rectangles denote the safe handling envelope, bounded by slip angle and yaw rate limits.

At low speed (5 m/s), trajectories converge rapidly and directly to the equilibrium, indicating a strongly damped response and wide stability margins. As speed increases (10–15 m/s), trajectories become more elongated and approach equilibrium more gradually, reflecting higher inertia and reduced damping. At 20 m/s, convergence is noticeably slower and trajectories bend around the equilibrium due to front tyre saturation, showing that the system operates closer to its physical limits.

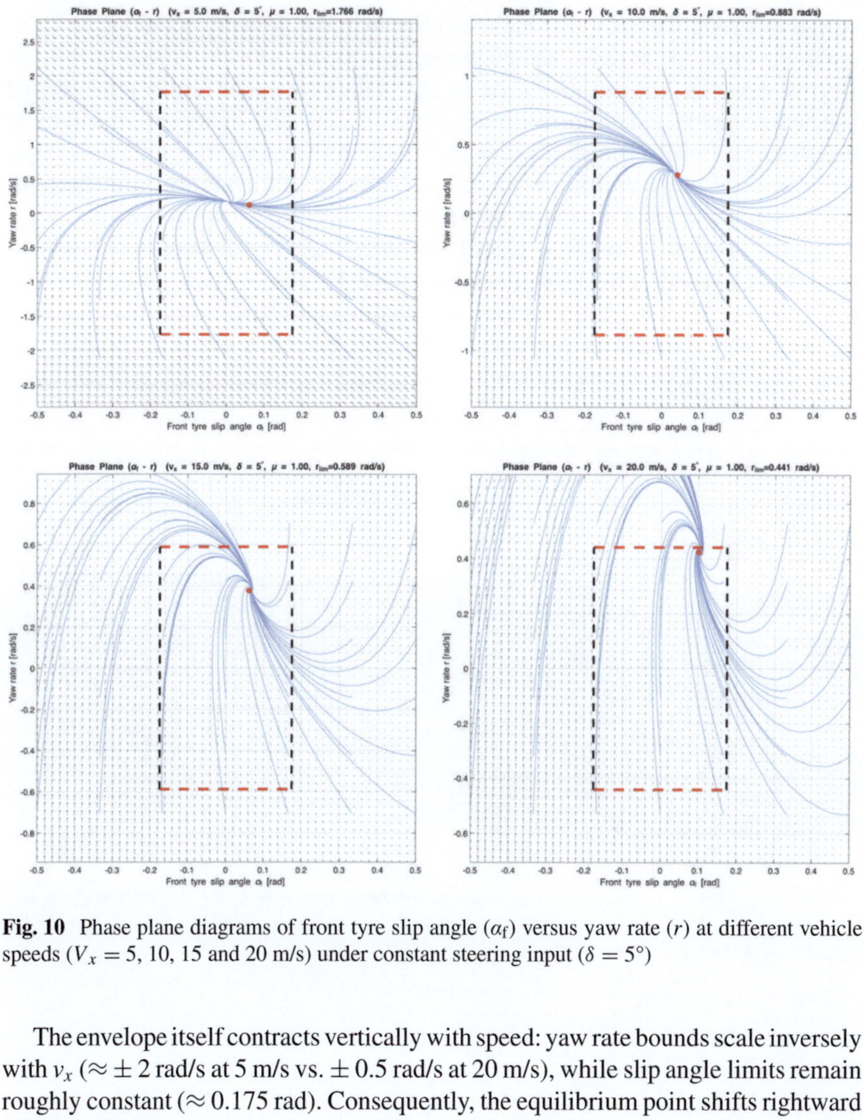

Fig. 10 Phase plane diagrams of front tyre slip angle (α_f) versus yaw rate (r) at different vehicle speeds ($V_x = 5$, 10, 15 and 20 m/s) under constant steering input ($\delta = 5°$)

The envelope itself contracts vertically with speed: yaw rate bounds scale inversely with v_x ($\approx \pm 2$ rad/s at 5 m/s vs. ± 0.5 rad/s at 20 m/s), while slip angle limits remain roughly constant (≈ 0.175 rad). Consequently, the equilibrium point shifts rightward (higher α_f) and downward (lower r) as v_x increases, moving closer to the envelope boundaries.

In summary, increasing speed reduces natural damping and shrinks the stable region, progressing from a comfortably stable regime at low speed to a near-limit regime at high speed. Exceeding the envelope at higher velocities would likely result in instability or require electronic stability intervention.

Figure 11 illustrates the phase-plane trajectories (front tyre slip angle, α_f, versus yaw rate, r) for different road conditions at a constant forward speed of $v_x = 10$ m/s and steering input of $\delta = 10°$. Under dry asphalt ($\alpha_{f,lim} = \pm 10°$), the handling envelope is widest, and all trajectories converge rapidly to the equilibrium point, indicating strong damping and high stability margins. In wet conditions ($\alpha_{\text{f,lim}} = \pm\, 6°$), the envelope shrinks, and convergence becomes slower, reflecting reduced effective damping. Under snow ($\alpha_{\text{f,lim}} = \pm\, 4°$), the envelope is further reduced, and trajectories approach equilibrium more gradually, showing increased sensitivity to perturbations. Finally, for ice ($\alpha_{\text{f,lim}} = \pm\, 2°$), the stable region is extremely narrow, convergence is sluggish, and even small deviations risk leaving the safe envelope, highlighting the critical instability of the system under very low friction.

Figure 12 compares the phase-plane trajectories under two distinct scenarios: low-speed dry road conditions ($V_x = 5$ m/s, $\delta = 5°$) and higher-speed icy conditions ($V_x = 15$ m/s, $\delta = 5°$). On the dry surface, the handling envelope is wide ($\alpha_{\text{f}} = \pm\, 10°$) and the system trajectories converge rapidly and directly towards the equilibrium point. This indicates strong damping, robust stability margins, and sufficient tyre

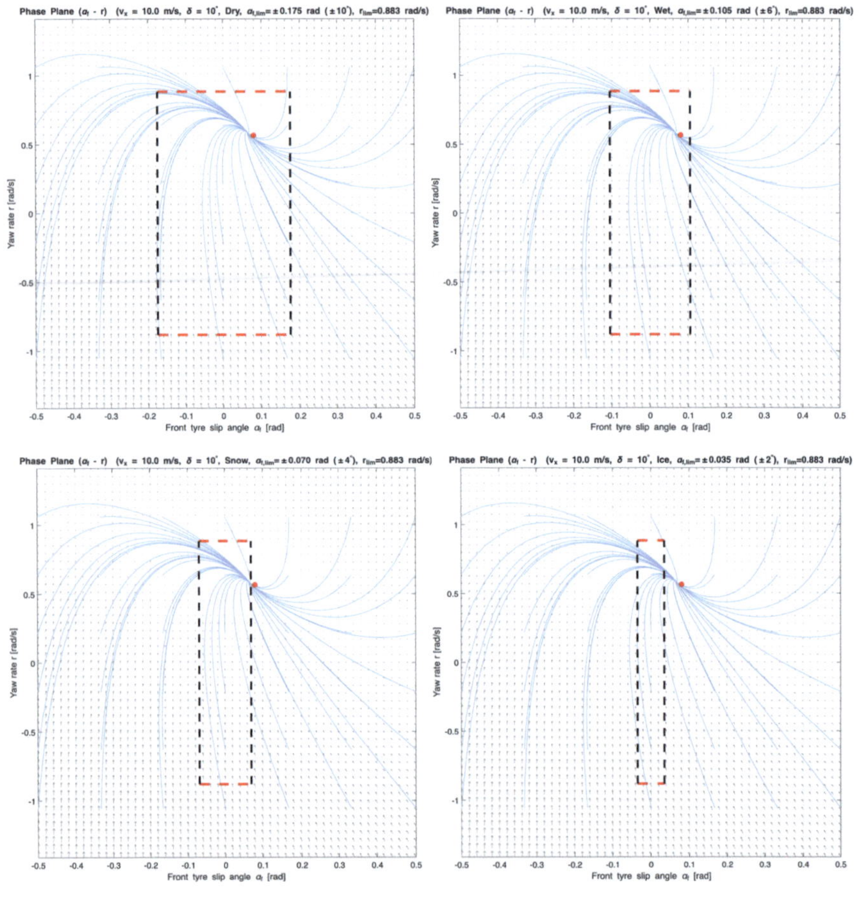

Fig. 11 Phase plane graphs of front tyre slip angle (α_{f}) versus yaw rate (r) at $V_x = 10$ m/s and $\delta = 10°$, illustrating handling envelopes under different road conditions: dry ($\pm\, 10°$), wet ($\pm\, 6°$), snow ($\pm\, 4°$), ice ($\pm\, 2°$)

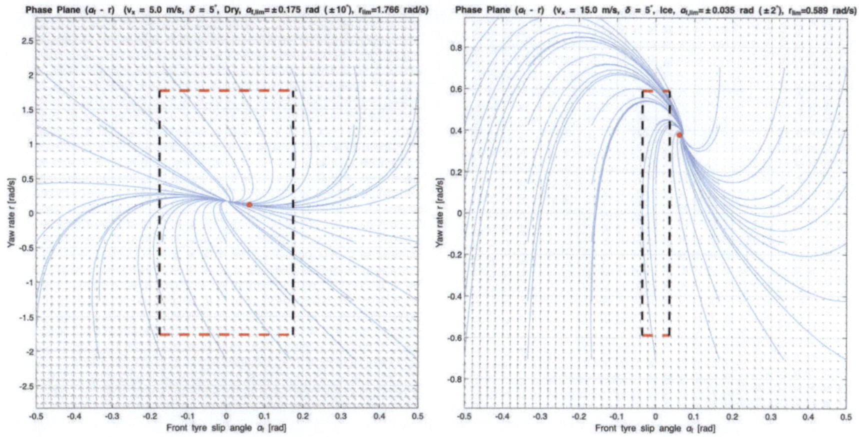

Fig. 12 Phase plane graphs of front tyre slip angle (α_f) versus yaw rate (r) under different conditions. (Left) dry road, $V_x = 5$ m/s, $\delta = 5°$ with slip angle limit $\alpha_f = \pm 10°$. (Right) ice road, $V_x =$ *15* m/s, $\delta = 5°$with slip angle limit $\alpha_f = \pm 2°$

force reserve to accommodate perturbations without departing from the safe region. In contrast, the ice condition exhibits an extremely narrow slip angle bound ($\alpha_f = \pm 2°$) and a lower yaw rate limit. The trajectories in this case converge more slowly and follow longer, skewed paths towards equilibrium, reflecting weaker damping and reduced stability. Even small deviations risk leaving the safe envelope, demonstrating how increased speed combined with low friction substantially degrades the vehicle's natural stability. The comparison highlights that while dry, low-speed driving allows stable and forgiving behaviour, icy, higher speed conditions impose very tight stability margins, requiring external interventions (e.g. ESC) or careful driver correction to prevent loss of control.

As vehicle speed increases or steering inputs become more aggressive, the size of the stable region in the phase plane typically decreases. In fact, at certain extreme combinations of speed and steering (high demand conditions), the stable equilibrium can disappear entirely, leaving only an unstable equilibrium (saddle point)—a scenario indicating the vehicle's critical handling limits. This phenomenon has been noted by [22], who observed that beyond a specific steering angle in their model, the previously stable focus equilibrium vanished and the vehicle could no longer sustain a steady-state turn without instability. Such analysis highlights how pushing the vehicle beyond its handling envelope (for example, very high speed cornering or sudden large steering inputs) can result in no available stable state, necessitating electronic stability interventions or driver correction.

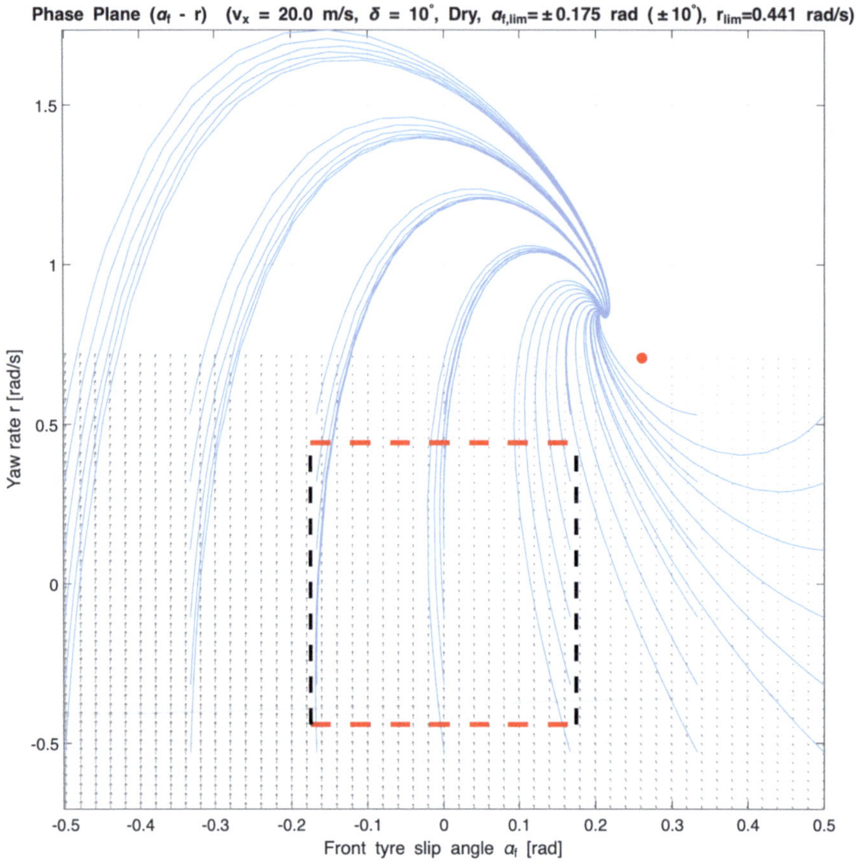

Fig. 13 Phase plane graphs of front tyre slip angle (α_f) versus yaw rate (r) for $V_x = 20$ m/s, steering angle $\delta = 10°$ (0.1745 rad), dry road condition. The red marker indicates the steady-state equilibrium, which lies outside the safe envelope defined by $\alpha_f < 0.175$ rad and $|r| \leq 0.441$ rad/s

3.4 Results with Controllers

The dynamic responses of the bicycle model under a fixed steering input of $\delta = 10°$ (0.1745 rad) and initial longitudinal speed of $V_x = 20$ m/s on a dry road surface were investigated with and without control. The phase-plane representation of the front tyre slip angle (α_f) versus yaw rate (r) is shown in Fig. 12. The equilibrium point of the open-loop system (red marker) lies outside the predefined stability envelope, which is bounded by $|\alpha_f| \leq 0.175$ rad ($\pm 10°$) and $|r| \leq 0.441$ rad/s. This result indicates that without intervention the vehicle cannot remain within the safe handling region under these operating conditions, as both the yaw rate and slip angle exceed their permissible limits (Fig. 13).

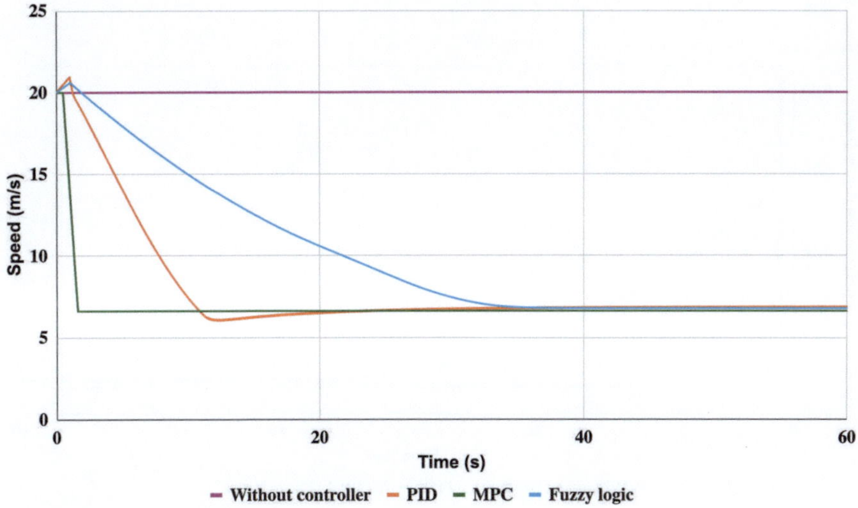

Fig. 14 Vehicle longitudinal speed at $V_x = 20$ m/s, steering angle $= \delta = 10°$ (0.1745 rad), under four cases: Uncontrolled, PID, MPC, fuzzy logic

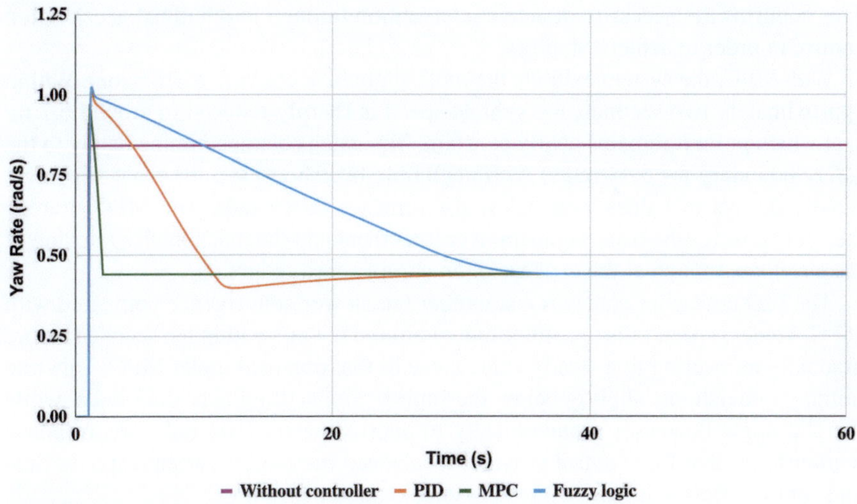

Fig. 15 Yaw rate response at $V_x = 20$ m/s, steering angle $\delta = 10°$ (0.1745 rad), for four different conditions: Uncontrolled, PID, MPC, fuzzy logic

The closed-loop results using three different controllers—PID, model predictive control (MPC), and fuzzy logic—are compared in Figs. 14, 15 and 16. In the uncontrolled case, the vehicle maintains its initial speed of 20 m/s, while both yaw rate and slip angle settle above their respective limits, consistent with the phase-plane analysis.

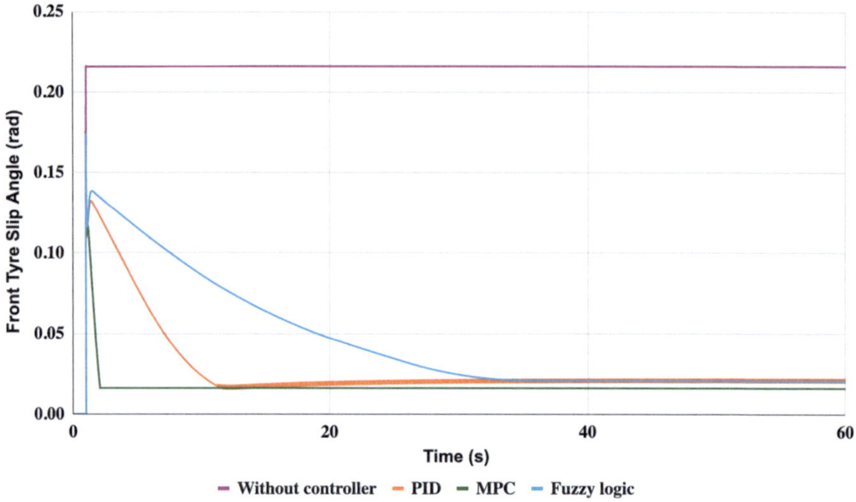

Fig. 16 Front tyre slip angle (α_f) response at $V_x = 20$ m/s, steering angle $= \delta = 10°$ (0.1745 rad) for four different conditions: uncontrolled, PID, MPC, fuzzy logic

This confirms the necessity of active intervention through longitudinal acceleration control in order to achieve stability.

With MPC, the system exhibits the most aggressive corrective behaviour. Within approximately two seconds, the vehicle speed is sharply reduced to around 6.7 m/s, at which point it remains nearly constant. Yaw rate is clamped immediately to the safety boundary ($\sim$ 0.44 rad/s) with negligible overshoot, and the front slip angle rapidly decays to values well below the limit ($\sim$ 0.015 rad). The MPC strategy therefore ensures the fastest constraint enforcement and the most precise tracking of the envelope, though at the expense of an abrupt speed reduction.

The PID controller achieves a smoother but slower convergence compared with MPC. Vehicle speed falls to a minimum of around 6.3 m/s within ten seconds before gradually recovering to a steady value close to that obtained under MPC. Yaw rate initially undershoots slightly below the limit before settling near 0.43 rad/s, while the slip angle decreases monotonically to approximately 0.02 rad. This response demonstrates that PID control provides a balanced trade-off between response time and control effort, with less aggressive deceleration than MPC.

The fuzzy logic controller, by contrast, produces the smoothest and most gradual transients. Speed decreases approximately exponentially from 20 m/s to about 6.7 m/s in 35 s. Both yaw rate and slip angle decline monotonically, converging to values just within the stability limits without overshoot. Although the fuzzy controller achieves constraint satisfaction later than PID and MPC, it minimises abrupt changes in control action, which may be advantageous for ride comfort and actuator longevity.

In summary, all three controllers succeed in bringing the system into the stable region defined by the phase-plane envelope, whereas the open-loop vehicle remains unstable. MPC ensures the fastest and tightest regulation but requires the largest sacrifice in speed. PID yields a more moderate response with similar steady-state outcomes, while fuzzy logic offers the smoothest but slowest corrective behaviour. Importantly, in all controlled scenarios the final steady-state vehicle speed converges to approximately 6.7 m/s, with yaw rate and slip angle constrained within the safe boundaries, thereby validating the use of the phase-plane limits as an effective reference for constraint-based vehicle stability control.

Additionally, in the scenarios evaluated in this dissertation, the yaw rate was generally observed to reach its boundary earlier than the front slip angle, consistent with the phase-plane limits shown. This indicates that, under the tested operating conditions (high-speed steering inputs on dry asphalt), yaw rate represents the dominant constraint. Nevertheless, it is important to note that in alternative conditions, such as low-μ surfaces or during sudden steering transients, the front slip angle may become the more restrictive factor, as it can saturate before the yaw rate develops due to vehicle inertia [1, 22]. Therefore, while the presented results highlight yaw rate dominance, comprehensive stability control design should account for both yaw rate and slip angle constraints.

4 Conclusion

In conclusion, the Model Predictive Control (MPC) strategy developed in this thesis has proven to be highly time-efficient in maintaining vehicle stability during extreme manoeuvres. Simulation results demonstrated that the MPC-based driver model brings the vehicle's state back into the safe stability envelope much faster than conventional controllers. For example, in a demanding high-speed steering scenario (initial speed 20 m/s with a 10° steering input), the MPC controller clamped the yaw rate to its safety limit and reduced the front tyre slip angle below its threshold within roughly 2 s. By contrast, the PID controller required about 10 s and the fuzzy logic controller about 35 s to achieve comparable stability. This swift intervention by MPC minimises the time the vehicle operates at the edge of its handling limits, underscoring the effectiveness of the MPC approach for real-time stability control in critical situations.

Beyond these performance findings, the thesis provided several key contributions to the field of vehicle dynamics and control. First, a nonlinear planar "bicycle" vehicle model (incorporating a Pacejka tyre model) was developed and rigorously validated against an ADAMS™ simulation, ensuring that the simplified model accurately captures the vehicle's yaw and lateral dynamics. Second, a phase-plane analysis framework was introduced, mapping yaw rate against front tyre slip angle to define a clear safe stability envelope for extreme manoeuvres. This framework established quantitative boundaries for stable operation and served as the reference for designing and evaluating the controllers. Third, using the validated model and the phase-plane envelope, an MPC-based stability controller was designed to modulate longitudinal acceleration (throttle/brake) and keep the vehicle within the defined envelope. A comprehensive comparative study was then conducted, evaluating this MPC strategy against a classical PID controller and a fuzzy logic controller under identical challenging manoeuvres. This comparative analysis highlighted important trade-offs: all three controllers ultimately succeeded in bringing the vehicle back inside the stability envelope, but the MPC provided the fastest and most precise regulation of yaw rate and slip angle, whereas the PID and fuzzy controllers achieved stability more gradually with smoother control actions. In particular, the MPC's aggressive corrections required a larger sacrifice in speed to regain stability, while the PID and FLC approaches produced gentler deceleration and thus a slower return to stable conditions. These findings validate the efficiency of the phase-plane envelope as a basis for constraint-based control and demonstrate the superior time-response of the MPC approach in enforcing vehicle stability.

Despite the promising results, several challenges and avenues for future work remain. One practical limitation is the difficulty of measuring the front tyre slip angle directly in a real vehicle. In practice this critical state must be inferred via sensor fusion or observers, which introduces uncertainty; improving slip angle estimation (for example, through advanced observers or additional sensing like steer angle measurement) will be important for real-world implementation. Another area for extension is the inclusion of actuator dynamics in the model and controller design. The current MPC controller assumes ideal, instantaneous throttle, and brake responses, whereas real actuators have response delays and limits, incorporating realistic actuator dynamics into the control.

References

1. Abe M, Kano Y, Satoh Y, Stephant J (2000) An experimental validation of side-slip control to compensate vehicle lateral dynamics for a loss of stability due to nonlinear tire characteristics. In: Proceedings of the 5th international symposium on advanced vehicle control (AVEC), Ann Arbor, Michigan
2. Arab A, Yu K, Yu J, Yi J (2024) Motion planning and control of autonomous aggressive vehicle maneuvers. IEEE Trans Autom Sci Eng 21(2):1488. https://doi.org/10.1109/TASE.2023.3245948
3. Balkwill J (2018) Performance vehicle dynamics: engineering and applications. Butterworth-Heinemann, Oxford
4. Beal, C.E. (2011) Applications of model predictive control to vehicle dynamics for active safety and stability. Ph.D. dissertation, Stanford University
5. Boada BL, Boada MJL, Díaz V (2016) Vehicle sideslip angle measurement based on sensor data fusion using an integrated ANFIS and an Unscented Kalman Filter algorithm. Mech Syst Signal Process 72–73:832–845. https://doi.org/10.1016/j.ymssp.2015.11.003
6. Chen B, Tomizuka M (2001) Experimental evaluation of steering angle measurement for vehicle lateral control. IEEE/ASME Trans Mechatron 6(4)
7. Chu D, Li H, Zhao C, Zhou T (2023) Trajectory tracking of autonomous vehicles based on model predictive control with PID feedback. IEEE Trans Intell Transport Syst 24(2). https://doi.org/10.1109/TITS.2022.3150365
8. Di Cairano S, Tseng HE, Bernardini D, Bemporad A (2013) Vehicle yaw stability control by coordinated steering and braking in the tire sideslip angle domain. IEEE Trans Control Syst Technol 21(4):1236–1248. https://doi.org/10.1109/TCST.2012.2200100

9. Falcone P, Borrelli F, Asgari J, Tseng HE, Hrovat D (2007) Predictive active steering control for autonomous vehicle systems. IEEE Trans Control Syst Technol 15(3):566–580. https://doi.org/10.1109/TCST.2007.894653
10. Huang X, Li L, Peng H, Wang Y (2020) A real-time trajectory reshaping strategy for reducing MPC computational burden in autonomous vehicles. Control Eng Pract 102:104537. https://doi.org/10.1016/j.conengprac.2020.104537
11. Lai CH, Wu YC, Chen YC (2019) PID control for vehicle stability: a survey and comparative study. J Dynamic Syst Measure Control 141(9). https://doi.org/10.1115/1.4043492
12. Li H, Wang X, Song S, Li H (2016) Vehicle control strategies analysis based on PID and fuzzy logic control. Procedia Eng 137:234–243. https://doi.org/10.1016/j.proeng.2016.01.255
13. MathWorks (2025) Tire-road interaction (magic formula). Available at: https://www.mathworks.com/help/sdl/ref/tireroadinteractionmagicformula.html. Accessed 20 July 2025
14. Pacejka HB (2006) Tyre and vehicle dynamics, 2nd edn. Butterworth-Heinemann, Oxford
15. Puşcul D, Lex C, Vignati M, Shao L (2024) A literature survey on sideslip angle estimation using vehicle dynamics based methods. IEEE Access. https://doi.org/10.1109/ACCESS.2024.3402429
16. Rajamani R (2011) Vehicle dynamics and control. Springer Science & Business Media
17. Rokonuzzaman M, Mohajer N, Nahavandi S, Mohamed S (2021) Model predictive control with learned vehicle dynamics for autonomous vehicle path tracking. IEEE Access 9:129493–129503. https://doi.org/10.1109/ACCESS.2021.3112560
18. Schwenzer M, Ay M, Bergs T, Abel D (2021) Review on model predictive control: an engineering perspective. Int J Adv Manuf Technol 117(1). https://doi.org/10.1007/s00170-021-07682-3
19. Uzunsoy E (2018) A brief review on fuzzy logic used in vehicle dynamics control. J Innov Sci Eng 2(1)
20. Weng W, Hu C, Li Z, Su H, Xie L (2024) An aggressive cornering framework for autonomous vehicles combining trajectory planning and drift control. In: 2024 IEEE Intelligent vehicles symposium (IV), Jeju Island, Korea, 2–5 June 2024. IEEE
21. Wenzel T, Burnham K, Isermann R (2006) Estimation of tire and vehicle dynamics states for advanced vehicle control using standard sensors. IEEE Trans Control Syst Technol 14(3)
22. Zhang H, Li X, Shi S, Liu H, Guan R, Liu L (2011) Phase plane analysis for vehicle handling and stability. Int J Comput Intell Syst 4(6)
23. Zhang Y, Wang Z, Wang Y, Zhang C, Zhao B (2022) Research on automobile four-wheel steering control system based on yaw angular velocity and centroid cornering angle. Measur Control 55(1–2):49–61. https://doi.org/10.1177/00202940211035404

Development and Simulation of a Dual-Mode Autonomous Braking System for Human and AI Controlled Driving in an EV Formula Student Car

Lewis Milne and Aydin Azizi

Abstract With many domestic vehicles now starting to incorporate autonomous features (ADAS) it is now a requirement for engineers to be aware of these elements and begin learning how to take advantage of the possibilities of their usage. Formula Student (FS) competitions around the globe are driving the development of young engineers by exposing them to autonomous vehicles by mandating that vehicles have autonomous capabilities with the German competition FSG needing ADAS capabilities to score maximum points. At Oxford Brookes the aim is compete at the highest level of FS which requires autonomous implementation into the successful human driven counterpart. The current focus is on the implementation of braking capabilities to the vehicle. This paper focuses on development and simulation of a braking system which delivers 1 g of deceleration during on track running. This involves development of a Simulink model, design of multiple PID and Fuzzy Logic Controllers and implementation of a braking model which is based on numerical methods. This methodology allows for quantitative performance data to be paired with control system performance to understand the real-world effects of slight variations within the control system. Research has found that a 2 degree of freedom PD controller provides the best performance with an absolute error value of 3.0498% which results in braking performance which closely matches the ideal performance with a difference in 0.1 m/s at the end of the braking event. The research also looked into the effects of Fuzzy Logic Controllers which shown that the performance required from the control system cannot be reached with the FLC's. Tests showed significant instabilities when the performance was close to the set point and only showed acceptable stability with significant undershoot in the performance which is not acceptable for autonomous vehicle performance.

L. Milne (✉) · A. Azizi
School of Engineering, Computing and Mathematics, Oxford Brookes University, Oxford, UK
e-mail: lewismilne55@gmail.com

A. Azizi (ed.), *Intelligent Control and Autonomous Systems*, Emerging Trends in Mechatronics, https://doi.org/10.1007/978-981-95-6402-6_3

1 Introduction

The world of self-driving vehicles is expanding exponentially year on year with many companies now at a stage to deploy these autonomous vehicles (AV's) into general use for the public, this requires expertise from Engineering and Computing Science disciplines to make these vehicles safe and robust. To aid the education of young engineers, the various Formula Student (FS) competitions around the world are beginning to require the cars to drive autonomously within certain events. Formula Student Germany (FSG) is considered the leading competition with teams from all over the globe competing at Hockenheim each season, FSG is now structured as such that without driverless (DV) capabilities it is impossible to earn the maximum points from the competition (Table 1).

Formula Student UK (FSUK) is the primary competition for Oxford Brookes University, and FSUK has introduced the DV aspect differently compared to FSG with FSUK offering the ADS-DV's (Fig. 1) which are vehicles that all teams can use where they must provide the sensors, instrumentation and write the code to run the vehicle. This provides an entry to DV development whilst lowering the barrier to entry. This vehicle can pull 1 g of lateral acceleration through corners and has a terminal velocity of 20 m/s which provides good performance for teams starting to develop their AV systems. This vehicle is used in 4 dynamic events: Acceleration, Skidpad, Autocross and Trackdrive. Autocross and Trackdrive are the biggest challenges and are where the braking system gets the most use, these events require the vehicle to navigate an unknown track as quickly as possible.

Oxford Brookes University operates 2 Formula Student teams, firstly, Oxford Brookes Racing (OBR) develops and operates the university's human driven vehicle (Fig. 2). Secondly, Oxford Brookes Racing Autonomous (OBRA) develops and maintains the AV system, required instrumentation and uses the ADS-DV at FSUK. It is a long-term goal for the university to compete at FSG which will require an integrated human and AV drivable vehicle. This presents multiple challenges which have various difficulties. Primary of these is the braking capabilities of the vehicle.

Integration of AV capabilities into a human driven FS car provides significant challenges thus when designing the systems required for AV braking the goal is for

Table 1 FSG scoring structure (https://www.formulastudent.de)

Event	Maximum points (Human)	Maximum points (DV)
Skidpad	50	75
Acceleration	50	75
Autocross	100	100
Endurance	250	200
Efficiency	75	75
Engineering design	150	300
Total	**675**	**825**

Fig. 1 ADS-DV in use by Oxford Brookes Racing Autonomous

Fig. 2 Oxford Brookes Racing 2025 Challenger

the least invasiveness as possible, this would allow the quickest implementation and testing time which is crucial for proper functionality. Braking has 2 functions in a Driverless FS car, firstly normal actuation of the braking system must be possible (AIBS). Secondly, the team must have the capabilities to emergency stop the vehicle

(EBS) these 2 systems are complex and as such focus will be on the development of the AIBS to fit within the OBR Vehicle. This paper looks to propose a design which can be implemented that allows AV braking whilst remaining legal to FSUK and FSG rules, there are no rules around the performance of the AIBS with the desired performance set by each team.

The aim of this paper is to present a viable solution for implementation of autonomous braking capabilities into the OBR vehicle. A number of objectives have been created which represent milestones in project development. These are:

- Design a concept of a reliable braking solution which delivers 1 g of deceleration
- Create a parameterised Simulink model of the OBR brake pedal and the chosen braking solution
- Implement and tune an effective control system solution which uses the desired braking command and actuates the brake system
- Create a simulation environment which focuses on braking dynamics and is validated against the OBR vehicle to conduct control system analysis.

This paper looks to design and simulate a reliable braking solution which can deliver 1 g of deceleration. Development will include mechanical integration within the chassis with significant development time being spent on the control system which will compare the use of PID control against Fuzzy Logic control. To compensate for a lack of real-world testing the designed control systems will be simulated to compare their braking performance using a simulation designed using numerical methods which is validated against the OBR vehicle. This simulation will allow for analysis on how variety in control signal affects ultimate braking performance with a recommendation on the optimal control method stated.

2 Literature Review

Rules and regulations constrain the valid options for autonomous braking on FS cars. Primary of which is a full Brake-by-Wire (BBW) system which is used on many EV's on the roads today, implementation of a robust BBW system that conforms to ISO 26262 [6] and has sufficient functional safety is a large task for students and as such is excluded from the rules on safety grounds. The main requirement is for the AV system to be able to lock the brakes of the vehicle which is tested and has to be passed before any autonomous running is allowed.

Implementation of AV braking requires various disciplines of engineering to be used in tandem to create an effective system Ciocca [2] presents a system which uses an electric motor mechanically attached to the brake pedal which actuates the hydraulic system. This system used PID control which had a rise time of roughly 0.2 s which is deemed comparable to human reaction time. This paper further analysed differences between multiple approaches with the mechanical approach chosen due to its cost and simplicity Shi and He [15] presents an Electro-Hydraulic Braking (EHB) system which uses small electric motors to actuate the hydraulic braking system. This paper creates a model where a torque from a motor is converted into linear motion within a master cylinder. It lays out the mathematical expressions required to model this system in ideal conditions. This method further uses Model Predictive Control (MPC) to control braking response too; this was compared against results from a Sliding Mode Controller (SMC) on a test bed. It was observed that MPC has half the rise time of the SMC at 180 ms with it generally being noticed that the MPC can handle changes in pressure requirement a lot smoother than SMC. Redundancy of braking systems is a concern for manufacturers, Liu et al. [9] analyses EHB methods and discusses hardware which is in use on commercial vehicles using this system. It details the relationship between motor torque and the force on the master cylinder pushrod which help provide a framework for modelling these systems on MATLAB. It uses SMC's too with it controlling wheel rate rather than pressure in the hydraulic system. The implementation of this system includes consideration for losses within brake pipes due to friction, nonetheless results show strong tracking behaviour of the expected braking behaviour versus the recorded data.

Freidrich et al. [3] presents Elefant Racing's work on their Driverless Formula Student car with a focus on its software and hardware systems. This paper presents a method of braking using 4 outboard motors which are controlled via CAN signals. The team has designed an Emergency Brake System (EBS) which uses pneumatics to actuate the system. A key requirement is functional safety and redundancy which translate to an "always on" feature when the car has no power. This locks the brakes and prohibits movement. The paper created 2 independent pneumatic circuits which are paired with a hydraulic system. These use a shuttle valve to allow the brake pedal to be actuated independently; however, when the pneumatics are activated, this will overpower the human system and control braking. It uses pressure tanks which must be refilled after each use, it introduces functional safety into the system by not allowing the AV system to run if a pressure check shows less than the required pressure. Kazban [7] presents a paper about ETH Zurich's driverless system. This paper also uses the method of outboard motors to perform braking manoeuvres with the added benefit of torque vectoring to aid performance. They also use a pneumatic EBS however details are not provided. Politechnico di Torino provided substantial research into the AV system as part of their Formula Student program. Sorrentino [17] provides detail into the state machine used to control the vehicle when in driverless mode, the control systems used and it also touches on the EBS system the team uses. This research is added to by Sorrentino et al. [18] with a standalone paper in the EBS system. A similar approach to Elefant is used with dual pneumatic systems included, this approach actuates the system via a shuttle valve embedded on the brake line itself to give greater flexibility in placement. It details the methods of controlling the EBS logic which are rule compliant too.

Vehicle modelling is a large part of understanding braking dynamics and creating a controller which can handle the quirks of the system which is inherently non-linear Savaresi and Tanelli [14] presents multiple models which have been used in traditional brake models. A single corner and dual corner model. The key difference between the 2 is the dual corner model considers transient behaviour of the braking system such as the weight transfer due to system actuation. It further goes on to detail the control systems behind the models using control theory first principles to explain dynamics. A key issue with these models is that they ignore certain environmental effects and are deemed too simple, to tackle this issue Saraiev and Gorb [13] presents a model which is vastly improved compared to traditional methods by including aerodynamics which ensure a non-linear deceleration, implements a speed dependent grip coefficient to model the effects of vehicle lift or downforce at high speeds and also models the combined inertias of rotating components such as flywheels. The result of this is a model which has 5 percent error consistently when compared to real telemetry data of a F1600 vehicle around Brands Hatch.

Considering the control methods used within autonomous braking systems the PID controller is commonly used with it being implemented in the papers by Ciocca [2]. PID tuning does not have a set defined method of tuning which is discussed by Somefun et al. [16]. It discusses how issues such as poor mathematical modelling, poor behaviour over time and computational efficiency can all affect performance of the PID controller. MPC control is often discussed as an upgrade to PID control, MPC is an optimisation-based approach which solves finite horizon problems Köhler et al. [8] discusses this in detail with details of a MPC which can be used as a foundation that provides a stable response using a terminal cost compared against a terminal set. MPC is incredibly versatile with usage within high level autonomous control systems however it has also been used within the built environment, Saloux et al. [12] details how MPC is used within buildings to control heating. It takes many variables into account such as ambient temperature, occupancy and time of day to minimise energy usage with results showing a 25–30% reduction in usage when this method is implemented. Fuzzy Logic control is commonly used with autonomous vehicle development with usage seen in areas such as collision avoidance and lateral control. Usage for longitudinal control is further discussed by Mattas et al. [10] which uses Fuzzy Logic control to control vehicle longitudinal velocity with comfort, stability and safety being maximised. This system used a Sugueno FIS with 5 membership functions to control velocity with the distance between vehicles and velocity error being considered by its inputs with acceleration being determined. This system produced good results with smooth accelerations and braking being observed in simulation and real-world testing.

The safety of AV braking systems is paramount within Formula Student and the wider motoring world. There have been numerous standards introduced which provide guidelines for designing these systems so consumers can have confidence in their reliability and robustness. ISO 8800 [5] created a safety framework for integration of autonomous systems into road vehicles which covers the electronic integration and also the machine learning components. Part of its purpose is to extend the safety standards already published in other standards such as ISO 26262 and ISO 21448 [4] which cover standards for hardware/software failures and edge case testing respectively. ISO 8800 introduces an AI Safety Lifecycle which is made to fit around automotive needs and includes factors such as the safe development of AI systems and the rigorous risk assessments which are required for these systems to be used by the public.

The literature reviewed shows that the implementation of autonomous braking functionality is best done using outboard motors if possible or failing that a mechanical approach. The current implementations of mechanical actuation show promising results which could further be built upon. The EBS systems are always heavily discussed and the trend is to use a pneumo-hydraulic system. There is a gap in research regarding the comparison of control approaches for mechanical braking integration too. Furthermore, integration into existing Formula Student vehicles is scarcely discussed and challenges such as packaging are not openly discussed. It is expected that these challenges are considered heavily within teams when designing these systems.

In this research, to fulfil the aim of creating an effective autonomous braking implementation within the OBR vehicle a novel approach is developed in regards to the control approach of the brake system. The control approach is based upon pedal actuation percentage rather than a pressure based system which is seen in literature. Creation of the Simulink model take inspiration from work completed by Ciocca [2] with novel components added to determine pedal actuation. Finally, the braking simulation which is to be used to quantify performance uses methods developed by Saraiev and Gorb [13] with slight changes being made to the numerical system to work with pedal percentage actuation commands. After a brief introduction and literature review, this paper is structured as such that the comparison of viable braking approaches is followed by chosen methodologies being discussed. The paper then moves to modelling the chosen system and braking simulation within MATLAB. These simulations are used to implement PID and Fuzzy Logic control with these 2 methods using the braking simulation to determine numerical data to be used in comparison. These results are then discussed, conclusions presented and future work detailed.

3 Methodologies

This paper follows a simple methodology (Fig. 3) which requires sequential development of distinct milestones. Initially, the structure allows for concept development with the requirements of the system being set and the relevant rulesets being considered to understand the limitations of the project. This moves into deciding what actuation method should be developed. This stage forms a large part of the literature review with each considered method being researched alongside reviewing literature around control approaches which are employed in different projects. With the actuation choice being confirmed, it is possible to consider theoretical performance against the performance targets set at the start of the project.

The chosen method can now be integrated within the OBR vehicle chassis to ensure fitment, this ensures that the chosen method is feasible method physically within the chassis before the control system can be created. With this method being successful, the model can be created in Simulink with the physical geometry being parameterised. This model should be created to firstly show realistic behaviour and

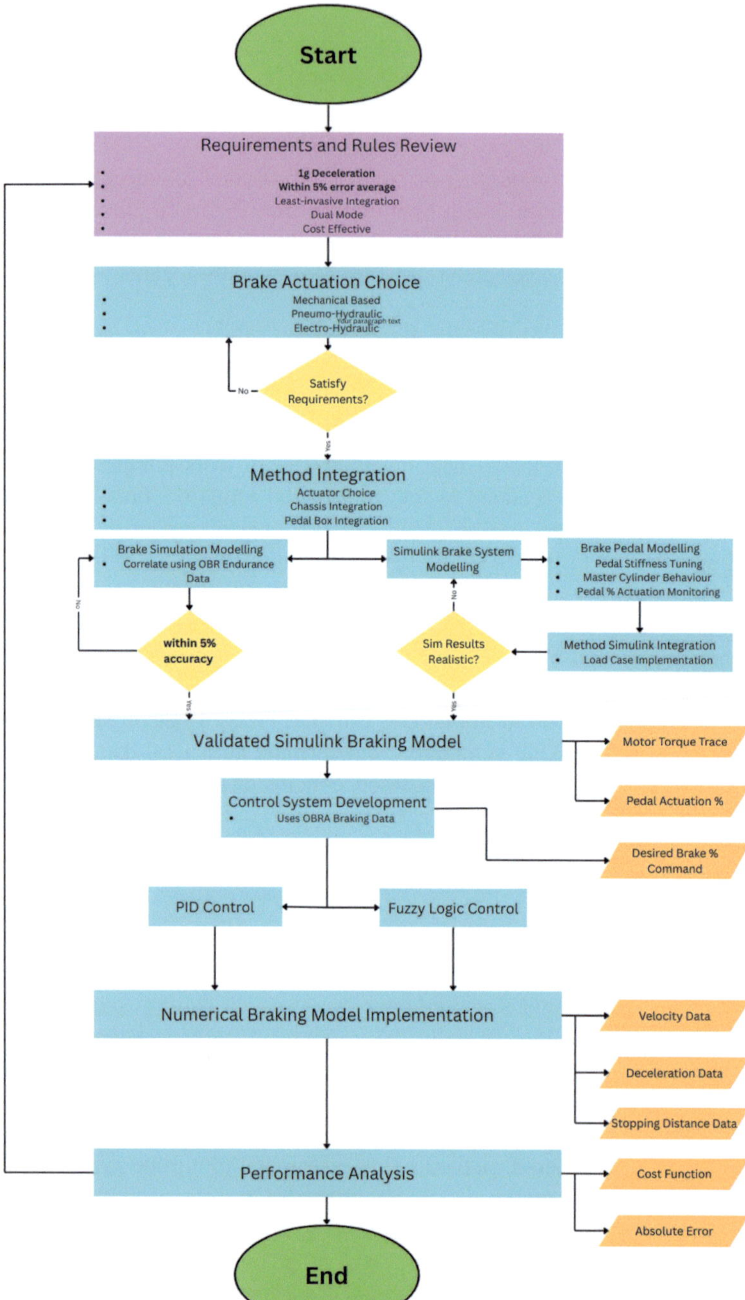

Fig. 3 Methodology flow chart

outputs of the model be within 5% accuracy. If these criteria are not met the model can be further developed. The desired control approach can now be developed, these control systems use the error of the pedal actuation to determine the torque output.

The control system development produces data which contains pedal actuation percentage which is used to calculate quantitative data which allows for realistic comparison of performance. The braking simulation is based on numerical methods. Research of these methods further tie into the literature review. Using the data generated from this model allows for discussion of real-world performance with comparison of differences in control system performance against realistic braking performance.

3.1 Actuation Method Comparison

3.1.1 Mechanical Actuation

The goal of this method is to actuate the brake pedal which the human driver would use. To maximise track time, the system should be as least invasive as possible to ensure quick turnaround between human and driverless modes. Two methods are presented as viable approaches; the first method is the use of a linear actuator to push the pedal. This is a method used by OBRA on their small scale testing buggy which replicates the ADS-DV. The system works by attaching the actuator to the brake pedal using a rocker. This method uses basic geometry to output the desired force with the rocker's motion ratio driving the dynamics. There are issues with this approach especially when driving at high speed which requires responsiveness, linear actuators cannot actuate quickly which means another method of actuation is required. The second method is using a servomotor with a pulley to pull the pedal backwards. Both methods use different dynamics to actuate the pedal however these have similar complexity. The choice of method is driven by the performance of the actuator itself. The AIBS is required to operate at high frequency and be able to mimic human braking inputs as closely as possible. Alongside this the actuator should have sufficient output to actuate the pedal so that the brakes lock. The peak force for human actuation is 500N which slows the car at 1.6 g. In driverless mode this would be enough to lock the brakes if need be. To understand the behaviour of using a linear actuator versus a servomotor qualitative analysis has been conducted taking into account the requirements these actuators have within the brake system (Fig. 4).

Based on the results in Fig. 4, it is clear that the use of a servomotor is best for implementing mechanical actuation if chosen. The combination of low volume, quick response and acceptable torque output make it feasible for integration.

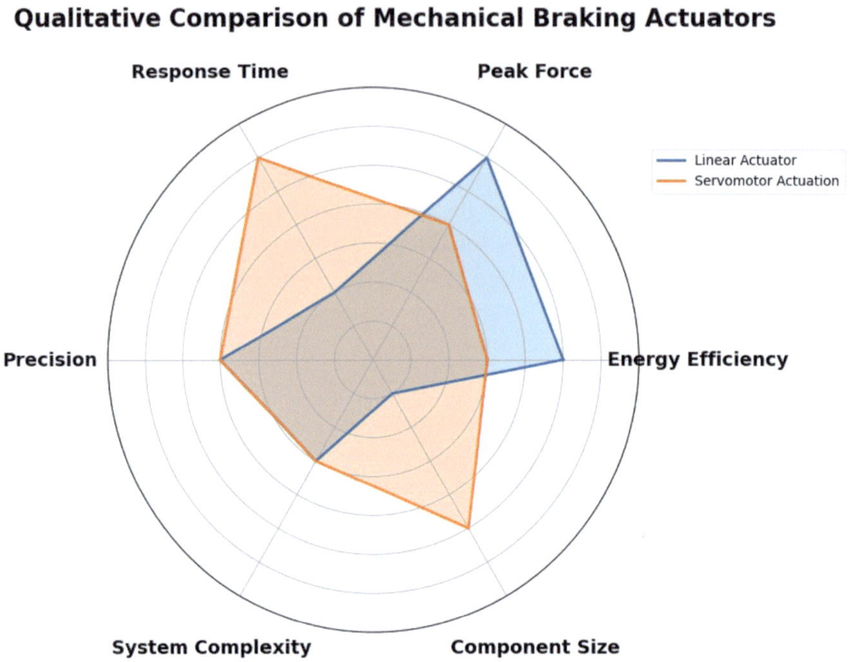

Fig. 4 Linear actuator and servomotor control compared using a spider diagram

3.1.2 Electro-Hydraulic Actuation

Electro-Hydraulic Actuation (EHB) is the principle upon Brake-by-Wire systems (BBW) are built on. BBW systems on road vehicles measure the percentage of pedal actuation and use that command to actuate the hydraulic brake system. This method is prohibited on FS cars which increases the components needed in the footwell area which suffers from issues with space. Housing the motors which actuate the system will thus be difficult. There is a scenario where an EHB system can function which would use a motor to actuate a 2nd master cylinder plumbed into the hydraulic circuits. This motor could then be actuated by a set control signal this approach has numerous flaws including such as packaging, in an FS car the space required is limited. These master cylinders could be placed independently from the pedal box and even in the rear of the car however then there would be extra mass added which would hampered vehicle performance in human and driverless modes too.

3.1.3 Pneumo-Hydraulic Actuation

Pneumatic braking systems have a place within the automotive sector and are commonly used on large vehicle such as trucks and lorries (Fig. 5). These vehicles

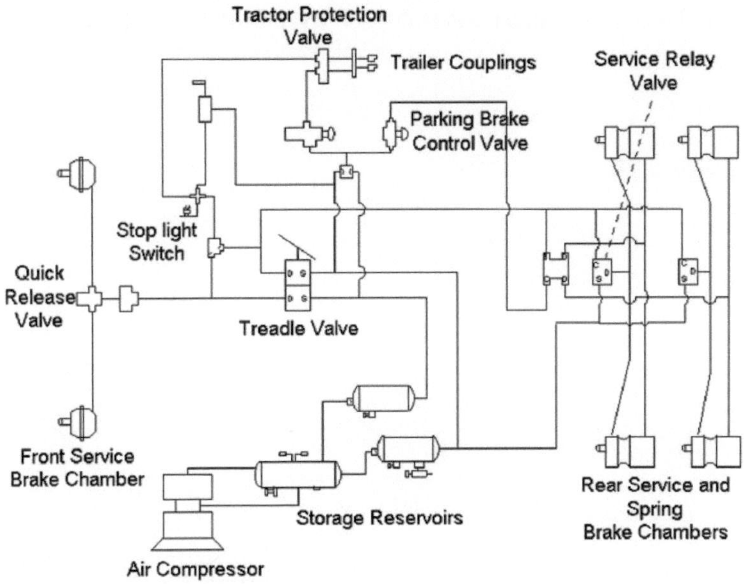

Fig. 5 Typical schematic of an air brake system

require large braking forces which can be delivered by pneumatics. These pneumatic systems require more components to deliver function which add weight and complexity to the vehicle, for example to ensure the reservoirs can provide air they need a compressor to constantly fill them. This increases mass and uses space which is not available on an FS car. Pneumatics are commonly used by European FS teams for their EBS systems which have to stop the vehicle once the Emergency Stop is triggered. They can function without compressors due to this being a one-time requirement with an external compressor being connected to the system in the pits. Pneumatic systems required for Formula Student applications benefit from the fact that required components are relatively cheap when compared to other systems.

3.1.4 Approach Selection

With each approach being explored using literature and document characteristics of each approach it is possible to conduct a qualitative analysis of the 3 approaches to determine which is most feasible for development within the constraints of the OBR Formula Student vehicle. The key features of the AIBS must be quick response, low weight and complexity and reliability whilst running on the low voltage circuit.

Using the analysis in Fig. 6, the mechanical actuation of the existing brake pedal using a servomotor. The use of pneumatics for this purpose is not viable due to the slower response and inability to deal with quick changes of request; furthermore, whilst electro-hydraulic systems provide benefits in terms of brake performance

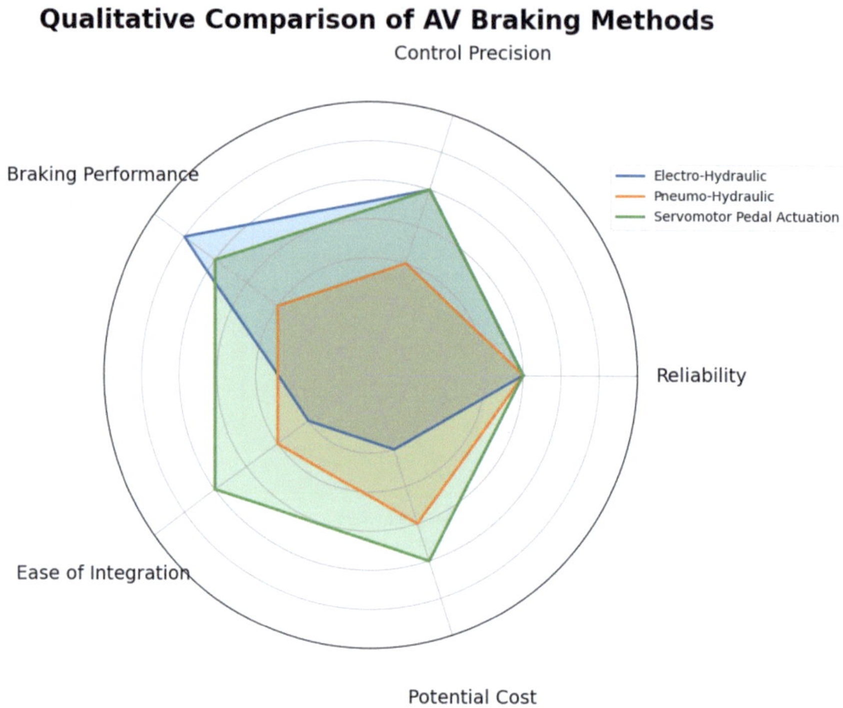

Fig. 6 Qualitative analysis of AV braking integration methods

with a comparable control precision the cost required to implement this system is deemed excessive. This leads to using the servomotor which provides comparable brake actuation forces, provides easier integration into the existing vehicle and is cheaper for its components.

3.2 Mechanical Actuation Development

3.2.1 Motor Selection

The goal for the braking performance was to consistently perform braking manoeuvres with a force of 1 g. This choice is set upon consultation of OBRA's data from FSUK where the peak braking force required averaged 0.5 g this choice allows for growth in the future too. The force required to stop at 1 g is 300N applied to the pedal. Servomotors are commonly used for low power applications with high energy density motors are generally expensive. A pulley is used on the end of the servomotor to convert the torque to a linear force. Initial placement of the motor showed

Table 2 Servomotor comparison

Motor name	Voltage (V)	Peak torque (Nm)	Dimensions (mm)	Cost (£)
T-Motor AK70-10	24/48	24.8	89 OD × 50.2	517.60
Dynamixel PM54-060-S250-R	24	38	54 × 126 × 54	1415.13
Harmonic Drive RSF-14B	24	28	50 OD × 168.5	740
Hitec HS-1005SGT	11–15	10.78	64.01 × 33.02 × 72.9	460.99

the pulley could be 25 mm in diameter which works well with the space constraints whilst providing the required force to actuate the pedal (Table 2).

The peak torque of the HS-1005SGT of 10.78 Nm provides 862.4 N, this is determined by applying $F = T/r$ with r being the radius of the pulley attached to the motor. The HS-1005SGT fits all the needs of the braking system with the motor capable of generating the 300 N needed to produce 1 g of deceleration with the 25 mm pulley attached. This provides performance which can actuate the full pedal whilst running on the 12 V low voltage circuit and having a lower cost relative to the other considered motors.

3.2.2 System Construction

Integration of the selected servomotor is constrained by the master cylinders and existing hydraulic components already in place. It has been chosen that the motor be mounted on the inside of the chassis, this allows for the pulley on the servomotor to align with the centre of the pedal whilst not obstructing the movement of the pedal which travels 25 degrees through its arc. A CAD model (Fig. 8) was created to test fitment before creating a parametric model within MATLAB. This included the motor, the 25 mm pulley and the mount which holds the motor in place and is bolted to the chassis. For successful integration of the motor, it is essential for the loads being transmitted into the chassis to be understood and considered. This is especially important for the carbon monocoque chassis which uses a large number of unidirectional fibres. This is out with the scope of this dissertation. The only other component which requires modification is the lever arm of the brake pedal. The current design uses a small tab on the topmost part which the metal cable attaches too (Fig. 7).

3.3 Braking System Modelling

To conduct analysis, modelling and simulation of the autonomous braking system MATLAB Simulink is used. The Simulink ecosystem allows for 3D modelling via

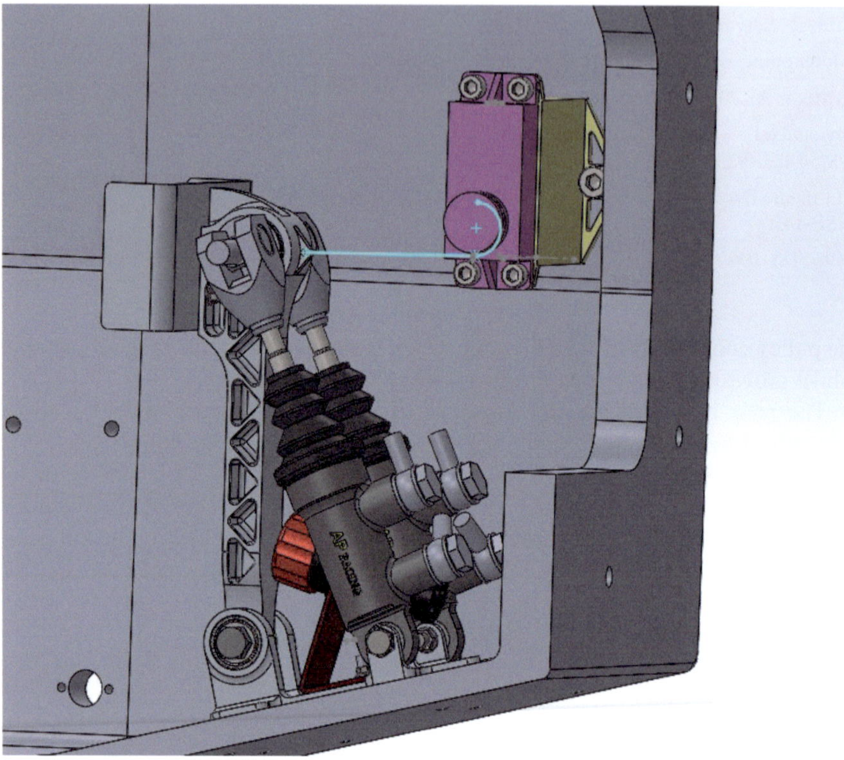

Fig. 7 Packaging of HS-1005SGT withhin the OBR chassis whilst remaining rules compliant and allowing full pedal travel

the Simscape package which is used to create a parametric model of the brake pedal seen within the OBR vehicle. Simulink has further benefits when it comes to designing control systems with many control methods having blocks that make design, integration and testing smoother. The autonomous system releases a desired pedal percentage signal which is used within this system. More information on the OBRA Autonomous System is within the Appendix. Information on the adjustments made to the Braking Model by Saraiev and Gorb [13] is further detailed in the Appendix too.

3.3.1 Simulink Model Construction

The first stage of simulating the braking is creating a Simscape model within Simulink. Using dimensions and geometry from CAD made the system simple to construct. The pedal body and face are direct imports that simplify the modelling required to place the axis where the master cylinders mount. The work completed here is influenced by the work done by Ciocca [2] in that paper they compare the system

Fig. 8 Simulink model of an OBR brake pedal using Simscape multibody

based on pressure within the system. In that case, the pressures of both hydraulic systems were simulated and combined into one reading which was augmented by the control method. The approach chosen in this paper is to create a Simscape model which outputs pedal actuation percentage which is then processed and used to determine braking forces and vehicle behaviour.

The master cylinder is modelled as 2 different physical bodies with the piston being excluded. It creates a model which captures real-world dynamics with the component rotating and staying in plane with the upper body which mounts on the balance bar. The pedal stroke is determined via the distance between the 2 faces which always stay parallel to each other. This is then compared to the known stroke of the master cylinder which is converted to a value between 0 and 1. In this model, a key assumption is that bearings are perfect and do not provide any friction, this keeps the model simple and helps minimise simulation run time.

The master cylinders themselves are effectively springs which push the pedal back to its original position. To model the systems stiffness, it was considered that the brake pedal has linear stiffness through its full travel until it reaches its bump stop, this further assumes that the brake system is bled perfectly with no air in the system and that the fluid is in a good condition with the viscosity stated by the supplier. The brake pedal is designed to depress the master cylinder by 30 mm when 500 N is applied by the driver. Using Hooke's Law, the linear stiffness can be derived by $K = F/x$ which is expressed in N/mm. For this system the linear stiffness is 16.667 N/

mm. Implementation of this stiffness into the model used a simple approach using the internal dynamics of the revolute joint which controlled the brake pedals lever arm. Multiple other methods were considered that did not work, these are discussed briefly in the Appendix. The linear stiffness which was initially calculated was converted to a rotational stiffness using;

$$K = Fr/\Theta$$

$K =$ Rotational Stiffness (Nm/deg).

$r =$ Pedal Arm Length (m).

$\theta =$ Pedal angle at maximum actuation (deg).

This rotational stiffness models the brake mechanism as one rotational spring that discounts effects of compliances within the hydraulic system such as pipe expansion or wear to components. For the purpose of this report these simplifications are deemed acceptable. The result of this simplification is a model which actuates fully through 100% of the brake travel consistently whilst also reducing computational load as the model is simplified.

The next step of modelling, the brake system is the forces generated by the master cylinders themselves.

OBRA uses a similar model on their vehicle simulator which was able to used and adapted for this model. The methods shown by Yang [19] are followed which results in a Simulink model (Fig. 9) which correlates with hand calculations. This circuit was then duplicated for the rear with the appropriate values changed to ensure realistic behaviour, these are namely the master cylinder diameter, brake bias and number of pistons per calliper.

The final stage is applying the force from the servomotor onto the pedal assembly, this applied directly onto the foot plate identically to how humans actuate the system. This approach is computationally less complex as the wire tension is not simulated. This forms a key assumption that the wire has no slack which would be the case during normal operation. This approach is simplified however the system represents realistic behaviour which allows for development of the control system to begin.

4 Model Control

The control of the braking system is based upon measuring the percentage of actuation against the desired actuation. The autonomous system gives braking commands in % of desired pedal actuation. Before designing the control system, several goals are set to quantify its performance. The response of the controller is most important; the autonomous system determines braking response based on velocity at that moment in time. This translates to preferring slight overshoot in braking response rather than undershoot. In a scenario where brake response greatly undershoots the car is liable

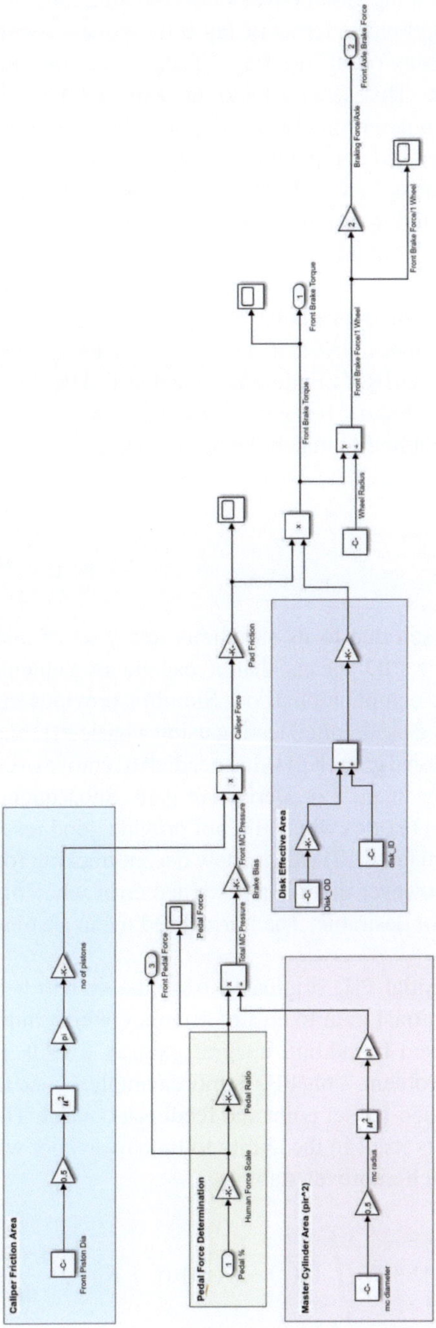

Fig. 9 Simulink model of the front hydraulic circuit of an OBR car

to go off track which can result in an event being over or large time penalties at a minimum. Conversely, braking system overshoots then the car gets to it target speed quicker and whilst not optimal in terms of lap time it does ensure safety which is what the OBRA team focus on at this stage. Factors such as rise time are tuned within OBRA's code too. This section looks to compare the control response of PID and Fuzzy Control to determine the best approach considering factors such as raw performance, stability and computational load. Both controllers function using brake pedal percentage actuation with small variations between the 2 of them in how that signal is translated to a torque which is used to actuate the revolute joint which simulates the pedal. Furthermore, the Simulink model is designed in a way that both controllers are integrated and can be switched between using a binary switch with the parameter script which runs in the background. To compare each controller effectively the set point will be consistent. This is defined using a lookup table which houses braking data from OBRA's Endurance event at FSUK 2025, this data is the best representation of the braking behaviour of the autonomous system with quick response and accuracy required to match the desired control.

4.1 PID Control

PID control [1] was chosen due to its simplicity and ease of understanding. The relatively basic nature of PID means that it can be implemented on low-level controllers with a minimal computational cost. Simulink provides an extensive library of controllers which can be used, initial testing using a basic PID showed large instabilities due to the proportional gain (Kp) value needed to remove overshoot, this could not be rectified by adding Integral or Derivative gain. Subsequent tuning attempts used the Zeigler Nichols [1] rules which did not provide good results. Results from the 1 DOF PID are seen in (Fig. 10) which show decent tracking for the lower quick jabs of the brake pedal however there is a sustained error when higher brake pedal actuation is required is not desirable. The normalised mean absolute error is 7.26% for this control method.

The issues with the initial PID originated from the system being unstable and requiring a lower proportional gain to ensure accuracy which induced steady state error that couldn't be fixed by adding integral gain. A 2 DOF PID (Eq. 1) was introduced to solve the problem. This PID is more complex since then it is possible to separate the gains applied for set point and feedback control. This affords higher flexibility which manifests itself in the ability to tune overshoot whilst maintaining disturbance rejection which improves stability.

$$u(t) = K_p(\beta r(t) - y(t)) + K_i \int_0^t (r(\tau) - y(\tau))\mathrm{d}\tau + K_d\left(\gamma \frac{\mathrm{d}r(t)}{\mathrm{d}t} - \frac{\mathrm{d}y(t)}{\mathrm{d}t}\right) \quad (1)$$

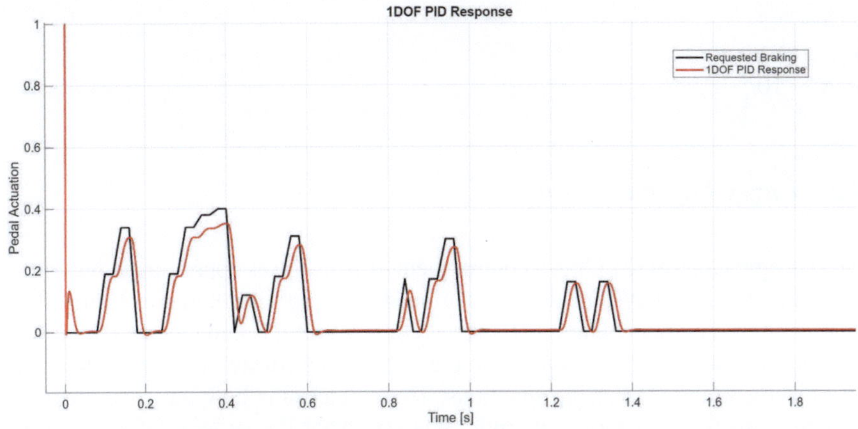

Fig. 10 Control response of a 1DOF PID controller

Introducing 2 DOF PID control resulted in several small changes which improved areas such as removing steady state error, improved tracking of the set point and also reducing oscillations in the response. This results in a reduction of the normalised mean absolute error to 5.87% with most of the error originating from the slight overshoot behaviour which is preferred to undershoot. This response also has a rise time which closely matches the set point with negligible error. The gains used within the 2DOF PID turn it into a 2DOF PD controller with the integral gain being unused. The gains and set point weighting coefficients are (Fig. 11).

- KP = 0.8
- KI = 0
- KD = 0.012

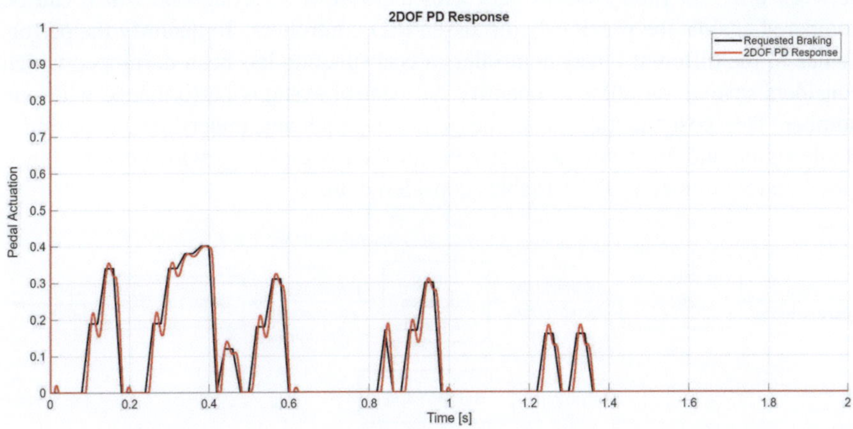

Fig. 11 Control response of a 2DOF PD controller

- $b = 2.5$
- $c = 0.25$
- $N = 10$.

4.2 Fuzzy Logic Control

Fuzzy Logic Control (FLC) is a control method which uses approximate reasoning to control a defined system without the need for a specific numerical modelling however it uses numerical inputs from the system which it is implemented. It mimics human decision making by converting crisp inputs into fuzzy inputs and fuzzy sets by implementing membership functions. These determine how the crisp input are changed into fuzzy" degrees of truth" this sets what the system defines as high, medium and low for example. Rules are then set using IF-THEN rules, these determine the response of the system. Once a response is determined the fuzzy output is "defuzzied" and converted to a crisp output. The fuzzy inference system described is a Mamdani system which uses membership functions on its inputs and outputs. Sugueno fuzzy inferences systems are an alternative approach which uses membership functions on the inputs; however, it does not use them for the outputs, its outputs are determined via a function with scaling coefficients assigned to the inputs which can be tuned (Fig. 12).

The Fuzzy Logic Controller used within the braking system utilises the error and error rate of the system with the torque request being the single output. To stabilise the inputs the error rate is filtered at 100 Hz and the output torque rate is capped at 231 degrees/second which matches the motor specification (Fig. 13).

Designing a robust Fuzzy Inference System (FIS) provided challenges, development was hindered by fuzzy controllers which were unstable and oscillating which meant effective tuning could not take place. Focus then switched to a comparison between different Fuzzy Controllers with the aim of selecting one which can be compared against the previously discussed PID controllers. To quantify the performance of the different Fuzzy controllers a cost function has been developed which considers several variables to quantify the control systems performance with one number. This cost function takes the goals of penalising undershoot, requiring a stable signal and error into account with different weightings which can be tuned based on requirements. The variables considered are:

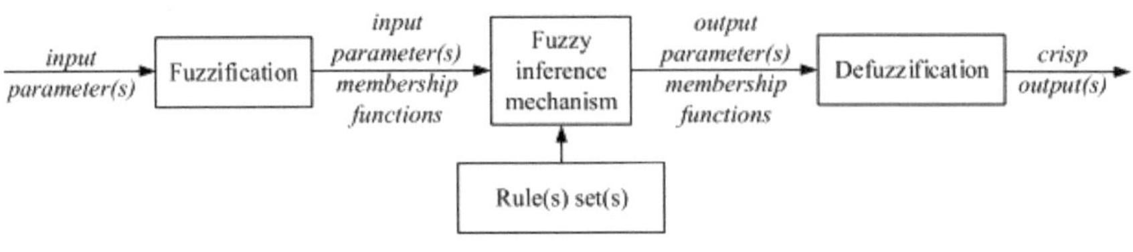

Fig. 12 Flow chart of a fuzzy logic system

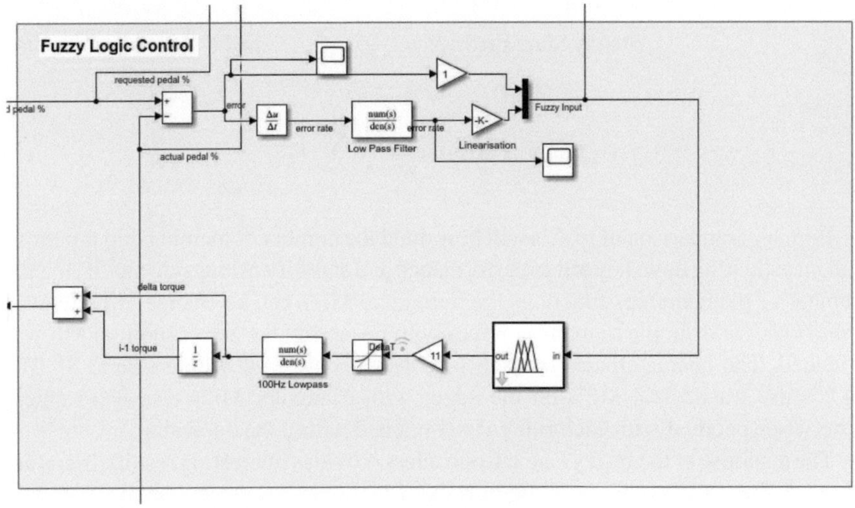

Fig. 13 Simulink implementation of the fuzzy logic controller

- Safety—Does the torque control react the correct way, Does the torque output move in the opposite direction to the error? This is weighted the highest.
- Stability—Does the output signal chatter or flip
- Accuracy—What is the total accumulated error, overshoot and steady state error?
- Efficiency—Is the controller making smart choices—this is weighted the lowest.

$$\text{Total Cost: } J = W + 50C + F + E + 0.5O + 1S + 0.5e \tag{2}$$

$$\text{Wrong Sign Torque : } W = \frac{1}{N}\sum_{k=1}^{N} 1\{e_k \cdot u_k < 0\} \tag{3}$$

$$\text{Charter: } C = \frac{1}{N-1}\sum_{k=1}^{N-1} |u_{k+1} - u_k| \tag{4}$$

$$\text{Flips: } F = \sum_{k=1}^{N-1} 1\{\text{sign}(u_{k+1}) \neq \text{sign}(u_k)\} \tag{5}$$

$$\text{Integral of Absolute Error: } E = \sum_{k=1}^{N} |e_k|\Delta t \tag{6}$$

$$\text{Overshoot: } O = \max\left(0, \max_k (y_k - r_k)\right) \tag{7}$$

$$\text{Steady State Error: } S = \frac{1}{M} \sum_{k=N-M+1}^{N} |e_k| \tag{8}$$

$$\text{Control Effort: } e = \frac{1}{N} \sum_{k=1}^{N} u_k^2 \tag{9}$$

Primary comparison of FLC's will be around the number of membership functions and investigating how they affect performance and stability. Alongside modifying the number of memberships functions the individual MF's can be changed to influence behaviour, for example to manage issues with instability the zero condition MF was widened. The characteristics of each FLC test is described within (Table 3). The FLC's use trapezoidal MF's on the edges with triangular MF's elsewhere unless otherwise specified with each rule map contained within the appendix.

The response of the Fuzzy Logic Controllers provides interesting results, literature suggests that FLC's with more membership functions have the potential for greater accuracy which comes at a cost of sensitivity. FLC 1 with 7 MFs on error, error rate and the output torque has the highest cost which makes it the most unsuitable for use. The key trend noticed with the fuzzy controllers is that there is a distinct trade-off between accuracy and stability. Common procedures for tuning FLC's state that to avoid instability you should widen the zero MF. This results in a lack of stability. To troubleshoot this issue the FLC's were modified to have trapezoidal zero MF's which add a deadband to error. On occasion this would help however it had the downside of inducing an acceptable error. Gaussian distributions were identified as a solution to this problem as the nature of these shapes promoted smooth transitions between rules. The parameters within the membership functions are incredibly sensitive to change too, changing the standard deviation (σ) from 0.08 to 0.12 makes a noticeable change to performance and stability. Figure 14 shows this peculiarity, whilst Table 4 shows the cost function performance of these 2 Fuzzy Logic Controllers with this slight change (Fig. 15).

The dynamics, quick reacting nature of the braking system is proving to be difficult for the Mamdani FLC, Sugueno systems are a possible alternative. These systems remove fuzzification of the output and generate a crisp output using the MFs for the input variables. The output can be modified via constants which can change performance. Sugueno systems were investigated as a solution however there was not a worthwhile change in performance, the ability to modify the output coefficients

Table 3 Tabulated descriptions of the fuzzy logic controllers designed

Name	Number of MF's	Notes
FLC 1	7	Gaussian zero MF, positive MF's angled towards 0, even spaced torque MF's
FLC 2	5	Evenly spaced MF's, dError wider zero MF
FLC 3	3	Outer MF's point to 0 for linearness, zero MF very wide Gaussian

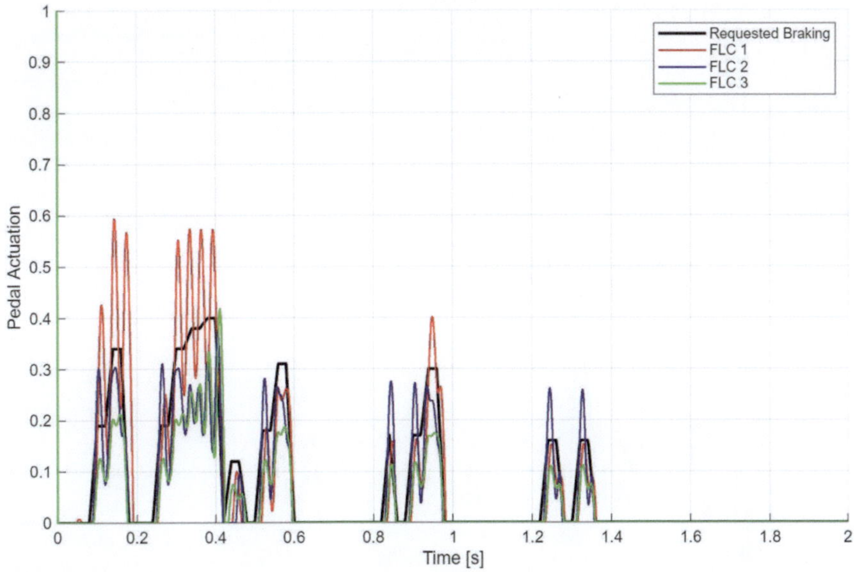

Fig. 14 Plot of pedal actuation control response for fuzzy logic controllers with varying membership functions

Table 4 Tabulated descriptions of the fuzzy logic controllers designed

Variable	FLC 1	FLC 2	FLC 3	FLC 1 (unstable)	2DOF PD
Performance cost (J)	29.5890	24.6540	27.1563	30.5605	25.4743
Wrong sign torque (W)	0.0777	0.1627	0.0774	0.1246	0.1913
Chatter (C)	0.0063	0.0104	0.0046	0.0095	0.0026
Flips (F)	22	17	18	24	21
Integral of absolute error (E)	6.0938	5.5491	7.8315	4.6930	3.0498
Overshoot (O)	1	1	1	1	1
Steady state error (s)	0	0	0	0	0
Control effort (e)	1.2072	1.8438	1.0350	1.5360	1.2078

was initially helpful in reducing error however it started introducing instability to the system when nearing the set point at the same stage as the Mamdani systems. As discussed above the tuning of membership functions is a precise art and the results help conclude the fact that the MF's are not functioning correctly and cannot reach the desired error.

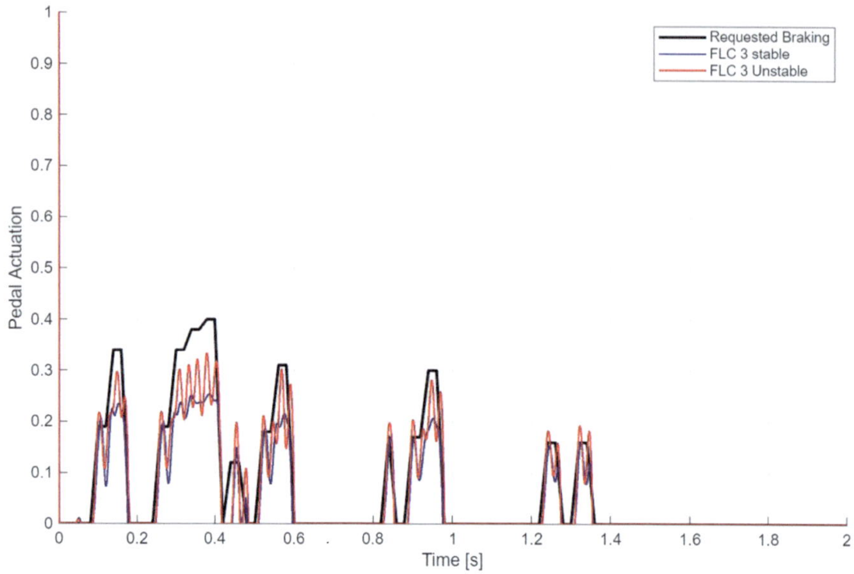

Fig. 15 Comparison of FLC 3 error and stability with the zero MF error Gaussian modified where stable $\sigma = 0.08$ and unstable σ is 0.12

4.3 Control System Comparison

Using the cost function detailed in Eq. 2, it is possible to quantify the performance of both the PID and FLC systems. The outputs of this function are seen within Table 4. The 1DOF PID has been discarded due to the discussed differences between them with the 2DOF PID providing better performance.

The performance of each of these control systems varies with each one having strengths and weaknesses. An interesting situation is that FLC 4 and 5 have similar performance when the cost function is considered. The next stage of analysis is fitting the control response into the numerical braking model which was presented earlier, this will provide quantitative data to understand how each controller affects performance.

5 Braking Performance

It is possible to use the braking simulator presented earlier to gain quantitative data and understand how control system performance affects the vehicles ability to stop as intended. The set value trace will be simulation initially which will be used to compare the different control approaches. It is expected that the brake pressure application

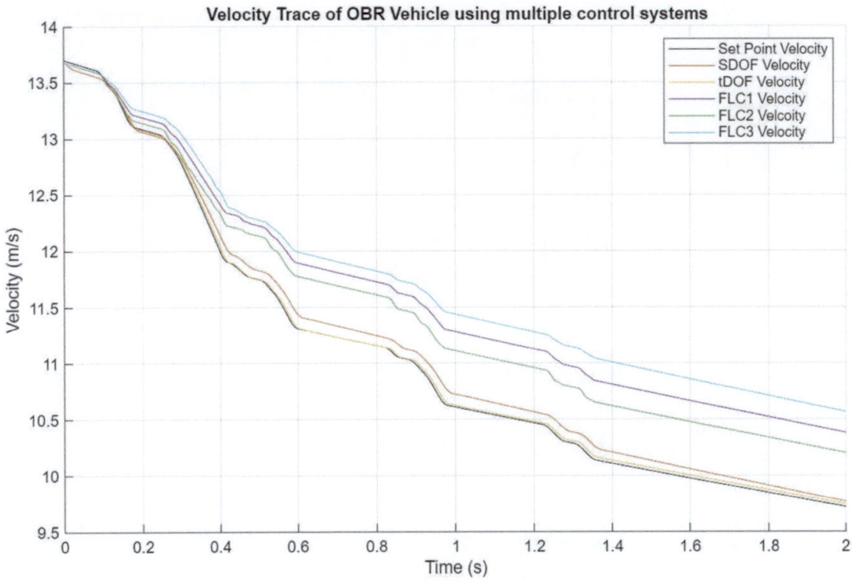

Fig. 16 Velocity trace of the 5 discussed control systems compared against the set point

will fluctuate and mirror the pedal actuation percentage which is simulated however it is important to look at how that affects the overall deceleration of the vehicle.

The velocity trace (Fig. 16) plots the theoretical ideal braking behaviour against the braking behaviour which results from the control response of the discussed controllers. The Fuzzy Logic Controllers are severely lacking in performance with a relatively large undershoot which is consistent with what was seen when analysing the control response independently. The difference in performance between the FLC's and the set value makes the use of these controllers hard to justify. Relating back to the initial goals of preferring overshoot rather than undershoot. The results are showing error in the areas with larger desired actuation greatly affect the deceleration more than error in small spikes. The PID control provides more promising results though, the 1DOF PID has slight offset due to control undershoot which is not as pronounced as the FLC counterparts. The strong performance is from the 2DOF PID controller which follows the set point strongly with a 0.05 m/s difference in velocity at the end of the braking event (Figs. 17 and 18).

6 Discussion

The choice to base all control on pedal position provide good accuracy whilst making the design easy to follow and understand. Modelling the stiffness of the braking system that is felt at the pedal took multiple iterations to have working due to Simulink

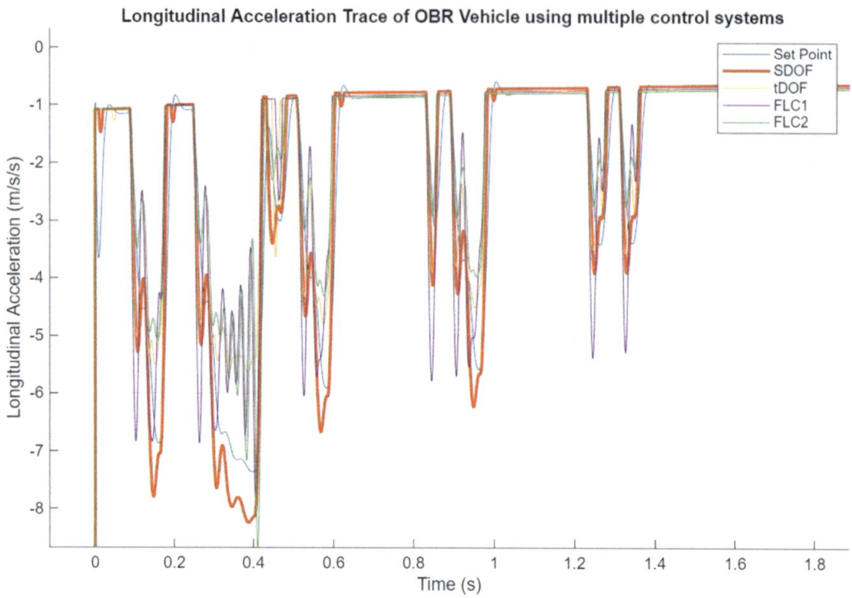

Fig. 17 Longitudinal acceleration trace of the 5 discussed control systems compared against the set point

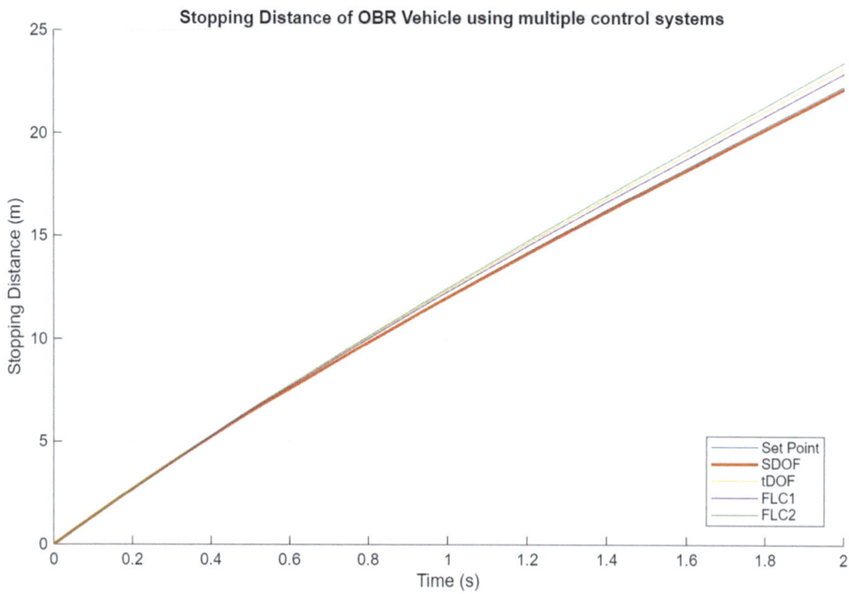

Fig. 18 Stopping distance of the 5 discussed control systems compared against the set point

blocks having weird behaviour, the choice to equate pedal position to percentage of maximum brake force works well for this paper however in reality there would be small "numb" zones at the start and end of the stroke due to the brakes first engaging the pads properly and then at the end due to compliances giving extra travel. The glaring omission in the simulations conducted is the lack of the wire pulling system being simulated. The lack of tension being taken by the servomotor skews the results slightly however as the first actuation of the pedal would not have the desired effect due to the motor taking the tension, once tensioned the cable would work similarly however it introduces more compliances to the system. The implemented solution works identically to the cable pulling though with the lack of tension and stress in the pedal not simulated. The integration of controllers into Simulink required the signals to be filtered, these were initially set up wrong with the output of the Fuzzy Logic block being filtered at 100 Hz which is orders of magnitude lower than what the control system would function at on low-level electronics. Increasing the frequency above 1 kHz stabilised results and aided performance. The simulation of braking dynamics has shown strong performance and correlation with OBR data. Implementation of this model presented challenges with integration with the Simulink model which led to the process of simulating control response and determining braking performance a split process with the data of pedal travel needing imported to the MATLAB script before running. Once the simulation runs it provides strong results. Considering the control approaches of PID first, PID was expected to provide strong results which it did. The dynamics and quick response required of the system did complicate the development with the 1DOF PID not being able to function as a stable control approach whilst maintaining stability. The choice of using the braking signals from OBRA aided development a lot as the braking response required is among the most technical seen by the team this season, by designing a system that can function on this response ensures that it would work on all expected responses. Moving to the 2DOF PID which allowed adjustment of the B and C values provided the responsiveness and accuracy which was desired. Initial efforts to use the Zeigler Nichols method of tuning was not working well so a trial-and-error approach was employed with one variable being modified at a time to understand response. The final values used to set the PID response are quite odd with a large B value which sets the weighting of the set point. This is required to be high to reach the set point with the proportional gain functioning to remove instability within the system. Using the cost function to quantify the performance of the controller is a quick metric that allows comparison with the resulting cost of each system, the 2DOF PD shows the 2nd lowest cost of 25.4743 which is only slightly higher than the cost of FLC 2. The 2DOF PD control has a relatively smaller absolute error when compared to the other tested controllers which is visually verified when looking at Fig. 11. The relatively low error translates to a velocity trace which matches the set point simulation incredibly closely with a velocity difference of 0.05 m/s difference at the end of the braking event which further proves this 2DOF PD implementation is fit for purpose and can function effectively as a control system for the AIBS functionality within the OBR car. The Fuzzy Logic Controllers have been developed in parallel to the PID approach, initial work and research into these controllers presented promising signs however effective implementation of these controllers proved difficult. Consulting the cost of each controller from the cost function (Table 4) shows that all 3 designs perform worse compared to the 2DOF PD controller in the key metrics of absolute error and chatter which aligns with the issues discussed about tuning which is that in this system the signal can either stable and have significant undershoot or unstable and have less undershoot. This was tested with the comparison of FLC 3 with the zero Gaussian MF widened and narrowed by 0.04 which made drastic changes to performance. The unstable FLC 3 has a low absolute error since the signal oscillates around and past the set point, this instability is shown in the higher chatter values. The issue all originates with the setup of the membership functions with them being extremely sensitive to change. Whilst the PD controller has better performance the development time of that compared to the FLC's is roughly a third. The results now are showing that based on comparison of error and behaviour the Fuzzy Logic Controllers are not suitable for use within this braking system. Looking at the performance analysis using the braking simulator backs up this conclusion too. The vehicle is usually 1 m/s faster than required which will cause a failure and cones to be hit during competition as a minimum. The closed loop nature of the autonomous system prefers the car to brake harder which will allow the brakes to be released earlier.

7 Conclusion

The work presented within this paper shows that integrating AI braking functionality into an OBR car is not only possible but can be integrated cleanly and for relatively low cost. The work required to integrate this system into the chassis is very low and most of the work required is tuning the control system. The presented PID and Fuzzy Logic control systems provide a comparison of performance for this system which requires high accuracy, quick response and consistent behaviour. These requirements have proved too much for the implementation of Fuzzy Logic Controllers with 3 distinctly different controllers designed, implemented and tested with the key difference being the number of membership functions. These systems require a trade-off between accuracy and stability based upon the requirements of the system which is disappointing. These controllers provided a minimum error of 0.1 at their peak during the harder braking event which is not satisfactory performance for this system. The PID approach proved a lot more effective, implementation of a 2DOF PD controller was relatively simple within Simulink and the tuning process was intuitive which has resulted in a controller that is stable and provides great performance with an absolute error value of 3.0498 which is mainly generated during slight overshoot whilst the controller settles. The development of a parameterised Simulink model using the Simscape package allowed for effective simulation of performance with the stiffness of the real brake pedal effectively simulated using key assumptions such as the brake system being perfectly bled. The choice of using pedal actuation percentage as the input allows flexibility for scenarios where the brake system has air in it or is reduced in performance when compared to approaches which use the desired brake pressure which would be skewed if the system was impure. Validation of controller behaviour was possible due to the numerical braking performance model which was validated to be comparable to the OBR vehicle. This braking model is effective in its use for this analysis however it has potential to be integrated into a wider vehicle model to allow for accurate analysis of braking dynamics. The future work which would develop this project focuses largely on the work required for implementation into the OBR vehicle and fine tuning the control performance. For example, this paper has examined the torque inputs required to actuate the pedal to the desired percentage. This would require further development to implement the required PWM signals required for actuation. Alongside this development of real-world use of the servomotor, the pedal assembly of the OBR vehicle would require small modification to fit a sensor which would provide the pedal actuation percentage. Finally, the braking system developed within this paper requires extensive testing to ensure its safety and that its response is acceptable which compared to the requested signal. The controllers themselves are effective in their current state however extra research could cover the effects a Sugueno FLC as Mamdani FLC's were primarily used in this paper.

A. **OBRA's Autonomous System**

The autonomous system functions using a collection of ROS 2 nodes which communicated between each other and are primary written in Python. The system is segmented into distinct sections namely Perception, Path Planning, Localisation and Mapping and Controls. The whole system runs on an innovative distributed computing system which is optimised for power efficiency whilst 4× the CPU performance per watt [11]. The Control Module is the primary node of concern, at the most basic level the AV system uses a Pure-Pursuit Controller (PPC) to determine steering angle with modifications being made to improve longitudinal control. This control is based on observed curvature with this value linearised and a desired velocity set via linear interpolation using predefined maximum and minimum velocities which the team can tune based upon the track characteristics. When the current velocity exceeds the desired velocity, the system determines a brake percentage value which the vehicle uses to apply the brakes. It is assumed that the stroke of the master cylinder and pedal travel is proportional so that 10% of pedal actuation request equals 10 % of the master cylinder stroke. This is the signal which the AIBS must uses to determine braking effort.

B. **Hydraulic Braking System**

The hydraulic braking system used by OBR is relatively simple using 2 master cylinders with diameter 15.8 mm and 15mm for the front and rear hydraulic circuits. AP Racing callipers are used with 2 pistons and 1 piston callipers on the front and rear, respectively. These components are connected with copper hard lines that run through the chassis with the brake lines that are outside being flexible to allow movement and compliance. The team uses a bespoke pedal box which is designed and manufactured by students at the university which houses the master cylinders. Figure 19 shows the brake pedal which will be connected to the servomotor. It is machined from high strength aluminium alloy with its geometry only needing slight modification to be compatible with the desired braking approach.

C. Simulink Model Construction Difficulties

Initially, the aim was to simulate both master cylinders independent as each one exerts a different stiffness due to their hydraulic construction. Implementing this in Simulink using the Simscape Spring and Damper Block proved difficult with the model having more movement than intended. To rectify this spatial contact forces were implemented to try and simulate the bump stops. Issues arise with the response of the system as the "bumpstop" would cause massive oscillations in the system and cause unrealistic behaviour like the pedal being stuck at 100 % actuation when forces greater than 400N were applied. The result of this was a system which had usable movement to around 60% of pedal stroke but was unusable above that threshold.

D. Saraiev Braking Model and Validation Against OBR Car

A sub-goal of this project was to utilise a simulation which captured the complex dynamics of braking. The initial approach was the implementation of the numerical model proposed by Saraiev and Gorb [13] which was discussed earlier. This model presents an ODE which describes the deceleration experienced during braking as function of grip, aerodynamic forces and road condition. This model was implemented within MATLAB and modified to function using a given percentage of brake pedal actuation. The brake actuation of the aforementioned Simulink model is saved and then used within this code.

$$\begin{cases} -\dfrac{dv}{dt} = \dfrac{g}{\delta_J}\left[\left(\cos\lambda - \dfrac{0.5c_y\rho F_y v^2}{G_a}\right)(\phi + f) + \dfrac{0.5c_x\rho F_x v^2}{G_a} \pm \sin\lambda\right], \\ v(0) = v_0 \end{cases} \tag{10}$$

Modifying the code for the OBR vehicle required simple values such as mass, aerodynamic coefficients and surface areas alongside grip coefficients. Extra consideration was made to include driver mass as a separate values so simulation of the effects of driverless operation can be studied in future. To validate the simulation real-world data was used, specifically the data gathered via an GNSS unit from FS Czech 2025 Endurance. An extra step was added as the team does not measure brake percentage actuation so a lookup table for percentage actuation had to be created via trial and error which created the desired longitudinal deceleration. A complete copy of the braking model code will be available within the appendix.

The comparison of data (Fig. 20) shows that the deceleration experienced by the vehicle matches the simulation strongly. The dips in the recorded data originated within in IMU. Looking at the velocity it matches well at higher velocities however as the brakes are released the final velocity has a relatively larger difference. Figure 21 plots the error over time, with a peak of roughly 17% at the end of braking. Over the complete braking event the average percentage error is 5.33% which is deemed satisfactory and whilst slightly over the specified error of 5% for the simulation

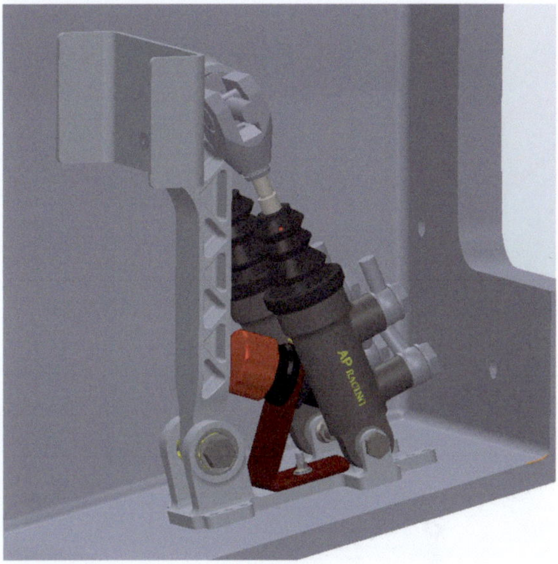

Fig. 19 OBR brake pedal

the low average as the vehicle decelerates means that confidence is high in this simulation.

E. **MATLAB Braking Model Code**

```
function F = brake_force(v, Fy, Fx)
% FS Car Braking Model with Time-Varying Pedal Input clear; clc; close all;

%% --- 1. Basic Physical and Aerodynamic Parameters –g = 9.81; % [m/s^2].

% Vehicle Data m_car =
236.63; m_driver=75; m =
m_car+m_driver; Ga =
m*g;

% Aerodynamic parameters.
p = 1.115;
c_x = 1.5; c_y = -
4.42; F_x = 1.9;
F_y = 2.0;

% Road slope lambda_deg = 1; lambda =
lambda_deg * pi/180;
```

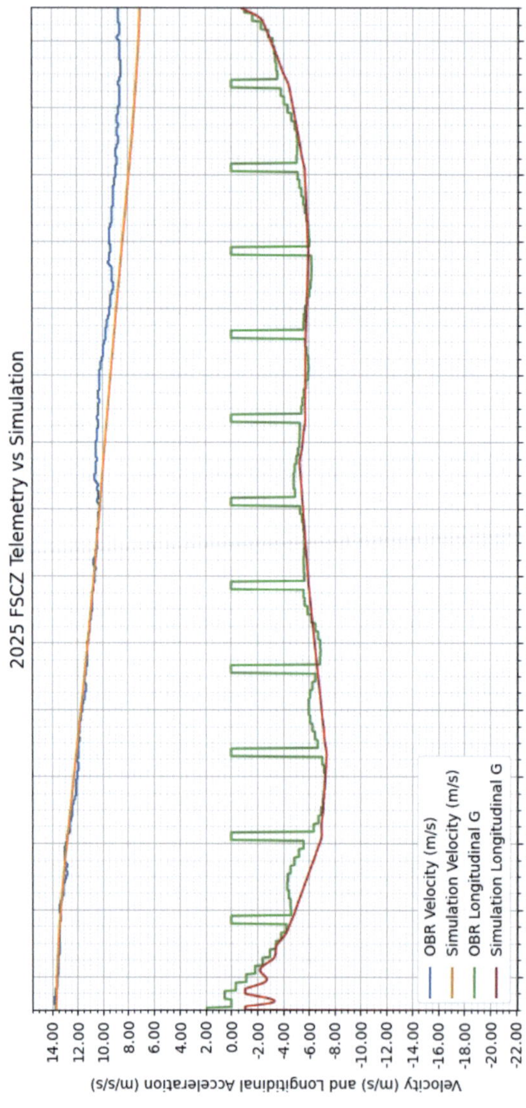

Fig. 20 FSCZ endurance data overlayed against numerical braking model outputs

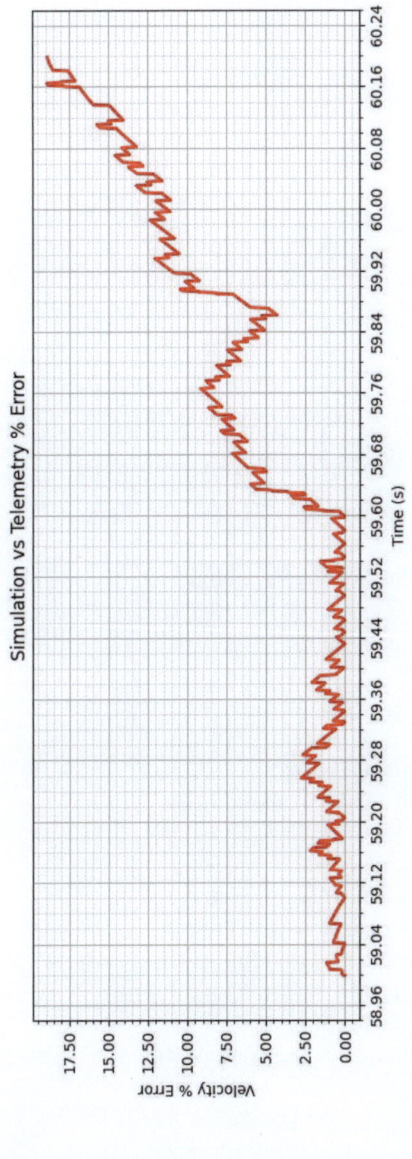

Fig. 21 Velocity error of recorded data versus simulation data

```
%% --- 2. Tire-Road Friction Parameters –q0 = 1.5; f = 0.1;

useSpeedDep = true; if
useSpeedDep.
A = 0.002;
else
A = 0;
end

%% --- 3. Rotational Inertias and Effective Mass Factor ---
Tk1 = 0.30;
Tk2 = 0.30;
Jk1 = 1.53;
Jk2 = 1.53; R_eff =
0.30;

wheelsCont = (Jk1/Tk1^2 + Jk2/Tk2^2) / m; gamma_target = 1.04;
engineCont_req = gamma_target - 1 - wheelsCont;
Je = engineCont_req * (R_eff^2 * m);

contrib_wheels = wheelsCont;
contrib_engine = Je/(R_eff^2*m);
gamma = 1+contrib_wheels+contrib_engine;

%% --- 4. Aerodynamic Constants ---
C_n = 0.5 * c_y * p * F_y / Ga;
D_n = 0.5 * c_x * p * F_x / Ga;

%% --- 5. Display Constants ---
% fprintf('--- Calculated Model Constants ---\n');
% fprintf('Vehicle weight, Ga = %.2f N\n', Ga);
% fprintf('Vehicle mass, m = %.2f kg\n', m);
% fprintf('Air density, p = %.3f kg/m^3\n', p);
% fprintf('Drag coefficient, c_x = %.3f\n', c_x);
% fprintf('Lift coefficient, c_y = %.3f\n', c_y);
% fprintf('Frontal area, F_x = %.3f m^2\n', F_x);
% fprintf('Lift area, F_y = %.3f m^2\n', F_y);
% fprintf('Road slope, lambda = %.2f deg (%.4f rad)\n', lambda_deg, lambda)
;
% if useSpeedDep
% fprintf('Tire friction: q0 = %.2f, speed-dependence A = %.4f, additional f
= %.2f\n', q0, A, f);
% else
```

```
%fprintf('Tire friction: constant q0+f = %0.2f\n', q0+f);
% end
% fprintf('Wheel parameters: Jk1 = %.2f, Tk1 = %.2f; Jk2 = %.2f, Tk2 =
%.2f \n', Jk1, Tk1, Jk2, Tk2);
% fprintf('Engine inertia: Je = %.2f kg·m^2, R_eff = %.2f m\n', Je, R_eff);
% fprintf('Rotational contributions: wheels = %.5f, engine = %.5f\n', contrib_
wheels, contrib_engine);
% fprintf('Total rotational mass coefficient, gamma = %.4f\n', gamma);
% fprintf('Aerodynamic constants: C_n = %.8f, D_n = %.8f\n', C_n, D_n);
% fprintf('----------------------------------\n\n');

%% --- 6. Load Pedal Data from Simulink --
load('pedal_perc.mat'); 80ts = ans;

% Convert Time to numeric (seconds)
if isdatetime(ts.Time).
t_numeric=seconds(ts.Time - ts.Time(1)); % shift start to 0.
elseif isduration(ts.Time).
t_numeric = seconds(ts.Time);
else
t_numeric = ts.Time; % already numeric
end

% Extract Data
data_numeric = ts.Data;

% Create interpolation function
pedalFun = @(t) interp1(t_numeric, data_numeric, t, 'linear', 'extrap');

%% --- 7. Define Braking ODE ---
frictionFun = @(v,t) (q0 + f - A*v).* pedalFun(t);

brakingODE = @(v,t) (g/gamma) * ( frictionFun(v,t) * (cos(lambda) - C_n*v
.^2) - D_n*v.^2 - sin(lambda));
ode_fun = @(v,t) -brakingODE(v,t);

%% --- 8. Numerical Integration via Euler ---
v0 = 13.7;
dt = 0.001; t_max = 10;
N_steps = ceil(t_max/dt);

time = zeros(N_steps,1);
v_num = zeros(N_steps,1);
```

```
a_num = zeros(N_steps,1);
N_effArr = zeros(N_steps,1);
F_frictionArr = zeros(N_steps,1);
F_dragArr              = zeros(N_steps,1);
F_gravityArr = zeros(N_steps,1);
pedalArr = zeros(N_steps,1);

v_num(1) = v0;
time(1) = 0;

for i = 1:N_steps-1 t =
time(i);

% Pedal input
pedal_perc_val = pedalFun(t); pedalArr(i) =
pedal_perc_val;

% Friction coefficient
q_eff = (q0+f - A*v_num(i)) * pedal_perc_val;

% Forces
N_eff = Ga*cos(lambda) - 0.5*c_y*p*F_y*v_num(i)^2;
F_frict = q_eff * N_eff;
F_drag = 0.5*c_x*p*F_x*v_num(i)^2;
F_grav =
Ga*sin(lambda);

% Store
N_effArr(i) = N_eff;
F_frictionArr(i) = F_frict;
F_dragArr(i) = F_drag;
F_gravityArr(i) = F_grav;

% Acceleration and update
a = (F_frict - F_drag - F_grav) / (m*gamma);
a_num(i) = a;
v_num(i+1) = v_num(i) + a*dt; 146time(i+1) = t + dt;

if v_num(i+1) < =0.1
v_num(i+1:end) = 0;
time(i+1:end) = t;
pedalArr(i+1:end) = pedal_perc_val;
break;
```

```
end.
end.

% Trim arrays
lastIndex = find(v_num > 0,1,'last');
if isempty(lastIndex), lastIndex = length(v_num); end.
time = time(1:lastIndex);
v_num = v_num(1:lastIndex);
a_num = a_num(1:lastIndex);
N_effArr = N_effArr(1:lastIndex);
F_frictionArr = F_frictionArr(1:lastIndex);
F_dragArr = F_dragArr(1:lastIndex);
F_gravityArr = F_gravityArr(1:lastIndex); pedalArr =
pedalArr(1:lastIndex);

%% --- 9. Compute Stopping Distance --x_num =
cumtrapz(time, v_num);

%% --- 10. Plot Results ---
figure;
subplot(4,1,1);
plot(time, v_num,'b-','LineWidth',2);
xlabel('Time (s)'); ylabel('Speed (m/s)'); title('Vehicle Speed vs. Time'); grid
on;

subplot(4,1,2);
a_model = -[diff(v_num)./diff(time); a_num(end)];
plot(time, a_model,'r-','LineWidth',2);
xlabel('Time (s)'); ylabel('Deceleration (m/s^2)'); title('Vehicle.
Deceleration vs. Time'); grid on;

subplot(4,1,3);
plot(time, x_num,'k-','LineWidth',2);
xlabel('Time (s)'); ylabel('Distance (m)'); title('Stopping Distance vs. Time');
grid on;

subplot(4,1,4); plot(time, -pedalArr,'m-','LineWidth',2); xlabel('Time (s)');
ylabel('Pedal Position (0-1)');
title('Pedal Input vs. Time'); grid on;

csvexportdata = [time(:), v_num(:),a_num(:)]; writem-
atrix(csvexportdata,"Braking_Sim_Correlation.csv");
```

```
%% --- 11. Output Final Values --stopping_time = time(end); stopping_
distance = x_num(end)
; fprintf('Estimated stopping time: %.3f s\n', stopping_time);
fprintf('Estimated stopping distance:
%.3f m\n', stopping_distance);

%% --- 12. Detailed Forces --figure; plot(time,F_frictionArr,'g-
','LineWidth',1.5);
hold on; plot(time,F_dragArr,'r--','LineWidth',1.5); plot(time,F_
gravityArr,'b-
.','LineWidth',1.5); xlabel('Time (s)'); ylabel('Force (N)');
legend('Friction','Drag','Gravity','Location','best'); title('Force Contribu-
tions vs.
Time'); grid on; end
```

F. **Fuzzy Logic Controller Rule Maps**

FLC 1 Rule Map

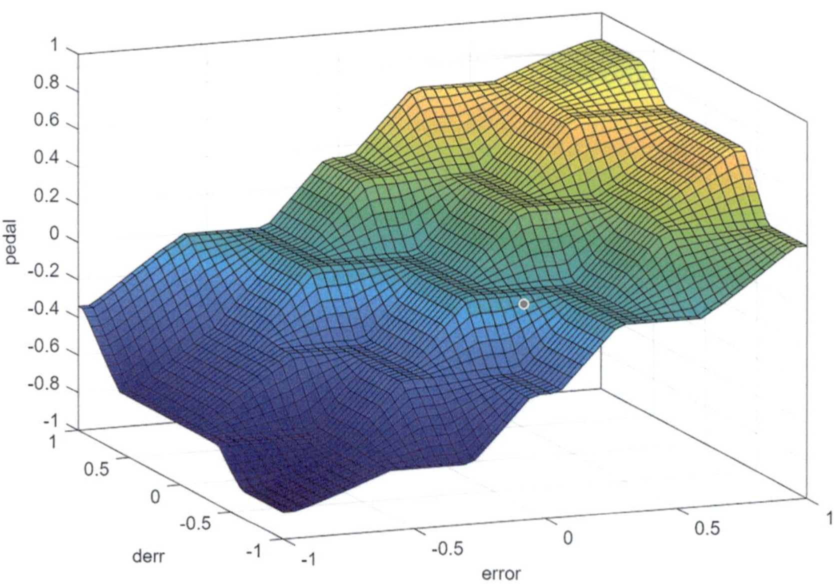

FLC 2 Rule Map

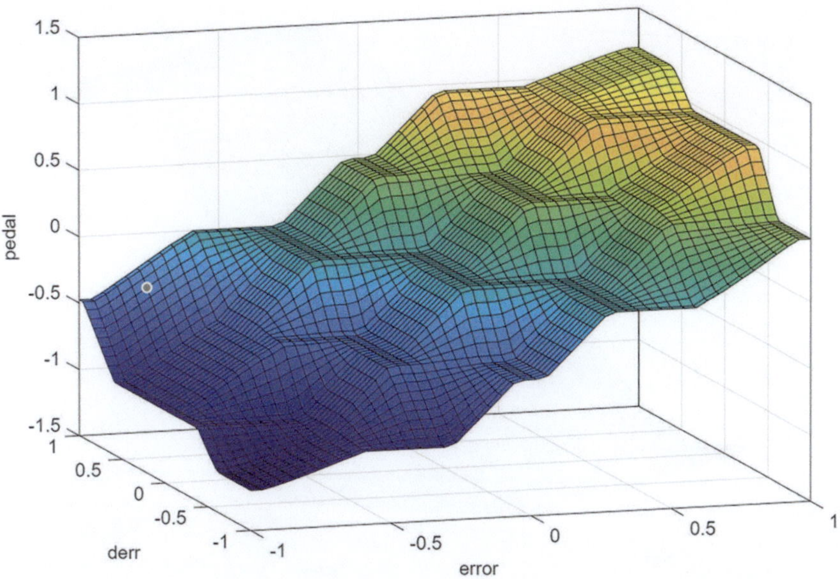

FLC 3 Rule Map

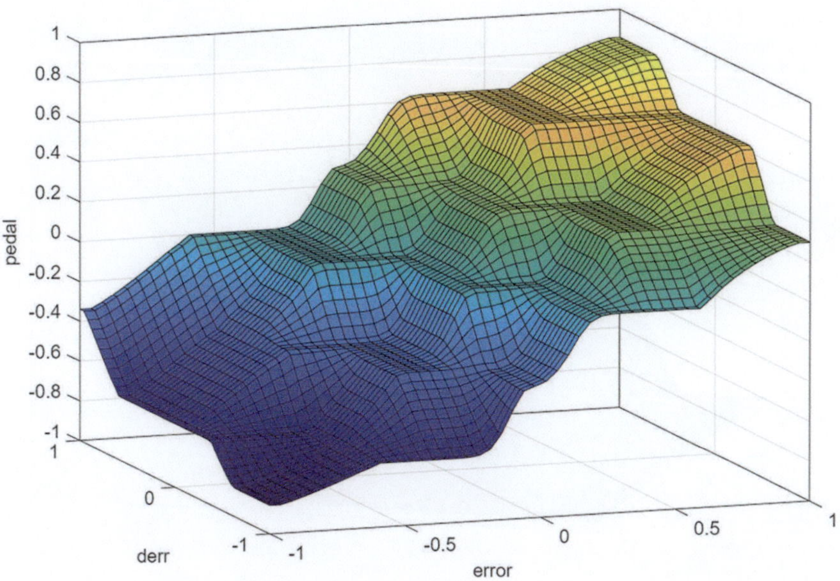

FLC 4 Rule Map

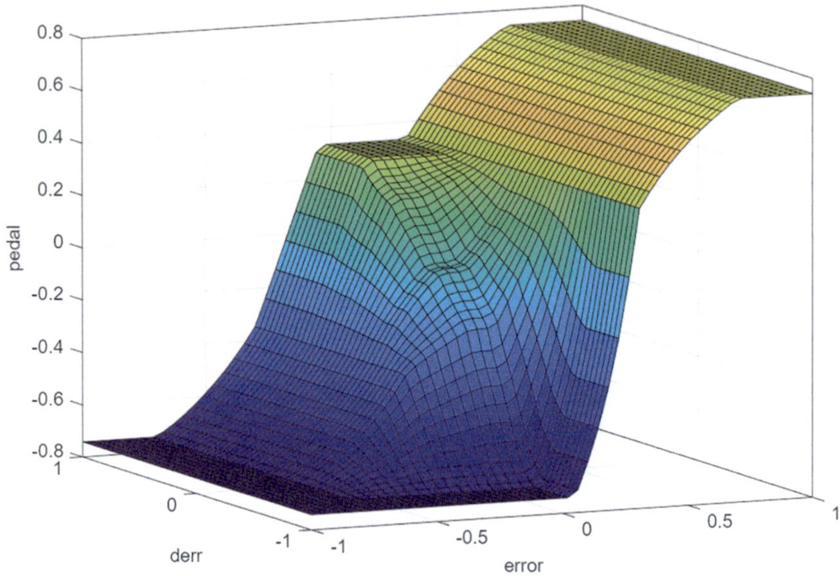

FLC 5 Rule Map

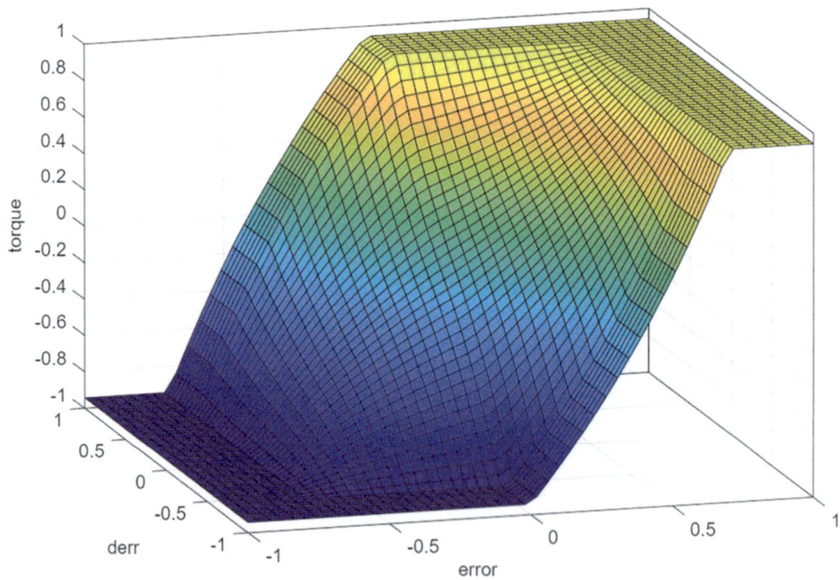

References

1. Åström KJ, Murray RM (2002) Feedback systems: an introduction for scientists and engineers, Chapter 6. https://www.cds.caltech.edu/~murray/courses/cds101/fa02/caltech/astrom-ch6.pdf. Accessed 27 Aug 2025
2. Ciocca S (2022) Design and implementation of autonomous braking system for formula student driverless application. https://webthesis.biblio.polito.it/secure/25008/1/tesi.pdf. Accessed 19 Mar 2025
3. Freidrich T, et al. (2021) Winning through simplicity: autonomous car design for formula student. https://arxiv.org/abs/2406.13256. Accessed 19 Mar 2025
4. International Organization for Standardization (2019) ISO/PAS 21448:2019—Road vehicles—Safety of the intended functionality. Publicly available specification. https://www.iso.org/standard/70939.html. Accessed 24 Aug 2025
5. International Organization for Standardization (2024) ISO/PAS 8800:2024—Road vehicles—Safety and artificial intelligence. Publicly available specification. https://www.iso.org/standard/83303.html. Accessed 24 Aug 2025
6. International Organization for Standardization (2018) ISO 26262:2018—Road vehicles—Functional safety. Standard. https://www.iso.org/standard/68383.html. Accessed 21 Aug 2025
7. Kazban J (2019) AMZ driverless: the full autonomous racing system. https://arxiv.org/pdf/1905.05150. Accessed 20 Mar 2025
8. Köhler J, Müller MA, Allgöwer F (2024) Analysis and design of model predictive control frameworks for dynamic operation—An overview. https://www.sciencedirect.com/science/article/pii/S1367578823000937. Accessed 01 Sept 2025
9. Liu Y, Pei X, Guo X (2022) Redundancy control of anti-lock braking system based on electro-hydraulic braking system. SAE Int J Veh Dyn Stabil NVH 7(1). https://doi.org/10.4271/10-07-01-0004.
10. Mattas F et al (2021) Safety aware fuzzy longitudinal controller for automated vehicles. https://www.sciencedirect.com/science/article/pii/S2095756421000635. Accessed 01 Sept 2025
11. Milne L et al (2025) OBR-autonomous: system design for an autonomous racing vehicle. Technical report. Oxford Brookes Racing Autonomous
12. Saloux E et al (2025) From theory to practice: a critical review of model predictive control field implementations in the built environment. https://www.sciencedirect.com/science/article/pii/S0306261925008219. Accessed 01 Sept 2025
13. Saraiev O, Gorb Y (2018) A mathematical model of the braking dynamics of a car. In: SAE technical papers on CD-ROM/SAE technical paper series. https://doi.org/10.4271/201801-1893.
14. Savaresi SM, Tanelli M (2010) Active braking control systems design for vehicles. Springer
15. Shi Q, He L (2023) A model predictive control approach for electro-hydraulic braking by wire. https://ieeexplore.ieee.org/stamp/stamp.jsp?arnumber=9736620. Accessed 20 Mar 2025
16. Somefun OA, Akingbade K, Dahunsi FM (2021) The dilemma of PID tuning. https://www.sciencedirect.com/science/article/pii/S1367578821000407. Accessed 1 Sept 2025
17. Sorrentino G (2020) Design of autonomous control systems for a driverless race car: remote emergency system and autonomous state machine. Master's thesis, Politecnico di Torino. https://webthesis.biblio.polito.it/16282/1/tesi.pdf. Accessed 19 Mar 2025
18. Sorrentino G et al (2021) Remote emergency braking for autonomous vehicles. https://www.researchgate.net/profile/Gennaro-Sorrentino/publication/356315865_Remote_Emergency_Braking_System_for_Autonomous_Racing_Electric_Vehicles/links/653a2ebd1d6e8a70704f2887/Remote-Emergency-Braking-System-for-AutonomousRacing-Electric-Vehicles.pdf. Accessed 19 Mar 2025
19. Yang Y (2018) Design of a FSAE braking system. S.B. thesis. Massachusetts Institute of Technology. https://dspace.mit.edu/handle/1721.1/119947. Accessed 26 Aug 2025

Trajectory Tracking and Stability Evaluation of a Formula Student Vehicle Using Model Predictive Control

Oliver Fragoso and Aydin Azizi

Abstract In this research, a Model Predictive Control (MPC) approach has been utilized to perform trajectory tracking and evaluate vehicle stability within a Formula Student application. The results show that the framework achieves accurate path following while respecting vehicle constraints, offering a reliable and adaptable tool for both simulation and real-world testing. This research has been developed with the purpose of supporting a Formula Student team, aiming to improve vehicle performance through the implementation of advanced control strategies grounded in vehicle dynamics. The central objective is to enhance the drivability of the car, achieve smoother dynamic behaviour, and strengthen the correlation between real-world data and simulation tools by incorporating a deeper understanding of the vehicle's behaviour. The work focuses on the design and implementation of a Model Predictive Control (MPC) system for trajectory tracking, using the nonlinear bicycle model as the foundation. The system is discretized through Euler integration, enabling real-time feasibility. The MPC framework relies on a quadratic cost function, where weighting matrices are applied to balance the importance of states and control inputs, ensuring optimal trade-offs between accuracy and control effort. Linear inequality constraints are formulated to represent the physical and safety limits of the car, maintaining realistic and safe operating conditions. A structured programming approach in Python was adopted, following object-oriented principles to ensure scalability, modularity, and straightforward integration into the team's existing environment. The program architecture was designed to allow rapid modification of parameters, weights, and constraints, enabling the team to perform efficient tuning and faster iterations during development and testing. This flexibility not only reduces the time required to evaluate different control setups but also provides a practical foundation for real-world implementation. Overall, the research establishes a robust platform that combines theoretical control concepts with practical usability, supporting ongoing improvements in Formula Student competition performance.

O. Fragoso · A. Azizi (✉)
School of Engineering, Computing and Mathematics, Oxford Brookes University, Oxford, UK
e-mail: aydin.azizi@brookes.ac.uk

A. Azizi (ed.), *Intelligent Control and Autonomous Systems*, Emerging Trends in Mechatronics, https://doi.org/10.1007/978-981-95-6402-6_4

1 Introduction

Autonomous driving has become one of the most active areas of research in modern transportation, with applications ranging from advanced driver-assistance systems to fully automated vehicles. A central challenge in this field is the accurate and robust tracking of reference trajectories while respecting physical and safety constraints. Path tracking is particularly complex because it involves the coordination of multiple objectives: maintaining accuracy and ensuring stability at the same time. Traditional controllers such as Proportional Integral Derivative (PID) have been widely used in automotive applications due to their simplicity, but their limitations become evident in multivariable and constrained systems [12]. Other approaches, such as the Pure Pursuit algorithm, have also been applied successfully in lightweight autonomous vehicles, but they are less effective in scenarios where high accuracy and stability are required [18].

Model Predictive Control (MPC) has therefore become a strong candidate in this area. MPC predicts the future behaviour of the vehicle using a dynamic model and optimizes the control inputs to minimize a cost function while handling constraints in real time. It has already been applied across different subsystems, from steering and yaw control to longitudinal dynamics and driver-assistance functions [4, 15]. In path tracking, MPC has demonstrated clear advantages over classical controllers, achieving improved accuracy and stability, although computational effort remains a key concern for real-time deployment [6, 26]. Recent work has even explored hybrid strategies, such as combining MPC with machine learning, to reduce computational demands while improving prediction quality [22].

A crucial design step for MPC-based vehicle control is the choice of model. The bicycle model, in both kinematic and dynamic forms, has been extensively adopted because it is simple enough for real-time use yet rich enough to capture the key dynamics of yaw rate, sideslip, and steering behaviour [5, 15]. Precise modelling requires accurate parameter identification, and recent studies show that the yaw moment of inertia and related vehicle parameters have a direct effect on prediction quality and stability [25]. Advanced formulations, such as Linear Parameter Varying (LPV) frameworks, have also been proposed to extend the performance of simpler models, allowing controllers to adapt more flexibly to changing vehicle dynamics [10].

Beyond the dynamic modelling, vehicle stability can be further improved by integrating additional control actions such as direct yaw moment. Hierarchical MPC frameworks that distribute steering and yaw moments have shown improved handling in low-adhesion conditions, both in simulation and hardware in the loop validation [9, 13]. Complementary strategies using nonlinear methods such as sliding-mode control also highlight the importance of addressing yaw stability under challenging driving conditions [21]. Similarly, real-time yaw rate and sideslip controllers implemented with torque vectoring confirm the importance of computational efficiency when moving from simulation to embedded deployment [23]. Studies on coordinated steering angle and yaw moment distribution further reinforce that coupling these control actions yields superior vehicle stability [3].

The scope of MPC is not limited to vehicle dynamics alone. Research has also demonstrated its usefulness in energy management and battery systems, showing its versatility as a predictive optimization tool in the automotive field [7]. This reinforces its potential as a unified approach for addressing multiple subsystems in autonomous vehicles.

In addition to control and modelling aspects, the way the software is implemented plays a decisive role in ensuring that research outcomes can be integrated into real systems. Following sound programming practices helps reduce technical debt and allows systems to evolve as new algorithms are introduced. In this context, the SOLID principles of software engineering have been recognized as effective guidelines for achieving modularity, scalability, and long-term maintainability [1, 19]. Applying these principles in control implementation makes the MPC framework not only theoretically robust but also practically usable for iterative testing and deployment.

1.1 Research Gap and Contribution

While MPC has been widely studied for autonomous vehicles, most research focuses on either theoretical models or specific subsystems, often neglecting integration with modular software architectures tailored for iterative development. Moreover, although the bicycle model has proven effective, there is limited work combining its real-time discretization with structured programming practices that emphasize flexibility and scalability. In this research, we address this gap by designing and implementing a trajectory tracking MPC based on the dynamic bicycle model, discretized with Euler integration, and programmed in Python with a modular architecture that follows SOLID principles. This framework ensures that control solutions are expressed directly in terms of steering and acceleration inputs, enabling efficient tuning, faster iterations, and straightforward integration with the team's existing simulation and perception systems.

Table 1 presents a summary of the main gaps identified between the literature and the present implementation. While previous studies have advanced the state of the art in vehicle modelling, control design, and path tracking using MPC, several aspects remain underexplored, particularly regarding real-time discretization strategies and software design practices. Highlighting these gaps provides the foundation for defining the specific contribution of this research.

Table 1 Gaps between literature and implementation

Topic	Literature	Limitation	Contributions
Actuation scope	MPC controls throttle, brake and steering	Current system only controls throttle, no braking or steering	Develop an MPC model (MIMO)
Real-time implementation	Simulated MATLAB implementation, not validated in physical platforms	Not implemented and tested in formula student competition	Validate on the simulator used in OBRA
Path knowledge	Full path preview is used to define the reference trajectory	Formula Student car lacks memory and depends on 5 m of visible cones	Formulate local MPC using immediate cone inputs
Real-world validation	Validated in simulation only	It depends on how much time before the competition the car is loaned to the team for testing	Implement on Formula Student vehicle for real-time validation
Cornering	Yaw control is mentioned and stable high-speed cornering	Current doesn't use yaw moment	Apply MPC to handle those constraints

1.2 Structure of the Paper

The rest of this paper is structured as follows:

- Introduction provides background information and motivation for MPC.
- Methodology explains the approach used in this work. It has three parts: The control formulation (MPC), vehicle model (the bicycle model), and program (the software implementation and design choices).
- Results presents the outcomes, focusing on tracking performance, stability, and computational behaviour.
- Conclusions summarizes the main findings and points to future work.

2 Methodology

Figure 1 illustrates the methodological flowchart that structures this research. Each block corresponds to one of the main sections of the paper, outlining the logical sequence followed from the identification of the problem to the modelling and discretization stages, the design of the MPC framework, the program implementation, and finally the evaluation of results and conclusions. This flowchart provides a clear overview of the workflow and highlights how each step contributes to the overall objective of the study.

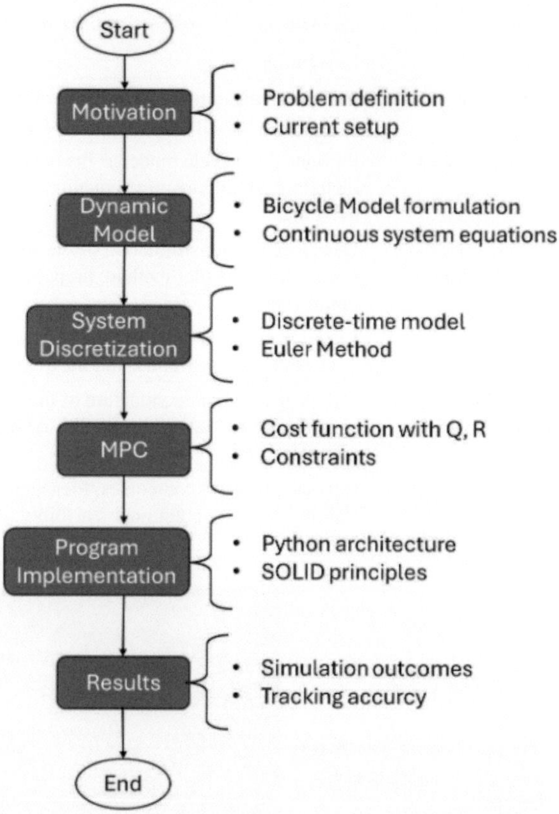

Fig. 1 Methodological flowchart of the research, showing the sequence of steps from problem, definition to implementation and results

Table 2 complements the methodological flowchart by summarizing the content of each section and highlighting the role of every step within the overall workflow.

2.1 Motivation

The current setup of the team relies on LiDAR cameras and GPS hardware, which are responsible for mapping the track and providing the coordinates that the vehicle must follow. Figure 2 shows the camera system used and an example of how the trajectory is represented in the onboard system.

At present, the team employs the Pure Pursuit algorithm to calculate the steering angle that should be applied to the vehicle. The algorithm uses a spline generated from the X and Y coordinates of the path ahead to determine the control action. Equation 1 illustrates the formula currently in use. The decision to rely on Pure Pursuit was motivated by its relatively low computational cost and its popularity in lightweight autonomous driving applications [18].

Table 2 Overview of the methodology and section content

Section	Description
Motivation	Introduces the current setup based on Pure Pursuit and interpolation, explains its limitations, and motivates the adoption of MPC
Dynamic model	Presents the bicycle model as the basis for predicting vehicle behaviour, including continuous-time equations for longitudinal, lateral, and yaw dynamics
System discretization	Describes the transformation of the continuous model into discrete-time form using the Euler method, preparing the system for real-time optimization
MPC	Explains the MPC formulation, including the cost function with weighting matrices Q, R constraints, and the quadratic programming representation
Program implementation	Details the Python architecture of the controller, highlighting modular class design and adherence to SOLID principles for scalability and maintainability
Results	Reports simulation outcomes, focusing on path tracking accuracy, stability, and computational feasibility of the proposed approach

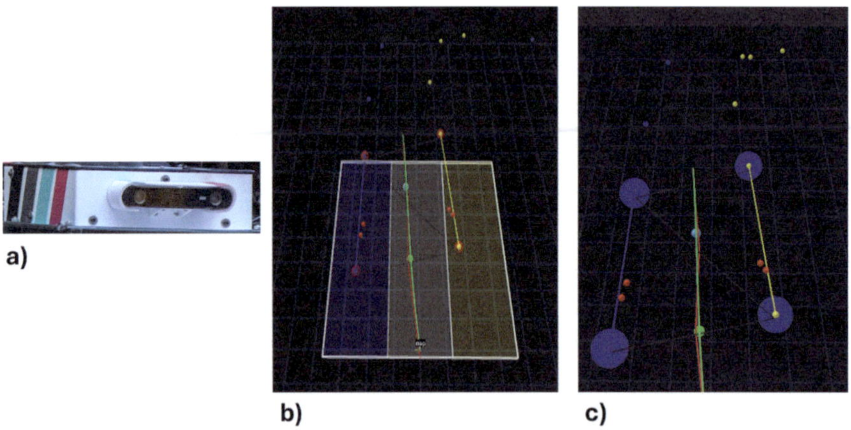

Fig. 2 **a**—OBRA LiDAR, **b**—it shows how the camera detects yellow and blue cones, **c**—track

$$\delta = \arctan\left(\frac{2 \cdot L \cdot \sin(\alpha_{pp})}{l_d}\right) \tag{1}$$

In addition to steering, the team uses linear interpolation to compute the longitudinal acceleration of the car. This interpolation takes into account the steering angle given by the Pure Pursuit algorithm and defines upper and lower velocity limits. By adjusting acceleration dynamically within these limits, linear interpolation allows the system to avoid errors that would occur if fixed values were applied instead [11].

2.2 Dynamic Model

To simulate the vehicle dynamics, the bicycle model was selected. This model provides a simplified yet effective representation of the car's motion while remaining compatible with the inputs to be controlled in this work: the steering angle and the longitudinal acceleration. The bicycle model has become a standard approach in control applications, particularly when combined with Model Predictive Control (MPC), because it allows the vehicle to follow a track while respecting constraints and minimizing a cost function. Its simplicity makes it attractive for real-time implementations, especially in MPC, where the model must be solved repeatedly within the optimization loop [9, 15]. Once the tuning parameters are adjusted correctly, this framework enables the car to be driven close to its handling limits [17].

Figure 3 illustrates the structure of the bicycle model and the main parameters that define it.

The state vector is defined as is shown in Eq. 2:

$$x = [\dot{x}, \dot{y}, \psi, \dot{\psi}, X, Y]^\top \tag{2}$$

The control inputs are shown in Eq. 3:

$$u = [\delta, a] \tag{3}$$

These states are obtained in real time from the LiDAR camera system developed by the team's electronics and engineering group. Using this setup provides accurate 3D perception of the environment and complements the vehicle's ability to generate reliable trajectories [16]. Figure 4 shows the advantages of having this system against other systems.

The kinematic relations describing position and heading are [15]:

$$\dot{X} = \dot{x} \cdot \cos\psi - \dot{y} \cdot \sin\psi \tag{4}$$

$$\dot{Y} = \dot{x} \cdot \sin\psi - \dot{y} \cdot \cos\psi \tag{5}$$

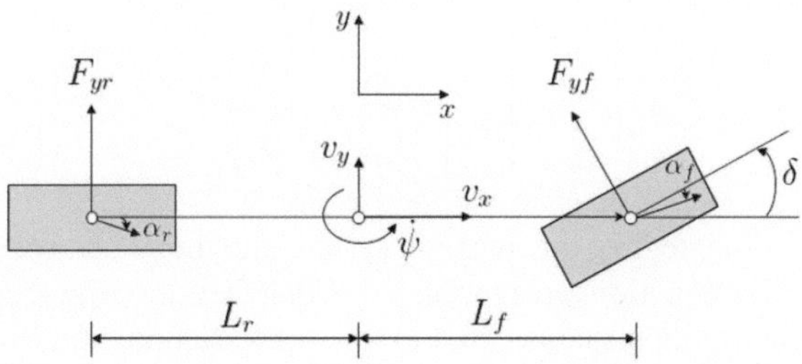

Fig. 3 Bicycle model [10]

Performance aspect	Human	AV			CV	CAV
		Radar	*Lidar*	*Camera*	*DSRC*	*CV+AV*
Object detection	Good	Good	Good	Fair	n/a	Good
Object classification	Good	Poor	Fair	Good	n/a	Good
Distance estimation	Fair	Good	Good	Fair	Good	Good
Edge detection	Good	Poor	Good	Good	n/a	Good
Lane tracking	Good	Poor	Poor	Good	n/a	Good
Visibility range	Good	Good	Fair	Fair	Good	Good
Poor weather performance	Fair	Good	Fair	Poor	Good	Good
Dark or low illumination performance	Poor	Good	Good	Fair	n/a	Good
Ability to communicate with other traffic and infrastructure	Poor	n/a	n/a	n/a	Good	Good

Fig. 4 Evaluation of perceptual systems in human drivers and advanced vehicle sensors [16]

For lateral and yaw motion, the classic linear single-track (2-DoF) form is used. In this representation, the system of forces is applied only at the front axle, which is steerable, while the rear axle is treated as a single wheel aligned with the vehicle's longitudinal axis [8].

$$F_x = F_{xf} \cdot \cos\delta + F_{xr} - F_{yf} \cdot \sin\delta \tag{6}$$

$$F_y = F_{yf} \cdot \cos\delta + F_{yr} - F_{xf} \cdot \sin\delta \tag{7}$$

$$M_z = l_f \cdot F_{yf} - l_r \cdot F_{yr} \tag{8}$$

Once the classical equations of the bicycle model are established, they must be expressed as a set of differential equations describing the behaviour of a dynamic system [8]. This representation makes it possible to discretize the model later, allowing numerical simulation of the vehicle's dynamics as a function of the chosen control inputs. In this way, the formulation produces updated state variables such as lateral velocity, longitudinal velocity, yaw angle, and yaw rate. Equation 9 presents the algebraic form of the resulting dynamic system.

$$\begin{bmatrix} \dot{y} \\ \dot{\psi} \end{bmatrix} = \begin{bmatrix} \frac{C_f+C_r}{mv_x} & \frac{-l_f C_f+l_r C_r}{mv_x} - v_x \\ \frac{l_f C_f-l_r C_r}{I_z v_x} & \frac{-l_f^2 C_f-l_r^2 C_r}{I_z v_x} \end{bmatrix} \begin{bmatrix} v_y \\ r \end{bmatrix} + \begin{bmatrix} \frac{C_f}{m} \\ \frac{l_f C_f}{I_z} \end{bmatrix} \delta \tag{9}$$

To integrate the dynamic system into the predictive control framework, the differential equations are rearranged so that they are expressed explicitly in terms of the two control inputs considered in this work: steering angle (δ) and longitudinal acceleration (a). This reformulation ensures that the solver computes trajectories directly as a function of these variables, aligning the optimization output with the vehicle's actuators. Equations (10)–(13) present the continuous-time state-space model expressed with δ and a as inputs, which forms the basis for the subsequent discretization stage.

$$\ddot{x} = -\frac{\mu g}{\dot{x}}\dot{x} + \frac{C_f \sin\delta}{m\dot{x}}\dot{y} + \left(\frac{C_f l_f \sin\delta}{m\dot{x}} + \dot{y}\right)\dot{\psi} - \frac{C_f \sin\delta}{m}\delta + a \tag{10}$$

$$\ddot{y} = -\frac{C_r + C_f \cos\delta}{m\dot{x}}\dot{y} + \left(-\frac{C_f l_f \cos\delta - C_r l_r}{m\dot{x}} - \dot{x}\right)\dot{\psi} + \frac{C_f \cos\delta}{m}\delta \tag{11}$$

$$\dot{\psi} = \dot{\psi} \tag{12}$$

$$\ddot{\psi} = -\frac{C_f l_f \cos\delta - l_r C_r}{I_x \dot{x}}\dot{y} - \frac{C_f l_f^2 \cos\delta + C_r l_r^2}{I_x \dot{x}}\dot{\psi} + \frac{C_f l_f \cos\delta}{I_x}\delta \tag{13}$$

2.2.1 System Discretization

When continuous-time models are adapted for predictive control, the choice of discretization method becomes a critical step. Three issues are typically considered in this process: first, selecting the most suitable discretization scheme; second, ensuring the convergence of the optimization algorithm applied to the finite-dimensional problem; and third, verifying that solutions converge to those of the original continuous problem as the time step is refined [14].

Common schemes include Euler, midpoint (or box), trapezoidal, and Runge–Kutta methods, all of which discretize state and control variables along the prediction horizon. Alternative approaches approximate the control trajectory using constant, linear, or spline functions while integrating the state equations more accurately, a strategy often referred to as control parameterization. In contrast, full discretization methods apply the scheme to both states and controls, with Euler being the most straightforward option. As shown in recent MPC studies, Euler discretization is often adopted because it offers low computational cost and straightforward implementation [5].

A summary of these discretization methods, together with their main features, is presented in Table 3.

Once the continuous-time bicycle model has been reformulated in terms of the inputs δ and a, it can be expressed in discrete form using a forward Euler approximation as is shown in Eq. 14:

$$x_{k+1} = x_k + T_s f(x_k, u_k) \tag{14}$$

Table 3 Common discretization methods for optimal control problems

Method	Key feature	Characteristics
Euler	Forward finite-difference approximation	Simplest; explicit update; widely used in MPC for efficiency
Midpoint	Uses average slope between intervals	Improves accuracy over Euler; moderate complexity
Trapezoidal	Approximates integral by trapezoid rule	Higher stability; implicit formulation
Runge–Kutta	Combines multiple evaluations per step for higher-order accuracy	Accurate but computationally more demanding
Control parameterization	Approximates control inputs (constant, linear, splines) and integrates states accurately	Reduces dimensionality; focuses on optimizing parameters

where $f(\cdot)$ represents the nonlinear dynamics of the bicycle model. This representation is widely adopted in predictive control studies as it provides a computationally efficient way to simulate system evolution under different input sequences [5, 15].

In addition to the state update, the system includes an output equation that links states and inputs to the measured or controlled variable. As is shown in Eq. 15:

$$y_k = C_d x_k + D_d u_k \tag{15}$$

For the present formulation, the direct feedthrough matrix D_d is set to zero, meaning that the outputs depend only on the current states and not directly on the inputs. This simplification is consistent with standard implementations of the bicycle model in predictive control, where variables such as lateral velocity, yaw rate, or global position are derived from the system states [4].

2.3 MPC

Path tracking in autonomous vehicles requires controllers that can handle multivariable dynamics and constraints while ensuring safety and comfort. Model Predictive Control (MPC) has emerged as one of the most promising approaches because it predicts future behaviour using a system model and optimizes control actions under constraints at each sampling instant. This makes it particularly suitable for autonomous driving tasks, where accuracy and robustness must be guaranteed simultaneously [4, 15].

Recent developments in the field confirm that MPC achieves superior performance compared with classical controllers in terms of stability and adaptability, especially under nonlinear and constrained conditions [16, 26]. In fact, path tracking is increasingly treated as a constrained optimization problem, where MPC minimizes trajectory error while also considering dynamic stability.

In contrast, classical controllers such as Proportional Integral Derivative (PID) control remain popular due to their simplicity and low computational requirements, but they are limited when dealing with multivariable, constrained, and nonlinear problems. Several studies highlight the advantages of MPC over PID in commercial and autonomous vehicle applications [12].

A summary of these differences is presented in Table 4, which compares the main features of PID and MPC for vehicle control.

Figure 5 shows the control flow adopted, where the planner provides the reference trajectory, the global MPC computes the optimal control increments considering system constraints, and the plant represents the vehicle model. An internal MPC layer is embedded to refine the control action, ensuring consistency between the planned trajectory and the actual vehicle response.

To implement predictive control in autonomous vehicles, MPC operates through a structured sequence of modules that interact in a closed-loop cycle. The framework begins with a planner that provides the reference trajectory, followed by the internal MPC where a vehicle model is used to predict future states and evaluate candidate control sequences. The global MPC layer then selects the first control action to be applied to the plant, and the resulting vehicle states are fed back for the next prediction cycle. This modular structure has been widely reported in the literature, highlighting MPC's ability to manage multivariable dynamics, constraints, and real-time feedback

Table 4 Comparison between PID and MPC in vehicle control

Aspect	PID	MPC
Prediction	Reactive, corrects error after it occurs	Predicts future states using a vehicle model
Constraints	Cannot explicitly enforce state input limits	Naturally incorporates constraints in optimization
Complexity	Simple to implement, low computational demand	Higher computational cost, requires solver
Adaptability	Limited adaptability to varying conditions	Adjusts to disturbances and nonlinearities
Performance	Effective in simple or linear systems	Superior tracking accuracy and stability in complex vehicle dynamics

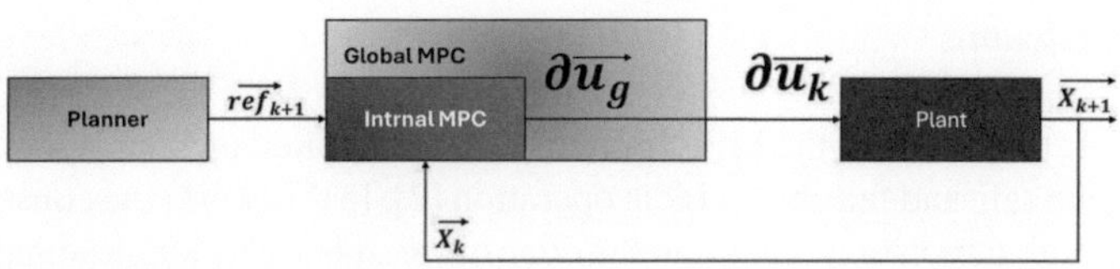

Fig. 5 Schematic representation of the MPC framework

Table 5 Step-by-step operation of the MPC framework

Step	Module	Operation	Inputs	Outputs
1	Planner	GPS and LiDAR provide the reference path, represented as a spline for smoother tracking	Road environment, GPS/LiDAR signals	Equation 2
2	Internal MPC	Uses the bicycle model with Euler discretization to predict vehicle states over horizon N. Generates candidate sequences of steering angle δ and acceleration a	Equation 2	$u = \begin{bmatrix} \delta_1 \\ a_1 \\ \delta_2 \\ a_2 \\ \vdots \\ \delta_n \\ a_n \end{bmatrix}^T \quad (16)$
3	Global MPC	Selects the first control actions from predicted sequence and formulates them as applied input	Equation 16	First step: $\begin{aligned} \delta_{k+1} &= \delta_k + \Delta\delta_1, \\ a_{k+1} &= a_k + \Delta a_1 \end{aligned} \quad (17)$
4	Plant	Vehicle model (simulation: bicycle model; real system: actual vehicle). Applies control inputs and generates next states	Equation 17	Updated vehicle states
5	Feedback	New vehicle states are sent back to internal MPC to close the loop, allowing evaluation and adjustment	Updated states from plant	Corrected state estimate for next iteration

in path tracking applications [6, 9]. Table 5 summarizes the operation of the MPC framework implemented in this work.

2.3.1 Constraints

An essential component of the MPC framework is the explicit inclusion of constraints that guarantee safe and feasible vehicle operation [2]. In this work, the constraints are defined for both the system states and the control variables. The longitudinal velocity is bounded within a minimum and maximum value to avoid unrealistic behaviours at very low or excessively high speeds. Similarly, lateral velocity is constrained as

Table 6 Constraints applied in the predictive control problem

Constraint type	Variable(s)	Bounds/Formulation
Fixed constraints	Longitudinal velocity ($\dot{x}$)	$\dot{x}_{\min} = 1$ $\dot{x}_{\max} = 30$
	Steering angle [24]	$\delta_{\min} = -\pi/4$ $\delta_{\max} = \pi/4$
	Acceleration increment (Δa)	$\Delta a_{\min} = -0.1$ $\Delta a_{\max} = 0.1$
Inequality constraints	Longitudinal acceleration (a)	$a_{\min} = -10 + \frac{F_{yf} \sin \delta}{m} + \mu g - \dot{\psi}\dot{y}$ $a_{\max} = 1 + \frac{F_{yf} \sin \delta}{m} + \mu g - \dot{\psi}\dot{y}$

a function of the longitudinal speed to maintain stability within admissible limits. The steering input δ is restricted by physical limits of the steering system, while the longitudinal acceleration is bounded to reflect the available traction and braking capacity.

Mathematically, the set of constraints can be expressed as a set of linear inequalities:

$$Gu \leq h \tag{18}$$

where G, h encode the limits on steering, acceleration, and their increments.

The MPC framework integrates two categories of constraints to ensure that the computed control inputs remain feasible and physically consistent. The first group corresponds to fixed constraints, which are defined by constant upper and lower bounds that limit the admissible range of a variable. The second group corresponds to dynamic constraints, which are expressed as inequalities that depend on the value of other system variables and therefore vary with the vehicle state. These two groups are summarized in Table 6.

2.3.2 Cost Function

The cost function in the MPC is expressed in a quadratic form so that it can be solved as a quadratic programming (QP) problem. At each time step, the sequence of control actions is grouped into a single vector (Eq. 16).

The objective is to minimize the difference between the predicted outputs and the reference trajectory, while also penalizing large control efforts and sudden changes in the inputs. This leads to the quadratic cost function:

$$J = \frac{1}{2} u^\top H u + f^\top u \tag{19}$$

where the matrix H depends on the weighting matrices for tracking accuracy, input magnitude, and input variation, and the vector f depends on the reference and the current state.

The penalty matrices play a central role in shaping the controller's behaviour. Q penalizes trajectory tracking errors, R penalizes the magnitude of steering and acceleration inputs, and an additional term. A higher weight in Q improves accuracy but may lead to sharper control actions, while higher weights in R generate smoother responses at the cost of larger tracking errors. Importantly, the choice of these weights is not trivial, as it represents a trade-off between conflicting objectives such as path accuracy. In future, the team might improve this task by approaching this task as a multi-objective optimization problem, where different weight combinations correspond to Pareto-optimal compromises between these objectives [20].

Together with the inequality constraints on steering and acceleration described earlier, the full optimization problem solved at each sampling step is:

$$\min_{u} \frac{1}{2} u^{\top} \mathrm{H} u + f^{\top} u \text{ subject to } Gu \leq h \tag{20}$$

This quadratic form is widely used in MPC formulations for autonomous driving, as it balances tracking performance with smooth and safe control actions [15].

2.4 *Python Code*

The Python implementation of the MPC framework was developed with the purpose of formulating the quadratic programming problem such that the solutions depend explicitly on steering angle and longitudinal acceleration. At each control step, the program constructs the matrices H, f, G, and h, which define the quadratic cost and the inequality constraints, and passes them to the solver to obtain the optimal input sequence. This ensures that the solver output is directly compatible with the vehicle actuators and aligns with the design goal of controlling the car based only on steering and acceleration.

The code structure follows the SOLID principles of modularity and separation of concerns. Each module is responsible for a single function: building the cost function, defining the constraints, preparing the solver interface, or updating the vehicle states. This modular organization not only increases readability but also simplifies debugging and future extensions. The design adheres to the SOLID principles of software engineering, which are widely recognized for improving code quality, scalability, and long-term maintainability in complex engineering applications [19].

By applying these principles, the implementation remains adaptable: new models, solvers, or tuning strategies can be integrated without restructuring the entire framework [1]. This flexibility is particularly relevant in MPC-based control systems, where modifications to cost weights or constraints are common during the iterative process of tuning and validation.

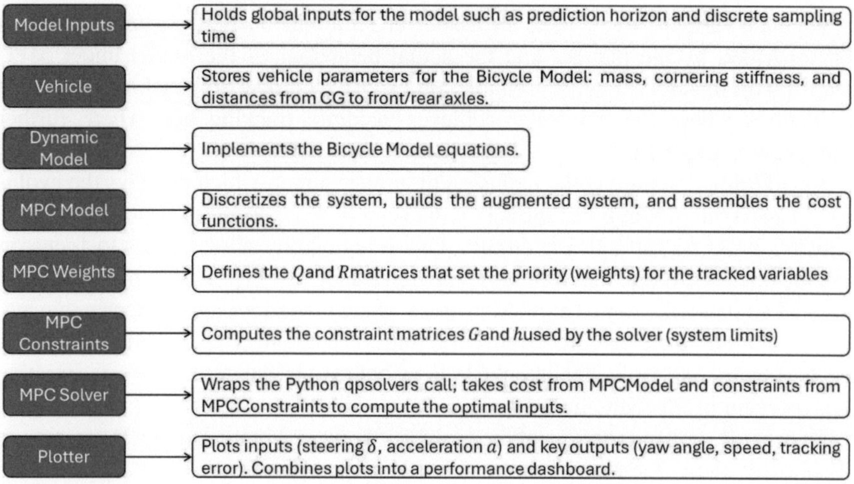

Fig. 6 Code structure (classes and responsibilities)

Figure 6 presents the classes that compose the Python program, together with a brief explanation of the role and responsibilities of each one. This structure reflects the modular design adopted in the implementation, where each class has a single responsibility to improve readability, maintainability, and future extensibility.

3 Results

The performance of the proposed MPC framework is evaluated through a digital dashboard, referred to as the *System Performance Dashboard*. This interactive environment was designed to present the most relevant variables for the team in a clear and accessible way, allowing both a detailed understanding of the vehicle's behaviour and more efficient tuning of the control parameters. By gathering all outputs in one integrated platform, the dashboard provides an immediate overview of how the model responds across different operating conditions.

The dashboard contains seven plots, each corresponding to a variable considered critical for assessing path tracking and dynamic performance:

- yaw angle,
- longitudinal velocity,
- steering angle,
- acceleration,
- total position error,
- error distribution,
- track position.

For each variable, two curves are displayed simultaneously: the reference trajectory provided by the planner and the actual behaviour of the model under MPC control. This dual representation enables quick visual comparison, making it possible to identify where the controller achieves accurate tracking and where further refinement is required.

An additional feature is the inclusion of a range slider, which displays the evolution of the system along the time axis, measured in seconds corresponding to the vehicle's progress through the circuit. This functionality allows users to focus on specific track segments, facilitating sector-based analysis. In practice, this means that targeted adjustments can be applied in areas where deviations are more pronounced, while maintaining the overall system configuration unchanged elsewhere. As a result, the dashboard not only documents system behaviour but also supports iterative improvement of the MPC framework.

To provide a clear overview of the variables displayed in the System Performance Dashboard, Table 7 summarizes the seven graphical outputs and explains the relevance of each one. These indicators were selected as they capture the most significant aspects of vehicle dynamics and tracking performance under MPC control.

Table 7 Description of the dashboard plots

Plot	Description	Purpose
Yaw angle	Comparison between the reference yaw profile and the predicted yaw response of the vehicle	Evaluates how well the controller maintains orientation along the track
Longitudinal velocity	Displays reference speed against actual velocity obtained from the model	Assesses speed regulation and the effect of acceleration input
Steering angle	Shows commanded steering input and its evolution over time	Monitors actuator demand and validates feasibility of steering actions
Acceleration	Represents longitudinal acceleration profile applied by the MPC	Ensures acceleration commands remain within comfort and safety limits
Total position error	Accumulated deviation between the reference trajectory and the actual path	Provides a global measure of path tracking accuracy
Error distribution	Displays position errors along the circuit	Highlights whether deviations are systematic or concentrated in specific regions
Track position	Overlay of reference track and simulated vehicle path	Visualizes spatial accuracy and identifies sectors requiring further tuning

3.1 Comparison of Poor and Optimized Tuning Outcomes

To illustrate the influence of the weighting matrices on vehicle behaviour, two representative scenarios are presented. Figure 7 shows a case of poor tuning, where the selected values of Q and R result in significant tracking errors and unstable responses in the dashboard plots. By contrast, Fig. 8 presents an optimized configuration of the same matrices, leading to improved tracking accuracy, smoother inputs, and overall better alignment with the reference trajectory. In both cases, the dashboard visualizations are displayed alongside the corresponding Q and R matrices, allowing a direct comparison between the tuning choices and the resulting system performance.

In the poor tuning scenario, the model initially follows the reference line with acceptable accuracy, but once it reaches the first curve the performance deteriorates. The yaw angle and longitudinal velocity plots exhibit irregularities, indicating that the car is unable to rotate adequately while maintaining the reference heading. At this point, the total position error plot shows a sharp peak, reaching deviations of up to 15 m, while the error distribution plot highlights in red the exact section of the track where this deviation occurs. The root cause is related to excessive entry speed, which prevents proper cornering and causes the system to lose alignment with the desired trajectory. To address this, the tuning strategy focuses on increasing the weights associated with yaw angle and acceleration in the Q and R matrices. This adjustment encourages the controller to prioritize steering during cornering, even at the cost of reducing velocity, while also paying closer attention to acceleration commands in order to avoid sudden spikes when exiting curves.

The optimized tuning scenario demonstrates the effectiveness of this adjustment. The updated values in the Q and R matrices produce plots where yaw angle and velocity show almost no deviation from their reference profiles. Although the overall vehicle speed is reduced, the trade-off results in a significant improvement in cornering capability, allowing the model to remain aligned with the track. The error plots confirm this improvement: total position error remains close to zero throughout the lap, and the error distribution indicates that large deviations have been eliminated. This scenario highlights how targeted adjustments of weighting matrices can transform the system from unstable to highly reliable performance, balancing accuracy and stability in line with the objectives of the Formula Student application.

3.2 Tuning of the Weighting Matrices

Motorsport environments are characterized by the need to perform rapid adjustments to engineering tools in order to increase the number of iterations during testing and simulation. Every additional iteration provides an opportunity to evaluate the system under slightly different conditions, improving adaptability to a specific track or race scenario. For this reason, the tuning process in this research is deliberately designed to

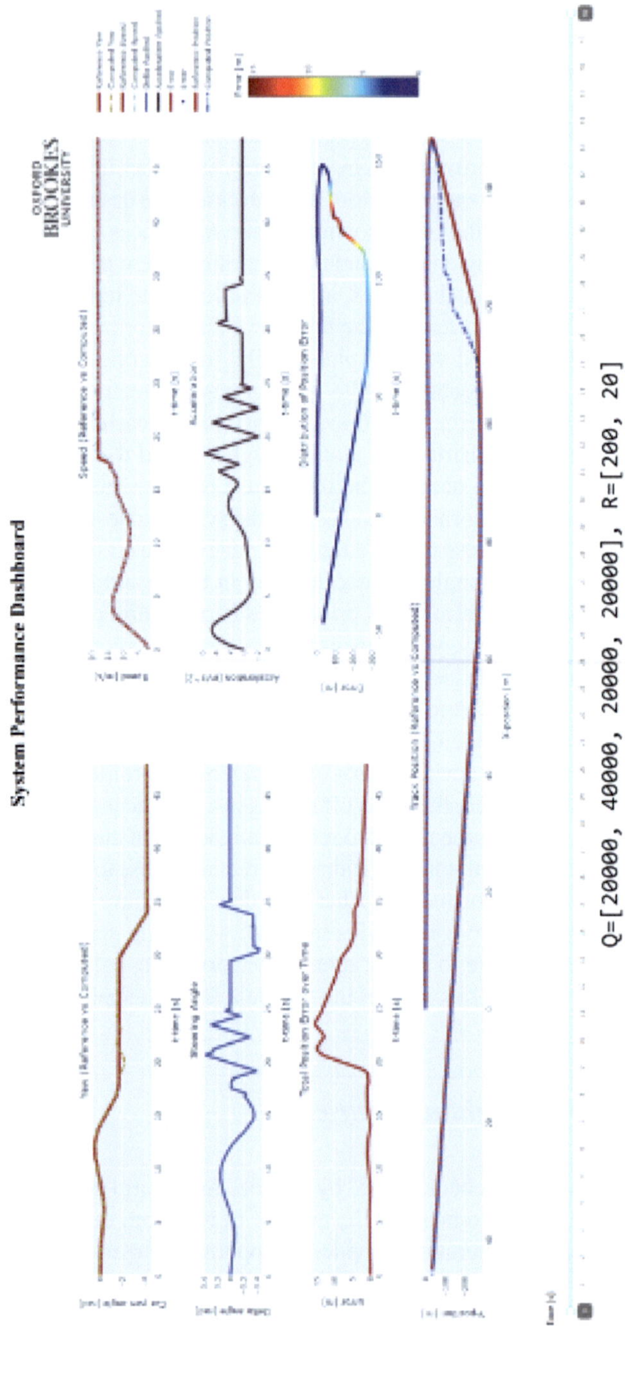

Fig. 7 Poor performance scenario

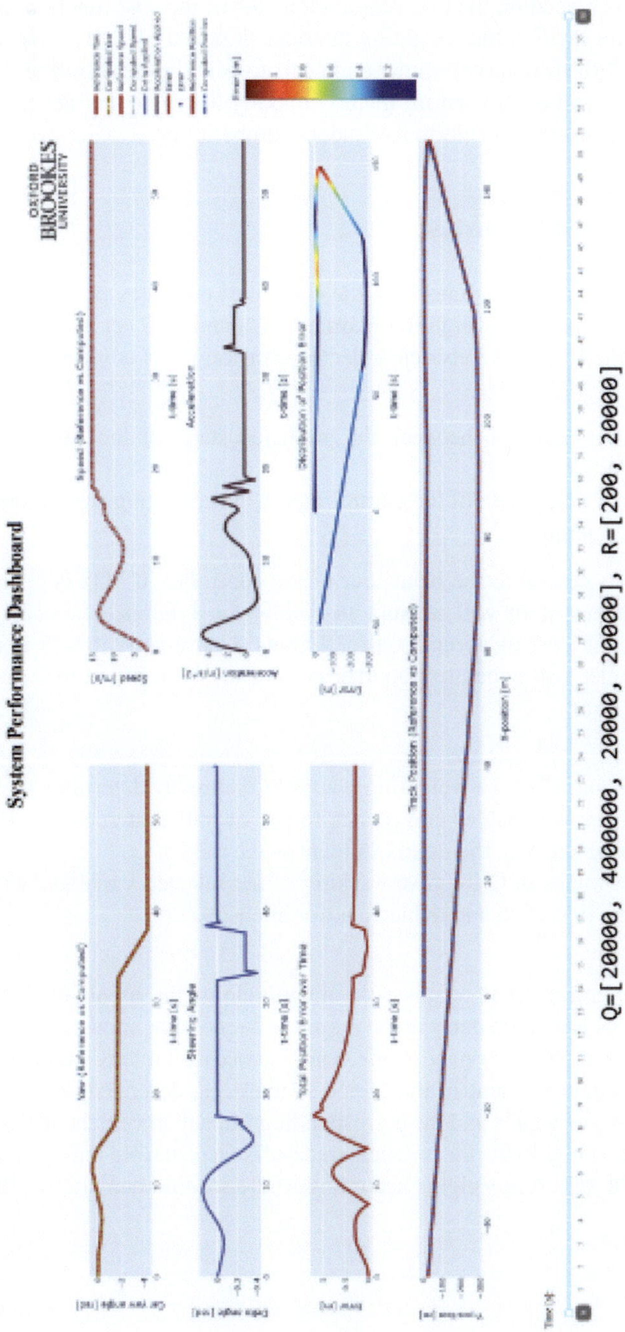

Fig. 8 Good performance scenario

remain straightforward and transparent. Instead of adjusting a wide range of parameters, the focus is placed on the two central elements of the cost function in Model Predictive Control (MPC): the weighting matrices Q and R. By concentrating on these matrices, the framework allows engineers to quickly shift priorities between tracking accuracy and control effort, thereby making the system easier to adapt in the fast-paced environment of Formula Student competitions.

3.2.1 Role of Q and R Matrices

In MPC, the cost function defines how the controller evaluates performance over the prediction horizon. The weighting matrices Q and R act as penalty terms that determine the trade-off between trajectory tracking and control smoothness. Specifically:

- Q penalizes the deviation between the predicted states of the vehicle and the reference trajectory.
- R penalizes the magnitude of the control inputs, discouraging overly aggressive steering or acceleration.

This balance is central to the behaviour of the controller. If Q is given much higher values, the system will attempt to follow the reference trajectory more closely, even at the cost of abrupt control actions. Conversely, if R is emphasized, the controller will generate smoother commands, possibly at the expense of some trajectory error.

The Q matrix

The diagonal entries of the weighting matrix Q directly determine how much error in each of these variables contributes to the overall cost function. Figure 9 illustrates how each entry in the matrix influences the model.

In short, high values in Q improve tracking of the selected variables, while low values allow more deviation and reduce control intensity.

The R matrix

The diagonal entries of R penalize the magnitude of the inputs of the model. Figure 10 explains how each entry in the matrix influences the model.

To provide a concise overview of the tuning process, Table 8 summarizes the role of each diagonal element in the Q and R matrices, describing the effects of assigning high or low values and the corresponding impact on vehicle performance. This representation highlights the practical trade-offs faced during tuning and serves as a guideline for selecting weights according to the specific demands of different circuits.

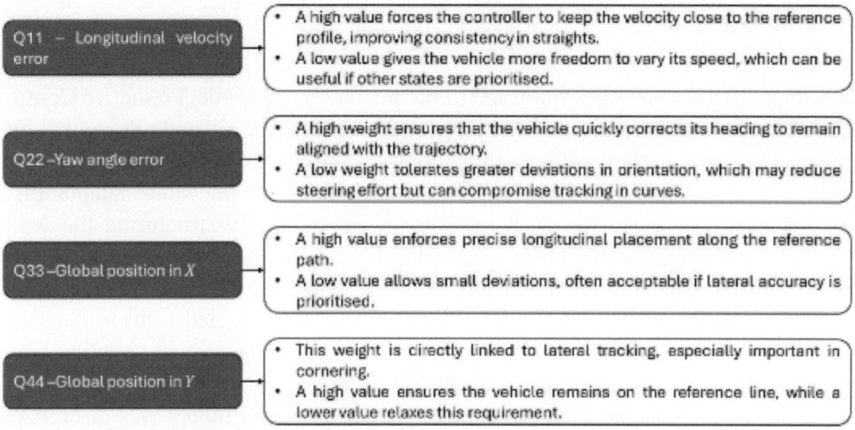

Fig. 9 Correspondence between Q matrix values and model behavior

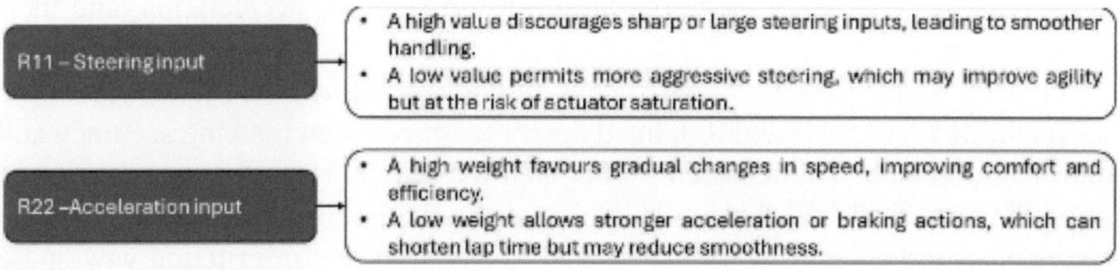

Fig. 10 Correspondence between R matrix values and model behavior

Table 8 Summary of tuning effects of Q and R matrices

Variable	High value effect	Low value effect	Impact on performance
Q_{11}	Keeps speed close to reference, improving consistency	Allows larger speed deviations	Balance between maintaining target velocity vs. flexibility in straights
Q_{22}	Strong correction to align heading with path	Tolerates orientation error	Higher stability in curves vs. reduced steering effort
Q_{33}	Accurate placement in longitudinal direction	Small longitudinal deviations tolerated	Important in lap-time optimization, especially in straights
Q_{44}	Precise lateral tracking, critical in curves	Permits lateral deviation	Higher cornering accuracy vs. reduced control effort
R_{11}	Avoids sharp steering actions, smooth response	Allows aggressive steering	Comfort and actuator safety vs. agility in tight turns
R_{22}	Smooth acceleration and braking	Stronger acceleration/ braking allowed	Ride comfort and efficiency vs. aggressive performance

4 Conclusion

This research presented the design and implementation of a Model Predictive Control (MPC) framework for trajectory tracking in the context of a Formula Student application. The study combined theoretical control concepts with practical implementation considerations, aiming to support a team environment where adaptability, fast iteration, and clear system behaviour are essential. By structuring the work through modelling, discretization, controller design, software implementation, and dashboard-based evaluation, a robust methodology was achieved.

At the modelling level, the dynamic bicycle model was selected as the foundation. This choice provided the necessary balance between computational simplicity and dynamic fidelity, enabling real-time feasibility while still capturing the essential aspects of yaw, lateral motion, and velocity dynamics. The model was discretized using Euler integration, which offered a straightforward yet effective method for implementing the prediction structure in Python. Constraints were formulated to ensure realistic behaviour, incorporating limits on steering, acceleration, and their increments.

The MPC framework was constructed around a quadratic cost function, where the weighting matrices Q and R defined the trade-off between tracking accuracy and control smoothness. The research highlighted the critical role of these matrices in tuning the system, demonstrating how changes in their values directly affected the behaviour of the vehicle. A dedicated analysis showed that prioritizing yaw angle and acceleration stability improved cornering capability, while smoother trajectories were achieved by increasing the penalties on control inputs. The process illustrated the practical importance of interpreting Q and R not only as mathematical weights but as design tools for achieving specific performance objectives.

Implementation in Python followed object-oriented programming principles and adhered to the SOLID guidelines. This ensured that the codebase remained modular, scalable, and easy to maintain. The structure allowed seamless adjustments of model parameters, horizons, and weighting matrices without the need to alter the underlying logic, aligning well with the ular software design, and visual analysis, providing the team with a tool that is both effective in simulation and adaptable for real-world implementation.

5 Future Work

While the present work demonstrated the effectiveness of the MPC framework based on Euler discretization, there is room for improvement in the numerical integration method used to predict vehicle dynamics. Euler's method was chosen because of its simplicity and ease of implementation, making it a practical choice for rapid prototyping and real-time feasibility. However, future research could explore replacing this approach with a higher-order method such as the classical fourth-order Runge–Kutta (RK4).

The main advantage of RK4 lies in its superior accuracy compared with first-order methods. By computing intermediate steps within each integration interval, RK4 significantly reduces local truncation error and provides a more faithful approximation of the continuous-time dynamics [27]. This improvement may translate into more precise predictions of vehicle states, particularly in scenarios involving sharp manoeuvres or high dynamic loads, where Euler's approximation might underrepresent the system behaviour.

At the same time, higher-order methods such as RK4 involve additional computational effort because each integration step requires multiple function evaluations. In embedded systems or time-critical applications, this increase in cost may limit the achievable sampling frequency, potentially offsetting the accuracy benefits. As highlighted in the literature, even optimized versions of RK4 require greater storage and computational resources than Euler, and their efficiency must be carefully weighed against accuracy gains in the specific application context [27].

Therefore, a promising direction for future work is to conduct a comparative study between Euler and RK4 discretization in the MPC framework. The evaluation should consider not only accuracy improvements in tracking and stability but also real-time feasibility and implementation complexity. Such analysis will help determine whether the trade-off justifies the adoption of RK4 in Formula Student applications or if Euler remains the most practical choice.

References

1. Ali AJM (2022) The impact of solid principles on code quality and software lifecycle. Int J Core Eng Manag 7(2):1–11. ISSN: 2348-9510
2. Batkovic I et al (2023) Experimental validation of safe MPC for autonomous driving in uncertain environments. IEEE Trans Control Syst Technol 31(5):2027–2042. https://doi.org/10.1109/TCST.2023.3291562
3. Choi J et al (2022) Coordinated steering angle and yaw moment distribution to increase vehicle regenerative energy in autonomous driving. In: 2022 IEEE 25th international conference on intelligent transportation systems (ITSC), pp 3083–3088. https://doi.org/10.1109/ITSC55140.2022.9921910
4. Di Cairano S and Kolmanovsky IV (2019) Automotive applications of model predictive control. In: Handbook of model predictive control. Springer International Publishing, pp 493–527. https://doi.org/10.1007/978-3-319-77489-3_21. ISBN: 978-3-319-77489-3
5. Emma T (2017) Robust model predictive control for autonomous driving. Technical Report
6. Hengyang W et al (2019) Path tracking control for autonomous vehicles based on an improved MPC. IEEE Access 7:161064–161073. https://doi.org/10.1109/ACCESS.2019.2944894
7. Hu X et al (2021) Research directions for next-generation battery management solutions in automotive applications. Renew Sustain Energy Rev 152:111695. https://doi.org/10.1016/j.rser.2021.111695. ISSN: 1364-0321. https://www.sciencedirect.com/science/article/pii/S1364032121009692
8. Jazar RN (2008) Vehicle planar dynamics. In: Vehicle dynamics: theory and applications. Springer, p 601. ISBN: 978-0-387-74243-4
9. Jin L et al (2024) A direct yaw moment control frame through model predictive control considering vehicle trajectory tracking performance and handling stability for autonomous

driving. Control Eng Pract 148:105947. https://doi.org/10.1016/j.conengprac.2024.105947. ISSN: 0967-0661. https://www.sciencedirect.com/science/article/pii/S0967066124001072
10. Kapsalis D et al (2021) Design and experimental validation of an LPV pure pursuit automatic steering controller. IFAC-PapersOnLine 54(2):63–68. https://doi.org/10.1016/j.ifacol.2021.06.010. ISSN: 2405-8963. https://www.sciencedirect.com/science/article/pii/S2405896321004389
11. Kuffel P, Kent K, Irwin G (1997) The implementation and effectiveness of linear interpolation within digital simulation. Int J Electr Power Energy Syst 19(4):221–227. https://doi.org/10.1016/S0142-0615(96)00045-2. ISSN: 0142-0615. https://www.sciencedirect.com/science/article/pii/S0142061596000452
12. Kulkarni NC (2024) Comparison of proportional integral derivative control and model predictive control for commercial vehicles. Int J Res Appl Sci Eng Technol JIRASET 12(8):866–875. https://doi.org/10.22214/ijraset.2024.63882
13. Lee JE, Kim BW (2024) Improving direct yaw-moment control via neural-network-based non-singular fast terminal sliding mode control for electric vehicles. Sensors 24(13):4079. https://doi.org/10.3390/s24134079. ISSN: 1424-8220. https://www.mdpi.com/1424-8220/24/13/4079
14. Martínez JM, Kaya CY (2007) Euler discretization and inexact restoration for optimal control. J Optim Theory Appl 134(2):191–206. https://doi.org/10.1007/s10957-007-9217-x.ISSN:1573-2878
15. Nan J et al (2024) Model predictive control for autonomous vehicle path tracking through optimized kinematics. Results Eng 24:103123. https://doi.org/10.1016/j.rineng.2024.103123. ISSN: 2590-1230. https://www.sciencedirect.com/science/article/pii/S2590123024013781
16. Peide W (2021) Research on comparison of LiDAR and camera in autonomous driving. J Phys Conf Ser 2093(1):012032. https://doi.org/10.1088/1742-6596/2093/1/012032
17. Polack P et al (2017) The kinematic bicycle model: a consistent model for planning feasible trajectories for autonomous vehicles? In: 2017 IEEE intelligent vehicles symposium IV, pp 812–818. https://doi.org/10.1109/IVS.2017.7995816
18. Rahman S et al (2024) An improved pure pursuit algorithm for the automated steering control of road vehicles. In: Advances in dynamics of vehicles on roads and tracks III. Springer Nature Switzerland, pp 586–595. ISBN: 978-3-031-66968-2
19. Ramachandrappa NC (2024) SOLID design principles in software engineering. Int J Comput Trends Technol 72(9):18–23
20. Stenger D et al (2023) What is the best way to optimally parameterize the MPC cost function for vehicle guidance? Mathematics 11(2). https://doi.org/10.3390/math11020465. ISSN: 2227-7390. https://www.mdpi.com/2227-7390/11/2/465
21. Sun X et al (2021) NFTSM control of direct yaw moment for autonomous electric vehicles with consideration of tire nonlinear mechanical properties. Bull Pol Acad Sci Tech Sci 69(3):e137065
22. Tătulea-Codrean A, Mariani T, Engell S (2020) Design and simulation of a machine-learning and model predictive control approach to autonomous race driving for the F1/10 platform. IFACPapersOnLine 53(2):6031–6036. https://doi.org/10.1016/j.ifacol.2020.12.1669. ISSN: 2405-8963. https://www.sciencedirect.com/science/article/pii/S2405896320322722
23. Tristano M et al (2022) Real-time implementation of yaw rate and sideslip control through individual wheel torques. In: 2022 IEEE vehicle power and propulsion conference (VPPC), pp 1–6. https://doi.org/10.1109/VPPC55846.2022.10003363
24. Vu TM et al (2021) Model predictive control for autonomous driving vehicles. Electronics 10(21). https://doi.org/10.3390/electronics10212593. ISSN: 2079-9292. https://www.mdpi.com/2079-9292/10/21/2593

25. Yashchenko DM et al (2020) Determination the yaw moment of inertia of an SUV in laboratory conditions. IOP Conf Ser Mater Sci Eng 1002(1):012028. https://doi.org/10.1088/1757-899X/1002/1/012028
26. Yu S et al (2021) Model predictive control for autonomous ground vehicles: a review. Auton Intell Syst 1:161064–161073. https://doi.org/10.1007/s43684-021-00005-z.ISSN:2730-616X
27. Zingg DW, Chisholm TT (1999) Runge-Kutta methods for linear ordinary differential equations. Appl Numer Math 31(2):227–238. https://doi.org/10.1016/S0168-9274(98)00129-9. ISSN: 0168-9274. https://www.sciencedirect.com/science/article/pii/S0168927498001299

Modelling and Control of a Two-Degree-of-Freedom Robotic Arm Using PID and Fuzzy Logic Control

Pablo Perez Martinez

Abstract The modern industry has experienced notable technological improvements in terms of performance, efficiency, and adaptability, largely thanks to the emergence of robotics. Widely used in highly repetitive tasks and hazardous environments, robotics continues to expand its field of application, but there is still considerable progress to be made, particularly regarding control. Numerous controllers are available, and the type employed depends on the application field, such as PID, MRC, fuzzy logic, or AI-based approaches. In this research, we have utilised PID and fuzzy logic control to test and compare their effectiveness. The results show that the performance of PID is significantly superior to that of Fuzzy Logic.

1 Introduction

1.1 Background

Robotics lies in a privileged position in technological development, being one of the fundamental pillars of the so-called Fourth Industrial Revolution, or Industry 4.0 [1]. Its importance is based on the ability of robotic systems to perform repetitive, complex, or dangerous tasks with a level of accuracy, speed, and reliability that is consistently unattainable by human beings [2]. As a result, a wide range of sectors, such as manufacturing, automotive, medicine, space exploration, and agriculture, have experienced notable advances in terms of efficiency, safety, and quality [3]. For example, in industrial environments, robotisation has enabled more flexible production with fewer errors, in medicine, robotic arms have been developed to assist in minimally invasive surgeries with a very high degree of accuracy [4], in space exploration, manipulators allow operations to be carried out under conditions considered highly hostile to humans [5], and in agriculture, robots are employed to perform harvesting and classification tasks with high precision [6]. In all these

P. P. Martinez (✉)
School of Engineering, Computing and Mathematics, Oxford Brookes University, Oxford, UK
e-mail: 19327523@brookes.ac.uk; pablopema2001@gmail.com

A. Azizi (ed.), *Intelligent Control and Autonomous Systems*, Emerging Trends in Mechatronics, https://doi.org/10.1007/978-981-95-6402-6_5

cases, the advantages provided by robotics include cost reduction, increased productivity, and minimisation of risks to people, all with a degree of automation that would have been unimaginable a few years ago. Furthermore, robotics is increasingly being incorporated into emerging areas such as rehabilitation and assistive technologies, in which robotic devices support patients with motor impairments, demonstrating impact beyond industrial contexts [7].

Among robotic systems, manipulator arms represent one of the most distinctive and widely employed devices. These arms, inspired by the biomechanics of a human being's arm, can adopt very diverse configurations, depending on the task and the requirements of precision, scope or load. There exist robotic arms with Cartesian, cylindrical, spherical, parallel, or articulated architecture, each with its own advantages and limitations in terms of mobility and complexity of control [8]. In this context, one of the principal characteristics that distinguish them is the number of degrees of freedom, which determines the motion capacity of the manipulator, and thus the workspace it is able to cover [9]. Therefore, while a robotic arm of one or two degrees of freedom may be sufficient for relatively simple applications of manipulation in a plane, six-degree-or-freedom arms are required for complex industrial tasks in three dimensions, where it is necessary to control the position as well as the orientation of the end effector. This wide variety of configurations has made robotic arms highly versatile tools, adaptable to a wide range of environments and needs [31]. Recent studies also emphasise their role as educational and research platforms, where simplified structures are used to prototype and validate controllers before deploying them on more complex manipulators.

In the particular case of two-degree-of-freedom robotic arms, these can provide an intermediate solution between the simplicity of a basic manipulator and the complexity of multi-degree-of-freedom robotic arms. A two-degree-of-freedom robotic arm is generally constituted by two arms connected by two rotational articulations that allow the end effector to be displaced within a two-dimensional plane, making it an ideal system for studies and validation tasks for control algorithms, as well as for practical applications that do not require three-dimensional movements. Among its more relevant fields of application are included the teaching and research, where it is used as a reference model for the simulation and analysis of advanced controllers [10], the automation of processes that require highly precise movements in a plane, such as trajectory tracking or the positioning of pieces [11], and medical rehabilitation, in which exoskeletons are employed, or portable two-degrees-of-freedom devices are used to assist patients in the recovery of upper limb mobility [12]. The relative simplicity of its design allows the implementation and testing of different control techniques, as well as the derivation of conclusions that can be extrapolated to more complex systems [13]. This is why two-degree-of-freedom arms are frequently adopted as benchmarks in experiments that combine classical and intelligent control, enabling rigorous validation of kinematic and dynamic models [14].

The control of robotic arms has evolved over the last decades, incorporating a wide variety of approaches. Traditionally, the most widely used controller has been the Proportional-Integral-Derivative (PID), thanks to its simplicity, ease of implementation and efficiency in the reduction of tracking errors [15]. However, the nonlinear nature of the systems and the coupled dynamics of the manipulators have led to the appearance of more sophisticated techniques. One example is the Model Reference Control (MRC), in which the arm is trained to follow the behaviour of a predefined reference model, which allows improved precision and robustness against perturbations [16]. Fuzzy logic offers a heuristic approach that is able to manage uncertainties and nonlinear behaviours without the need for such an exact mathematical model [17]. Finally, neural networks constitute a modern alternative with great potential, as they allow approximating complex dynamics and learning from operation data, offering adaptability to changes in the system conditions [18]. Each of these approaches presents specific advantages, and the comparison between their performances represents a major topic in current research on robotics control [19]. In particular, neural network-based estimators have shown strong performance for forward and inverse kinematics while keeping computational costs manageable [30]. Likewise, hybrid and data-driven controllers have reported improved tracking and robustness compared with purely PID schemes, reflecting a broader shift towards adaptive, learning-based control [20].

The development of more advanced artificial intelligence techniques, together with the integration of machine learning algorithms and reinforcement learning, promises to provide the manipulators with a more autonomous capacity for adaptation and decision-making, even in unstructured or changing environments [21]. In this way, the trend towards producing smaller components, the incorporation of more precise sensors and the improving in the processing capacity, will make it possible to design lighter, more efficient and safer robotic arms [22]. In the social sphere, these improvements could result in more accessible assistance systems for people with reduced mobility [23], collaborative robots able to work alongside human beings to meet high safety standards, and manipulators specialised in precision tasks in fields such as surgery or space exploration [24]. Ultimately, the evolution of the robotic arms will not only continue transforming the industry, but will also have a direct impact on people's lives, extending the horizons of what can be considered possible in the interaction between humans and robots [25]. Recent work anticipates that progress will hinge on the synergy between robotics and AI, enabling more autonomous, safe, and socially integrated systems [26].

Based on the literature review introduced in this research, the approach adopted has consisted of combining classical and intelligent control strategies within a simulation framework. A Proportional-Integral-Derivative (PID) controller and a fuzzy logic controller have been developed and implemented in MATLAB/Simulink, using the dynamic model of a two-degree-of-freedom robotic arm derived through the Euler–Lagrange formulation.

1.2 Approach

The methodological approach followed during the development of this research has been structured carefully into a sequential series of steps each of them forming the basis for the next one. The main objective has been to design, implement and compare different control strategies applied to a two-degrees-of-freedom robotic arm. The controller must be able to receive a position point as an input, expressed in Cartesian coordinates, and from that calculate the torque to be applied to each of the manipulator's joints to make the end effector achieve the desired position in a precise and stable manner. This approach represents a common challenge in the field of robotics, as it implies translating a high-level spatial objective into low-level physical commands, while managing nonlinearities and the coupled dynamics of the multi-joint systems.

As can be seen in Fig. 1, the first stage consists of the study and definition of the robotic arm's geometry as the case under analysis. In this phase, the fundamental physical parameters are selected, such as the length of the bars, the mass of each element, and the configuration of the general structure. These data are essential, as they determine the workspace reachable by the arm as well as the properties of inertia that influence its dynamic behaviour. Therefore, it has been necessary to define a reference system and a set of coordinates to describe the manipulator's movement. This has allowed the generation of a precise kinematic model that has been the basis of the dynamic model.

Once the geometry definition has been completed, the kinematic model is developed. This model incorporates the mathematical relationships between the joint variables and the end effector's final position in the Cartesian space. Even though the resolution of inverse kinematics was not the main objective of the project, it has been essential to have these equations to verify the capacity of the arm to reach the desired position.

After defining the kinematics, the next step has been the formulation of the dynamics of the system, which represents the core of any analysis in control. To develop the model, the Euler–Lagrange equations have been used, a common tool

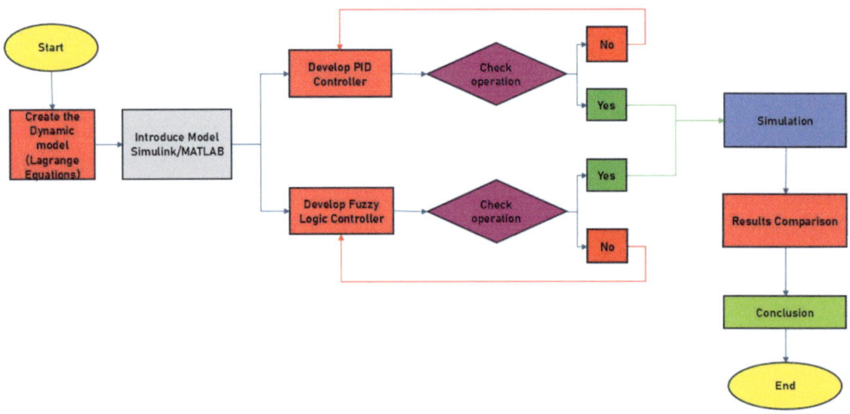

Fig. 1 Project flow chart

in multi-body systems due to their capacity to integrate systematically the contributions of kinetic and potential energy. Through this method, it has been possible to derive a set of nonlinear coupled differential equations that describe the dynamic behaviour of the two-degree-of-freedom manipulator. These equations define explicitly the relationship between the torques applied to the joints and the accelerations resulting from the links, constituting the theoretical basis for testing and validating the control strategies.

The motion equations have been implemented in MATLAB/Simulink, creating a block that can represent the behaviour of the arm and allow the evaluation of different controllers under equivalent conditions. The use of Simulink has been highly useful, as it offers a modular environment where it is possible to connect blocks that represent both the dynamics of the system and the control algorithms, thus facilitating the validation of each stage of the process. This model has been used as a common reference for the two control approaches that have been studied.

The fuzzy logic controller has been designed with the intention of managing the nonlinearities and possible uncertainties of the system through a heuristic approach. For that, membership functions associated with input variables (error and derivative of error) and a set of rules that produce the control action by applying torque have been defined. During this phase, different numbers and types of membership functions have been tested, evaluating their impact on the system response. The adjustment of these functions and the refinement of the set of rules have been carried out iteratively in Simulink, analysing the behaviour of the arm when trying to reach different target positions. These experiments have allowed the identification of the optimal configurations to provide a compromise between speed and stability of the response.

In parallel, a Proportional-Integral-Derivative controller has been implemented and widely used in industrial applications due to its simplicity and efficiency in the reduction of errors. In this case, the principal task has consisted of setting the parameters of the proportional, integral and derivative gains, to allow the end effector to achieve the desired position with minimal error and without excessive oscillations. To do so, manual iterative adjustments have been combined with automatic tools such as the PID Tuner in Simulink, which has permitted the identification of a parameter set with the sufficient robustness.

Once the controllers were configured, a validation process was carried out to assess their ability to both simulate and control the system response. This step involves evaluating whether the controllers can effectively minimise the error between the reference signal and the actual output, thereby confirming their suitability for the task. After this validation has been achieved, the controllers can then be reliably employed for simulation purposes, ensuring that the subsequent analysis was based on stable and verified control actions.

Finally, once both controllers have been implemented, the comparative phase is followed. In this final step, the results obtained with each technique have been analysed in terms of accuracy in tracking, response time, and stability of the system.

1.3 Summary of Literature Review

1.3.1 Mathematical Modelling

Mathematical modelling, particularly focusing on kinematics and dynamics, constitutes the foundational framework for effective robotic arm control. Reference [4] provide an extensive analysis of a two-degree-of-freedom robot arm using Denavit–Hartenberg parameters for the kinematic model and applying Euler–Lagrange equations for dynamics. Their contribution primarily addresses theoretical accuracy, establishing a robust basis for subsequent real-time control strategies.

Expanding upon similar methodologies, [8] present a comprehensive dynamic model derived via kinetic and potential energy calculations, subsequently linearised through feedback linearisation methods. While their work does not incorporate neural network or PID strategies explicitly, it underscores the necessity of rigorous dynamic modelling as a critical step in precise controller design, especially for Model Predictive Control (MPC) implementations.

Reference [26] explore a practical and human-centred application by developing a 2-DOF wearable robotic exoskeleton for stroke rehabilitation. Their approach integrates structural validation with hybrid control strategies, combining real-time PID control in joint space with simulations of sliding-mode compliance control. This highlights the dual need for precision and adaptability in human-interactive systems.

References [24] and [25] extend mathematical modelling into the data-driven domain through Multilayer Feedforward Neural Networks (MLFFNNs). Sharkawy develops neural architectures capable of accurately approximating both position and orientation in forward and inverse kinematics tasks. Bouzid et al. further investigate how variations in arm geometries and datasets affect the accuracy and adaptability of neural models, emphasising the role of data quality and neural architecture depth on performance outcomes.

Reference [10] further affirm the efficacy of neural network-based inverse kinematics solutions, revealing clear computational advantages over traditional regression methods. Such data-driven approaches provide enhanced flexibility and scalability, particularly suitable for real-time and highly variable robotic applications.

1.3.2 Classical PID Control and Enhancements

PID control remains widely used for its simplicity and reliability. Reference [4] apply PID controllers based on inverse kinematics, though without adaptation. Reference [3] improve upon this by using metaheuristic algorithms, Particle Swarm Optimisation (PSO), Artificial Bee Colony (ABC), and Chaos Game Optimisation (CGO) for tuning PID parameters, validating results in both simulation and hardware.

Reference [23] integrate fuzzy logic with PID and apply Social Spider Optimisation, achieving near-zero steady-state error. Fuzzy logic allows real-time adaptation to nonlinearities, boosting robustness.

Reference [32] further cemented the industrial relevance of PID control by conducting comparative analyses of several PID variants, P, PI, PD, and full PID in MATLAB Simulink environments. Their findings suggest that auto-tuned PID controllers offer superior overall performance, balancing fast response, minimal overshoot, and stability, providing explicit gain values ($K_\mathrm{p} = 2.1638$, $K_\mathrm{i} = 91.7476$, $K_\mathrm{d} = 0.0127$) that serve as benchmarks for similar applications.

These studies collectively affirm that while PID is effective, its adaptability is greatly improved through intelligent tuning (e.g. PSO, fuzzy logic). MATLAB-based simulation remains a core tool for controller testing, offering safety and flexibility prior to real-world deployment.

Despite advances in neural control, PID's dominance in industry reflects its interpretability, real-time performance, and low computational demand.

1.3.3 Fuzzy Logic Control

Fuzzy logic control (FLC) represents an increasingly prominent methodology in robotic arm control, particularly valued for its ability to manage system nonlinearities and parameter uncertainties without reliance on highly accurate mathematical models. By employing linguistic rules and membership functions rather than explicit equations, fuzzy controllers enable adaptive and intuitive decision-making, offering robustness and flexibility in real-time robotic applications.

Reference [23] demonstrate the integration of fuzzy logic with PID controllers enhanced through Social Spider Optimisation, achieving virtually zero steady-state error in robotic arm trajectory tracking. Their study highlights the capacity of fuzzy-based approaches to effectively compensate for nonlinear behaviours while maintaining stability. In a complementary comparative analysis, Reference [32] evaluate several PID variants against a fuzzy logic controller (FLC) in the control of a six-degree-of-freedom robotic manipulator. While conventional PID exhibited superior transient performance, the fuzzy approach produced reduced overshoot and smoother responses, underscoring its relevance in safety–critical or human-interactive contexts.

Earlier contributions further illustrate the adaptability of fuzzy paradigms. Reference [15] introduce a Neuro-Fuzzy Controller (NFC) for a two-degree-of-freedom robotic arm, where the integration of neural learning mechanisms enabled dynamic adjustment of fuzzy membership functions. Their results confirmed enhanced robustness and trajectory-tracking accuracy compared to classical PID strategies. Similarly, [17] propose a trapezoidal fuzzy two-degree-of-freedom PID algorithm for articulated manipulators, demonstrating improved stability and precision through optimised fuzzy inference rules.

Applications of fuzzy control extend beyond industrial settings into the domain of rehabilitation robotics, where adaptability to user-specific dynamics is essential. Reference [12] develop an intelligent fuzzy-neural control system for a two-degree-of-freedom rehabilitation robot, employing genetic algorithms to optimise fuzzy parameters and ensuring both safety and patient-specific adaptability. More recently, [26] presented a hybrid fuzzy-PID controller for a wearable robotic exoskeleton, validating its capacity to support stroke rehabilitation through both simulation and real-time experiments.

Collectively, these contributions underscore the advantages of fuzzy control in robotic systems, notably in contexts characterised by nonlinearities, uncertainties, or direct human interaction. The capacity of FLCs to combine with conventional PID architectures and to hybridise with neural or evolutionary optimisation techniques positions them as a versatile and robust strategy within modern robotic arm control.

1.3.4 Conclusion

The reviewed literature consistently highlights robotic arms as effective tools for enhancing automation, with significant benefits in terms of safety, precision, and operational efficiency, particularly in hazardous or repetitive environments. A recurring theme across the studies is the indispensable role of mathematical modelling, especially kinematics and dynamics, as the foundation upon which any reliable control strategy must be constructed. Whether for simulation, trajectory planning, or controller design, accurate modelling provides the framework for subsequent implementation.

Classical PID controllers remain dominant in industrial contexts due to their straightforward implementation, interpretability, and proven effectiveness in real-time applications. Numerous studies, however, emphasise that PID performance is significantly enhanced when coupled with intelligent tuning mechanisms, such as metaheuristic optimisation or fuzzy augmentation, which improve adaptability to nonlinearities and disturbances.

Fuzzy logic control (FLC) emerges as a particularly valuable complement and alternative to PID. Its capacity to manage uncertainties and nonlinear dynamics without explicit system models, together with its ability to integrate seamlessly into hybrid architectures, provides robustness in complex environments. This has been demonstrated both in industrial manipulators, where overshoot minimisation and smooth tracking are critical, and in rehabilitation robotics, where adaptability to human-specific dynamics is essential.

Taken together, the literature affirms that while PID controllers remain the cornerstone of industrial robotics for their reliability and simplicity, the integration of fuzzy logic substantially extends system adaptability and robustness. The growing use of hybrid approaches that combine PID and FLC reflects a clear trend towards controllers capable of balancing interpretability, precision, and resilience in both industrial and human-centred robotic applications.

Based on the reviewed literature, this research has adopted a dual-control approach, integrating a classical Proportional-Integral-Derivative (PID) controller and a fuzzy logic controller (FLC) for a two-degree-of-freedom robotic arm. This choice reflects the need to evaluate the comparative performance of a conventional, widely used method and a heuristic, model-free strategy, thereby providing a balanced perspective on robustness, adaptability, and accuracy in robotic control. Following this introduction, the remainder of the manuscript is organised as follows:

the methodology chapter develops the three main components of the work, namely the dynamic model, the fuzzy logic controller, and the PID controller, each of them explained in detail with their corresponding implementation in MATLAB/Simulink. The manuscript then presents the comparative analysis of the results, before drawing the final conclusions.

2 Methodology

In this section, the complete process undertaken in the research is presented in detail, from the dynamic model and its obtention step-by-step, to the design and implementation of each of the controllers that have been studied.

2.1 Dynamic Model

The dynamic model represents a crucial element when designing any type of system to be simulated and controlled. This model's main role is to define the relationships between the control actions (the torque in this case) and the response of the system (movement of the end effector). Therefore, the precision and performance of the controller depend highly on the accuracy of the dynamic model, and on its representation of the physical reality.

2.1.1 Geometry and Kinematics

The geometry of the robotic arm studied in this project is characterised by its simplicity, as its main role is to test and compare the performance of several controllers under the same conditions. The manipulator is composed of two links and two rotational joints that connect both bars and allow them to rotate.

Kinematics represent a key aspect in this project, because the desired position for the end effector will be introduced as a point in the Cartesian coordinate system (X, Y). However, as the motion of the arm bars is essentially rotational, the generalised coordinates will be the angles of the two joints, $\theta 1$ and $\theta 2$, which correspond, respectively, with the angles of rotation for each link, as can be appreciated in Fig. 2. As a result, we will define the x and y position of the arm's endpoint as a function of the angles $\theta 1$ and $\theta 2$; to do so, the forward kinematics will be used (Table 1).

The forward kinematics are:

$$x_{\text{end}} = x_2 = L_1 \cos(\theta_1) + L_2 cos(\theta_1 + \theta_2) \tag{1}$$

$$y_{\text{end}} = y_2 = L_1 \sin(\theta_1) + L_2 \sin(\theta_1 + \theta_2). \tag{2}$$

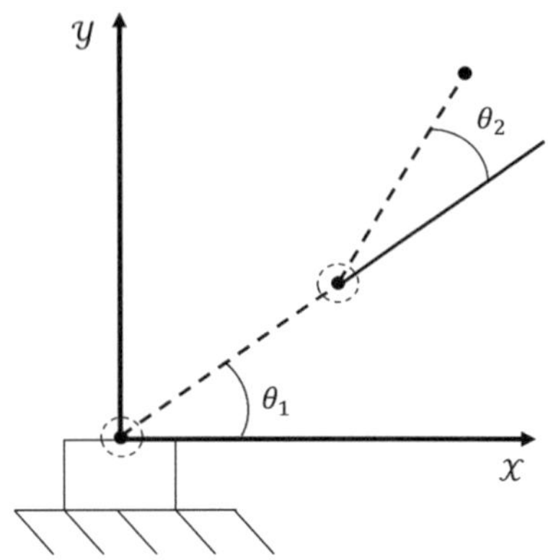

Fig. 2 Degrees-of-freedom robotic arm scheme

Table 1 Length parameters

Parameter	Value (m)
Length 1st bar (L1)	1
Length 2nd bar (L2)	1

The accuracy of the controller lies mainly in the appropriate modelling of the system. The more accurate the equations have, the better the motion will be optimised, and, as a result of this, the results will be better in terms of accuracy and lower oscillations.

2.1.2 Dynamic Parameters

In this study, the dynamic parameters of the robotic arm have been deliberately simplified to facilitate the modelling process. The masses of the two links have each been set to 1 kg, ensuring a uniform representation that eases the derivation of the dynamic equations. Furthermore, it has been assumed that the entire mass of each link is concentrated at its distal end. This assumption effectively eliminates the rotational inertia terms, thereby reducing the complexity of the governing equations while still capturing the essential dynamic behaviour. Such a modelling approach is consistent with methodologies found in the literature, where researchers investigating two-degree-of-freedom manipulators have adopted similar simplifications to enhance analytical tractability without significantly compromising the accuracy of dynamic response analysis [4].

By removing the need to account for distributed link masses and their associated inertia tensors, the resulting model becomes more tractable for controller design and simulation purposes. This simplification is particularly useful in preliminary studies where the focus lies on evaluating and comparing control strategies, rather than on achieving precise dynamic fidelity. Consequently, it provides a suitable balance

Table 2 System mass parameters

Parameter	Value (kg)
Mass 1st bar (M1)	1
Mass 2nd bar (M2)	1

between mathematical simplicity and physical relevance, enabling clearer insights into the performance of classical and intelligent control techniques (Table 2).

2.1.3 Equations of Motion

There are numerous methods to obtain the equations of motion of a two-degree-of-freedom dynamic system, such as the Newton–Euler method or the D'Alembert method. However, following the approach of Reference [21] the Euler–Lagrange method will be the one selected for this case. The method is based on a mathematical function known as the Lagrangian which is obtained from the kinetic and potential energies.

$$L = \text{Kinetic Energy} - \text{Potential Energy} \tag{3}$$

$$F = \frac{\mathrm{d}}{\mathrm{d}t}\left(\frac{\mathrm{d}L}{\mathrm{d}\dot{q}}\right) - \frac{\mathrm{d}L}{\mathrm{d}q}. \tag{4}$$

As mentioned, to obtaining the Lagrangian, it is necessary to calculate the kinetic and potential energies of the system. Once the parameters of the system have been defined, the first derivatives of the x and y coordinates will be obtained, for be used in the calculus of the kinetic and potential energies. Additionally, the coordinates of the end of the first link will be defined.

$$x_{\text{first bar}} = x_1 = L_1 \cos(\theta_1) \tag{5}$$

$$y_{\text{first bar}} = y_1 = L_1 \sin(\theta_1) \tag{6}$$

$$\dot{x}_1 = -L_1\,\dot{\theta}_1 \sin(\theta_1) \tag{7}$$

$$\dot{y}_1 = L_1\,\dot{\theta}_1 \cos(\theta_1) \tag{8}$$

$$\dot{x}_2 = -L_1\,\dot{\theta}_1 \sin(\theta_1) - L_2\left(\dot{\theta}_1 + \dot{\theta}_2\right)\sin(\theta_1 + \theta_2) \tag{9}$$

$$\dot{y}_2 = L_1\,\dot{\theta}_1 \cos(\theta_1) - L_2\left(\dot{\theta}_1 + \dot{\theta}_2\right)\cos(\theta_1 + \theta_2). \tag{10}$$

Now, with the coordinates of each bar and their derivatives calculated, the next step is calculating the kinetic and potential energies of the system:

The kinetic energy of the system is the sum of the kinetic energy for each bar

$$\mathrm{KE} = \mathrm{KE}_1 + \mathrm{KE}_2 \tag{11}$$

$$\mathrm{KE}_1 = \frac{1}{2} m_1 \left(\dot{x}_1^2 + \dot{y}_1^2\right) \tag{12}$$

$$\mathrm{KE}_1 = \frac{1}{2} m_1 \left(L_1^2 \dot{\theta}_1^2\right) \tag{13}$$

$$\mathrm{KE}_2 = \frac{1}{2} m_2 \left(\dot{x}_2^2 + \dot{y}_2^2\right) \tag{14}$$

$$\mathrm{KE}_2 = \frac{1}{2} m_2 \left[L_1^2 \dot{\theta}_1^2 + L_2^2 \left(\dot{\theta}_1 + \dot{\theta}_2\right)^2 + 2 L_1 L_2 \dot{\theta}_1 \left(\dot{\theta}_1 + \dot{\theta}_2\right) \cos(\theta_2) \right]. \tag{15}$$

After calculating the kinetic energy for each bar, the potential energy is calculated as well:

$$\mathrm{PE} = \mathrm{PE}_1 + \mathrm{PE}_2 \tag{16}$$

$$\mathrm{PE}_1 = m_1 * g * y_1 = m_1 g L_1 \sin(\theta_1) \tag{17}$$

$$\mathrm{PE}_2 = m_2 * g * y_2 = m_2 g [L_1 \sin(\theta_1) + L_2 \sin(\theta_1 + \theta_2)]. \tag{18}$$

When the kinetic and potential energies are already calculated for every bar of the system, is the moment to calculate the Lagrangian:

$$\begin{aligned} L = {} & \frac{1}{2} m_1 \left(L_1^2 \dot{\theta}_1^2\right) + \frac{1}{2} m_2 \left[L_1^2 \dot{\theta}_1^2 + L_2^2 \left(\dot{\theta}_1 + \dot{\theta}_2\right)^2 \right. \\ & \left. + 2 L_1 L_2 \dot{\theta}_1 \left(\dot{\theta}_1 + \dot{\theta}_2\right) \cos(\theta_2) \right] - m_1 g L_1 \sin(\theta_1) \\ & - m_2 g [L_1 \sin(\theta_1) + L_2 \sin(\theta_1 + \theta_2)] \end{aligned} \tag{19}$$

In order to define the equations of motion, the partial derivatives with respect to the angular velocities and their time derivatives will be calculated.

$$\begin{aligned} \frac{\mathrm{d}}{\mathrm{d}t} \left(\frac{\mathrm{d}L}{\mathrm{d}\dot{\theta}_1} \right) = {} & (m_1 + m_2) L_1^2 \ddot{\theta}_1 + m_2 L_2^2 \left(\ddot{\theta}_1 + \ddot{\theta}_2\right) \\ & + m_2 L_2 L_1 \left[-\sin(\theta_2) \dot{\theta}_2 \left(2\dot{\theta}_1 + \dot{\theta}_2\right) + \cos(\theta_2) \left(2\ddot{\theta}_1 + \ddot{\theta}_2\right) \right] \end{aligned} \tag{20}$$

$$\frac{\mathrm{d}}{\mathrm{d}t}\left(\frac{\mathrm{d}L}{\mathrm{d}\dot{\theta}_2}\right) = m_2L_2^2\left(\ddot{\theta}_1 + \ddot{\theta}_2\right) + m_2L_1L_2\left(\ddot{\theta}_1\cos(\theta_2) - \dot{\theta}_1\,\dot{\theta}_2\sin(\theta_2)\right) \tag{21}$$

In addition, the partial derivatives with respect to the angles of each joint will also be calculated.

$$\frac{\mathrm{d}L}{\mathrm{d}\theta_1} = -m_1gL_1\cos(\theta_1) - m_2gL_1\cos(\theta_1) - m_2gL_2\cos(\theta_1 + \theta_2) \tag{22}$$

$$\frac{\mathrm{d}L}{\mathrm{d}\theta_2} = -m_2L_1L_2\,\dot{\theta}_1\left(\dot{\theta}_1 + \dot{\theta}_2\right)\sin(\theta_2) - m_2L_2\cos(\theta_1 + \theta_2) \tag{23}$$

After having obtained every component of the Euler–Lagrange equations, it is substituted into Eq. 4:

$$\begin{gathered}\frac{\mathrm{d}}{\mathrm{d}t}\left(\frac{\mathrm{d}L}{\mathrm{d}\dot{\theta}_1}\right) - \frac{\mathrm{d}L}{\mathrm{d}\theta_1} = \\ (m_1 + m_2)L_1^2\ddot{\theta}_1 + m_2L_2^2(\ddot{\theta}_1 + \ddot{\theta}_2) + \\ m_2L_2L_1[-\sin(\theta_2)\dot{\theta}_2(2\dot{\theta}_1 + \dot{\theta}_2) \\ +cos(\theta_2)(2\ddot{\theta}_1 + \ddot{\theta}_2)] + m_1gL_1\cos(\theta_1) \\ +m_2gL_1\cos(\theta_1) + m_2gL_2cos(\theta_1 + \theta_2) = \tau_1\end{gathered} \tag{24}$$

$$\begin{aligned}\frac{\mathrm{d}}{\mathrm{d}t}\left(\frac{\mathrm{d}L}{\mathrm{d}\dot{\theta}_2}\right) - \frac{\mathrm{d}L}{\mathrm{d}\theta_2} &= m_2L_2^2\left(\ddot{\theta}_1 + \ddot{\theta}_2\right) + m_2L_1L_2\left(\ddot{\theta}_1\cos(\theta_2) - \dot{\theta}_1\,\dot{\theta}_2\sin(\theta_2)\right) \\ &+ m_2L_1L_2\,\dot{\theta}_1\left(\dot{\theta}_1 + \dot{\theta}_2\right)\sin(\theta_2) + m_2gL_2\cos(\theta_1 + \theta_2) = \tau_2\end{aligned} \tag{25}$$

Once the second-order differential equations are obtained, they are rearranged to obtain expressions for the angular accelerations. This is necessary for the implementation in Simulink that is explained later.

$$\begin{aligned}\ddot{\theta}_1 = {} & \frac{\tau_1 - m_1gL_1\cos(\theta_1) - m_2gL_1\cos(\theta_1) - m_2gL_2\cos(\theta_1+\theta_2) - m_2L_2^2\ddot{\theta}_2}{L_1^2(m_1+m_2) + m_2L_2^2 + 2m_2L_2L_1\cos(\theta_2)} \\ & \frac{-m_2L_2L_1\cos(\theta_2)\ddot{\theta}_2 + m_2L_2L_1(\sin(\theta_2)\dot{\theta}_2(2\dot{\theta}_1+\dot{\theta}_2))}{L_1^2(m_1+m_2) + m_2L_2^2 + 2m_2L_2L_1\cos(\theta_2)}\end{aligned} \tag{26}$$

$$\begin{aligned}\ddot{\theta}_2 = {} & \frac{\tau_2 - m_2gL_2\cos(\theta_1+\theta_2) - m_2L_1L_2\dot{\theta}_1\left(\dot{\theta}_1+\dot{\theta}_2\right)\sin(\theta_2)}{m_2L_2^2} \\ & \frac{-m_2L_1L_2\left[\ddot{\theta}_1\cos(\theta_2) - \dot{\theta}_1\dot{\theta}_2\sin(\theta_2)\right] - m_2L_2^2\ddot{\theta}_1}{m_2L_2^2}\end{aligned} \tag{27}$$

2.1.4 Simulink Model

The implementation of the dynamic model in Simulink has been carried out within a dedicated subsystem, hereafter referred to as the Dynamic Model, which is configured with two inputs and two outputs. Within this subsystem, two computational branches have been constructed, each corresponding to one of the previously derived equations of motion. These branches, however, are not entirely independent, since the dynamic equations are coupled and therefore require interconnections between them to accurately represent the system behaviour.

For the construction of this subsystem, basic mathematical operator blocks have been employed, such as addition, subtraction, multiplication, and division, as well as simple gain blocks to represent multiplication by constant values. In addition, blocks for the computation of trigonometric functions have been incorporated, as these are essential for capturing the nonlinear terms present in the model. Regarding the time evolution of the states, as shown in Fig. 3, integrator blocks have been used to obtain angular velocities and angular positions from the angular accelerations, ensuring a continuous representation of the system dynamics.

With respect to the system's interface, as shown in Fig. 4, the two inputs of the subsystem correspond to the torques applied to each joint, which are generated by the control action of the respective controller. Conversely, the subsystem provides as outputs the joint angles θ_1 and θ_2, which represent the angular positions of each link. These angles are subsequently used to compute the Cartesian position of the end effector, allowing for trajectory tracking and further performance assessment of the implemented control strategies.

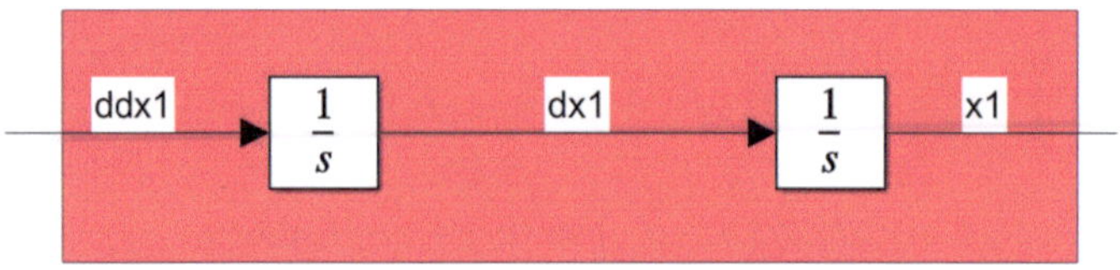

Fig. 3 Integrator blocks used for obtaining velocity and position

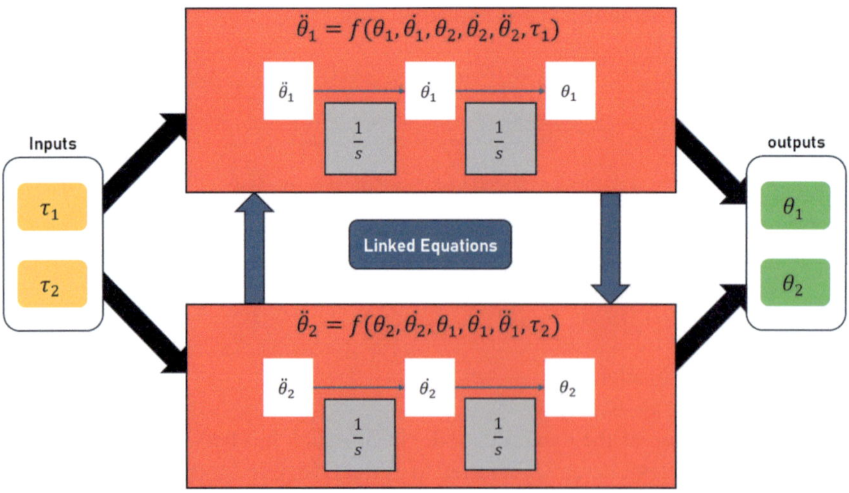

Fig. 4 Scheme of dynamic model subsystem block

```
function [theta1, theta2] = inverseKinematics2DOF(ref)
%#codegen
% inverseKinematics2DOF computes joint angles theta1 and theta2 for a 2-DOF planar robotic arm.
% Inputs:
%   x, y - Cartesian coordinates of the end effector
%   L1, L2 - lengths of link 1 and link 2
% Outputs:
%   theta1, theta2 - joint angles in radians
L1 = 1;
L2 = 1;

x = ref(1);
y = ref(2);


% Compute distance to the point

r2 = x^2 + y^2;

% Compute theta2 using the law of cosines
cos_theta2 = (r2 - L1^2 - L2^2) / (2 * L1 * L2);
% Ensure the cosine is within valid range to avoid NaN due to floating point errors
cos_theta2 = min(max(cos_theta2, -1), 1);
sin_theta2 = sqrt(1 - cos_theta2^2);  % Elbow-down configuration
theta2 = atan2(sin_theta2, cos_theta2);

% Compute theta1
k1 = L1 + L2 * cos_theta2;
k2 = L2 * sin_theta2;
theta1 = atan2(y, x) - atan2(k2, k1);
```

Fig. 5 MATLAB code for inverse kinematics

Apart from the dynamic model subsystem block, there are two other subsystems that are present in the Simulink models for both controllers. These blocks are for the kinematic transformation. As mentioned previously, the desired point to be reached by the end effector is introduced by the user in Cartesian coordinates, as an [*X*, *Y*] point. However, as also it has been mentioned, the equations of motion are designed for angular variables, in this case, the joint angles θ_1 and θ_2. This makes necessary to transform the coordinates of the objective point into values of both angles. To do so a MATLAB block with the inverse kinematics equations detailed above has been included, it can be seen in Fig. 5.

On the other hand, the opposite process must be carried out for the response. Once the controller has calculated the torque to be applied in order to reach the specific desired point, the joint angles resulting from the output of the dynamic model need to be transformed back into Cartesian coordinates. To do so, a subsystem block, shown in Fig. 6, is employed.

2.2 *Fuzzy Logic Controller*

Fuzzy logic control (FLC) is an intelligent control approach inspired by human reasoning, where decisions are made using linguistic variables without needing exact mathematical models. This makes it particularly suitable for highly nonlinear systems or those subject to uncertainty, such as robotic manipulators. Unlike classical PID

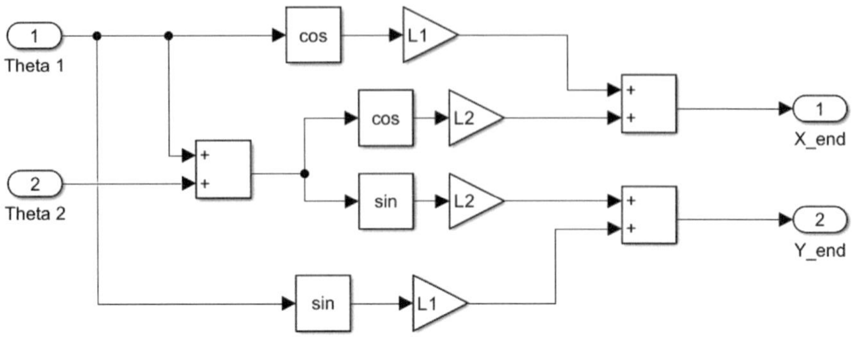

Fig. 6 Forward kinematics subsystem block

controllers, fuzzy controllers do not require an accurate mathematical model of the plant but instead rely on membership functions and a set of fuzzy rules. These rules provide the mapping between the input variables (e.g. error and rate of change of error) and the corresponding control action, which in this case is the torque applied to the joints. In this way, the controller can infer the appropriate response directly from qualitative relationships, avoiding the need for precise system identification. This flexibility allows fuzzy logic controllers to adapt better to varying operating conditions, offering robustness against disturbances and modelling inaccuracies [27]. In the context of robotic arms, where dynamic couplings, nonlinearities and external perturbations often compromise conventional control strategies, fuzzy logic provides an effective alternative, capable of delivering smoother responses and reduced overshoot [25].

2.2.1 Inputs and Outputs

The design of a fuzzy logic controller allows flexibility in the selection of input variables. Depending on the application, variables such as position, velocity, acceleration, or even external forces may be used, as these variables reflect the state of the system and guide the inference process [24]. This versatility distinguishes fuzzy logic controllers from more rigid structures, enabling them to adapt to a wide range of robotic applications and environments.

In robotic systems, it is very common to use the error in position and its derivative as controller inputs, as these variables properly capture the instantaneous deviation from the target and the dynamic tendency of the system. In the present study, as the robotic arm executes rotational movements, the selected variables have been the angular position error of each joint and the derivative of this error, as seen in Table 3. This choice aligns with the logic of classical PID controllers, which also rely on error-based inputs, but extends it through the fuzzy inference process. The output of the controller is the torque applied to each joint, which acts as the corrective action to drive the system towards the desired angular position.

Table 3 Input and output parameters for fuzzy logic controller

Controller	Input	Output
Fuzzy logic 1	$\theta_1 - \theta_{1\text{ref}}$ (error)	τ_1
	$\frac{d(\theta_1 - \theta_{1\text{ref}})}{dt}$	
Fuzzy logic 2	$\theta_2 - \theta_{2\text{ref}}$ (error)	τ_2
	$\frac{d(\theta_1 - \theta_{1\text{ref}})}{dt}$	

2.2.2 Membership Functions

Membership functions (MFs) are a key element of fuzzy logic controllers, as they define how numerical input values are translated into qualitative categories. Each input variable is divided into a set of linguistic terms such as Negative Large, Negative Small, Zero, Positive, Small or Positive Large. Each of these terms is represented by a membership function, which assigns a degree of membership to the input within a certain interval of values. For this project, if the input is the angular error, a value within a negative range may be classified as Negative Small or Negative Large depending on its magnitude. In this way, the controller interprets continuous signals by associating them with qualitative states, which are later used by the fuzzy rules to determine the final control action.

A wide range of membership function types can be employed, including triangular, trapezoidal, Gaussian, and bell-shaped functions. The choice of MF type has a significant influence on the controller's behaviour. Simpler shapes, such as triangular and trapezoidal, are computationally efficient and widely used in real-time control due to their straightforward implementation. More complex shapes, such as Gaussian or bell-shaped functions, provide smoother transitions between fuzzy sets but require more careful parameter tuning and can increase the computational cost. Importantly, increasing the number of membership functions does not always improve performance; while finer partitions may capture more detail in the input space, they also complicate the rule base and may lead to unnecessary sensitivity to noise or overfitting [28].

In this project, different configurations have been tested using five and seven membership functions per input variable, as can be seen in Tables 4 and 5. However, simulation results indicated that the configuration with seven membership functions does not achieve any improvement and, in fact, often led to poorer performance compared to the five-MF design. For this reason, the option with five membership functions has been retained as the most appropriate balance between complexity and results, despite being unable to fully control the system, as shown in Fig. 7.

Table 4 5 Membership functions states

5 membership functions				
Negative large	Negative small	Zero	Positive small	Positive large

Table 5 7 Membership functions states

7 membership functions						
Negative large	Negative medium	Negative small	Zero	Positive small	Positive medium	Positive large

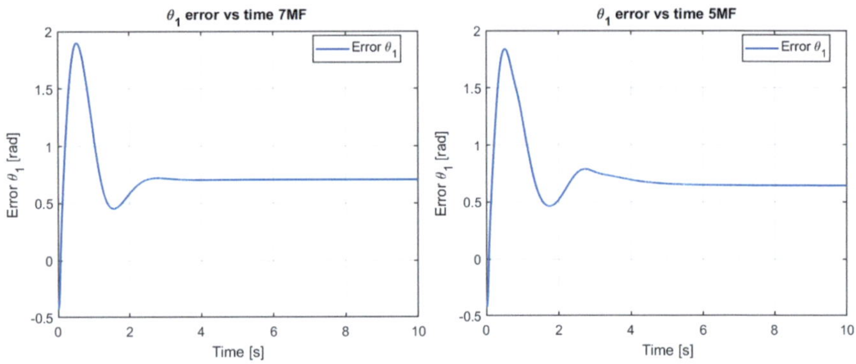

Fig. 7 θ_1 error comparison for 5MF and 7MF

It should be noted that the ranges of values for each of the previously defined states can be set independently and manually. However, in this case, the MATLAB Fuzzy Logic Designer tool of MATLAB has been used, allowing the number and type of membership functions to be defined and the functions to be created with uniform value ranges.

In this study, two types of MFs have been considered: Gaussian and triangular. Gaussian functions were initially explored for their smooth representation of input variations, but triangular functions have been ultimately selected due to their balance between accuracy, computational simplicity, and stability of results. The triangular MFs have enabled the system to provide consistent torque responses while keeping the design of the rule base manageable. This decision is in line with previous studies in robotic control, which emphasise the practicality and robustness of triangular membership functions in real-time robotic applications [29] (Figs. 8 and 9).

A comparative analysis has also been carried out between Gaussian and triangular membership functions. Although Gaussian functions allowed smoother transitions between states, their use did not improve the tracking accuracy and frequently resulted in less stable behaviour. Triangular functions, by contrast, demonstrated relatively more consistent results, with simpler tuning and lower computational demand. Nevertheless, neither approach succeeded in providing effective control of the robotic arm (Fig. 10).

In light of what these comparisons have revealed, the final configuration adopted in this work has been based on five triangular membership functions for each input variable, corresponding to the states Negative Large, Negative Small, Zero, Positive Small, and Positive Large. While this design does not achieve stable control, it has

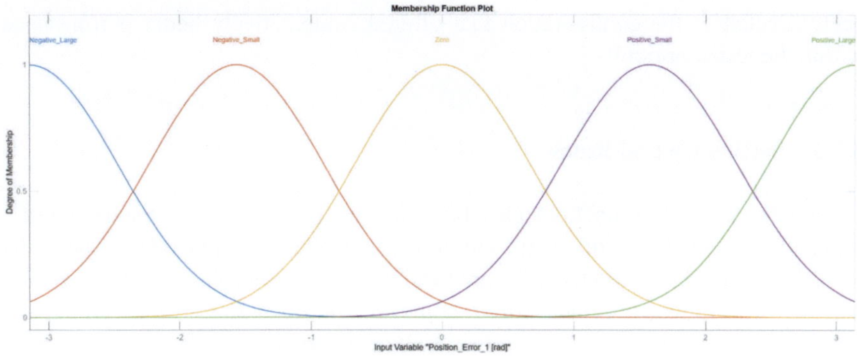

Fig. 8 Gaussian membership functions

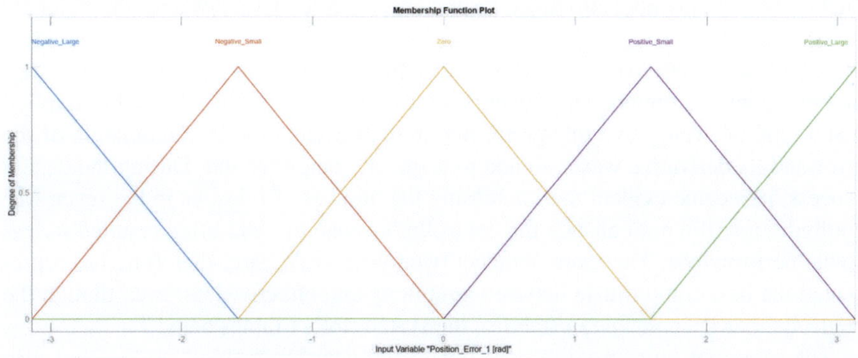

Fig. 9 Triangular membership functions

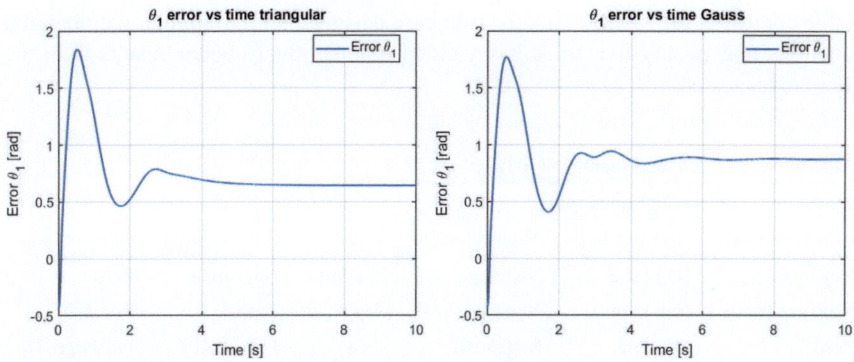

Fig. 10 θ_1 error comparison for triangular and Gaussian MFs

been selected as the configuration that offered comparatively better performance within the tested options.

2.2.3 Setting Up and Rules

Once the membership functions have been defined, the next step in the design of a fuzzy logic controller is the construction of the rule base. Fuzzy rules provide the logical framework that connects the input variables to the control action. They usually take the form in which the antecedent combines the fuzzy states of the inputs and the consequent specifies the appropriate output action. These rules are essential to the overall performance of the controller, as they directly determine how the system reacts to different operating conditions. The quality of the control depends not only on the choice of membership functions but also on the consistency and completeness of the rule base.

After testing both five and seven membership functions per input, the final configuration adopt five membership functions for each variable. This choice has implied that a total of twenty-five rules were required, as each possible combination of the error and its derivative was assigned to a specific output action. During the design process, it became evident that increasing the number of rules, as in the seven-MF configuration, did not enhance the controller's capability and often resulted in less stable performance. The more compact twenty-five-rule base, therefore, has represented the best compromise between simplicity and effectiveness, even though the controller has been unable to stabilise the system fully (Tables 6 and 7).

The rules not only provide structure to the decision-making process but also influence critical aspects such as overshoot, settling time, and robustness against disturbances. In this study, although the fuzzy logic controller has not achieved stable control of the robotic arm, the rule base designed with triangular membership functions and twenty-five rules has yielded comparatively better results than the alternatives tested.

Table 6 Set of rules for 5 membership functions

Rules for 5 membership functions					
Error\Derivative	Negative large	Negative small	Zero	Positive small	Positive large
Negative large	Neg large	Neg large	Neg large	Neg small	Zero
Negative small	Neg large	Neg large	Neg small	Zero	Pos small
Zero	Neg large	Neg small	Zero	Pos small	Pos large
Positive small	Neg small	Zero	Pos small	Pos large	Pos large
Positive large	Zero	Pos small	Pos large	Pos large	Pos large

Table 7 Set of rules for 7 membership functions

Rules for 7 membership functions							
Error\Derivative	Negative large	Negative medium	Negative small	Zero	Positive small	Positive medium	Positive large
Negative large	Negative Large	Negative Large	Negative Large	Negative Large	Negative Medium	Negative Small	Zero
Negative medium	Negative large	Negative large	Negative large	Negative medium	Negative small	Zero	Positive small
Negative small	Negative large	Negative large	Negative medium	Negative small	Zero	Positive small	Positive medium
Zero	Negative large	Negative medium	Negative small	Zero	Positive small	Positive medium	Positive large
Positive small	Negative medium	Negative small	Zero	Positive small	Positive medium	Positive large	Positive large
Positive medium	Negative small	Zero	Positive small	Positive medium	Positive large	Positive large	Positive large
Positive large	Zero	Positive small	Positive medium	Positive large	Positive large	Positive large	Positive large

2.2.4 Simulink Model

The Simulink model consists of the kinematic blocks (forward and inverse), the dynamic model subsystem, and the fuzzy logic-based controller blocks. In addition to these, derivative blocks are included to provide input signals to the controllers.

The model connections reflect the logical operation of the system. The process begins with the user defining the desired end effector position, which is introduced as a vector. This vector serves as the input for the inverse kinematics block, implemented in MATLAB code, which converts the Cartesian coordinates into angular values to be used as reference signals. The dynamic model subsystem block receives the torques as inputs and outputs the controlled angular positions. The reference signals generated by the inverse kinematics block are compared with these outputs within the controller blocks. Two fuzzy logic controllers are implemented, one for each joint, and their inputs are the errors computed as the difference between the reference values and the controlled variables. These errors are also differentiated and combined using a multiplexer, so that both the error and its derivative are provided as inputs to each controller.

The outputs of the controller blocks are the torques, which act as inputs to the dynamic model. Finally, a forward kinematics subsystem block converts the controlled angular positions back into Cartesian coordinates, enabling the user to observe the resulting end effector trajectory(Fig. 11).

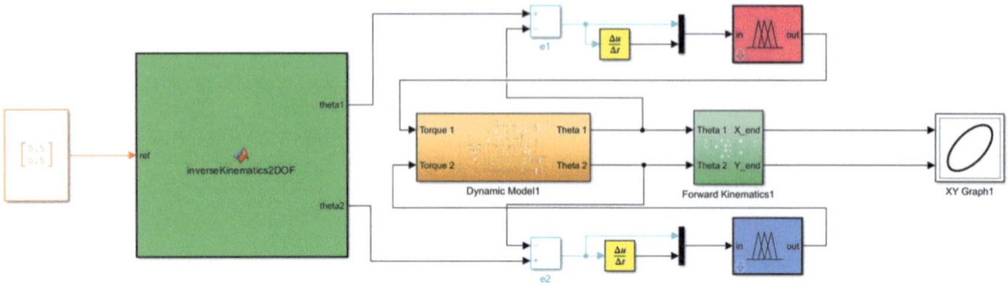

Fig. 11 Simulink model for fuzzy logic controller

2.3 *PID Controller*

Proportional-Integral-Derivative (PID) controllers remain one of the most widely adopted control strategies in industrial robotics due to their simplicity, robustness, and effectiveness in dealing with dynamic systems. In the context of robotic arms, PID control has been extensively applied to ensure accurate trajectory tracking, stability, and error reduction in real-time applications. Its widespread use in industrial manipulators is attributed to the ability to adjust system response through intuitive tuning of proportional, integral, and derivative gains, which directly influence tracking precision and transient behaviour [32]. Despite the emergence of advanced intelligent controllers, PID controllers continue to serve as a benchmark in robotics, providing a reliable basis for evaluating and comparing alternative techniques.

2.3.1 Inputs and Outputs

The inputs to the PID controller are the angular position of each joint and its corresponding reference value, which result from transforming the user-specified desired end effector position. The controller continuously calculates the error between the actual angle and the target angle, producing a corrective action that seeks to reduce the error. While in many control systems the action is expressed in terms of a force, in this case torque has been selected as the control variable; it is the rotational equivalent of force. The torque output of the PID controller is then applied directly as the input to the dynamic model of the robotic arm, as described earlier, driving the system towards the desired configuration (Table 8).

Table 8 Inputs and outputs for PID controllers

Controller	Input	Output
PID 1	θ_1	τ_1
	$\theta_{1\text{ref}}$	
PID 2	θ_2	τ_2
	$\theta_{2\text{ref}}$	

In the proposed control scheme, the Proportional-Integral-Derivative (PID) controller continuously evaluates the difference between the desired joint angle and the actual measured angle, producing the error signal that drives the control action. This error is decomposed into three components: the proportional term responds instantaneously to the magnitude of the deviation, the integral term accumulates the error over time to eliminate steady-state offsets, and the derivative term predicts the future trend of the error by assessing its rate of change. The weighted combination of these three contributions yields a corrective signal, which, in the context of the robotic arm, is expressed as a torque. Unlike a purely translational system, where the control action is typically a linear force, in a rotational system the torque constitutes the angular counterpart, directly influencing the joint accelerations according to the equations of motion. Through this mechanism, the PID controller transforms the error in the joint angle into the appropriate torque command, which is then applied to the dynamic model of the arm to reduce the error and achieve precise tracking of the reference trajectory.

2.3.2 Parameters

As mentioned earlier, PID controllers stand out due to their simplicity and ease of set up. However, it is essential to select the appropriate parameters to achieve effective control, and therefore, it is crucial to understand clearly the significance of each parameter that influences the performance of the PID.

The proportional term, denoted as K_p, provides an immediate response to the magnitude of the error between the desired and actual joint angles. Its role is to apply a corrective torque proportional to the current deviation, ensuring that the robotic arm reacts quickly whenever a discrepancy arises. A higher proportional gain increases the system's responsiveness, reducing the error more rapidly; however, if set too high, it can also cause excessive oscillations or even instability, as the controller may overreact to small deviations. In robotic arms, this translates into fast but potentially overshooting movements, where the end effector reaches the target position quickly but may overshoot before settling.

The integral term, K_i, addresses the accumulation of error over time, ensuring that any persistent deviation from the target position is corrected. By integrating the error, the controller generates an additional torque that compensates for steady-state offsets, which can arise from unmodelled dynamics, external disturbances, or gravity effects on the robotic arm. The presence of this term guarantees that the end effector will eventually reach the exact desired position without leaving a residual error. Nonetheless, excessive integral action can slow down the transient response and lead to integrator wind-up, causing sluggish behaviour or sustained oscillations if not managed properly.

The derivative term, K_d, predicts the future behaviour of the error by measuring its rate of change. Essentially, it provides a damping effect, applying a corrective torque that counteracts rapid variations and reduces overshoot. This predictive capability makes the system smoother and prevents the robotic arm from oscillating excessively around the target position, enhancing stability during fast movements. However, because it relies on error differentiation, this term is sensitive to measurement noise; an overly large derivative gain may amplify such disturbances, leading to unwanted vibrations. In robotic applications, a carefully tuned derivative component is crucial to achieve a balance between responsiveness and smoothness in the arm's trajectory tracking.

2.3.3 Setting Up

A wide variety of approaches have been proposed in the literature for the tuning of PID parameters, ranging from classical analytical techniques to more advanced computational tools. Traditional methods, such as the Ziegler–Nichols or Cohen–Coon rules, provide heuristic formulations based on the system's response to step inputs or sustained oscillations, offering a practical yet approximate means of setting proportional, integral, and derivative gains [15]. More recent strategies rely on automated or optimisation-based tools integrated into simulation platforms, such as MATLAB/Simulink, which enable real-time adjustment of controller parameters according to system dynamics [23]. These modern methods allow a better balance between speed, accuracy, and robustness, helping users significantly improve the performance of robotic arms in industrial and research contexts without requiring excessive time.

In this particular case, the tuning process has combined the use of the MATLAB PID Tuner application with an iterative refinement process. Initially, the automated tool was employed to generate baseline values of K_p, K_i, and K_d, based on the dynamic response of the robotic arm model within Simulink. These values were then progressively adjusted through manual iterations, evaluating the impact of each modification on error reduction, overshoot minimisation, and settling time. This hybrid methodology ensured that the controller parameters are not only consistent with the system's theoretical model but are also optimised through practical experimentation, allowing the user to achieve a configuration that offer a robust and stable performance of the controller.

Below, an example is shown of the initial PID controller values and the resulting end effector trajectory. The reference point introduced by the user is [0.5, 0.5] m (Table 9).

As can be seen in Fig. 12, with these values, the results do not achieve the desired point. The movement starts at the point [2, 0], which corresponds to both links of the arm being completely horizontal. However, the end effector of the arm does not follow a clear trajectory towards the desired point, exhibiting significant oscillations and following a much longer path before the simulation ends at 10 s.

Table 9 First iteration parameters

PID	K_p	K_i	K_d
1	10	10	10
2	10	10	10

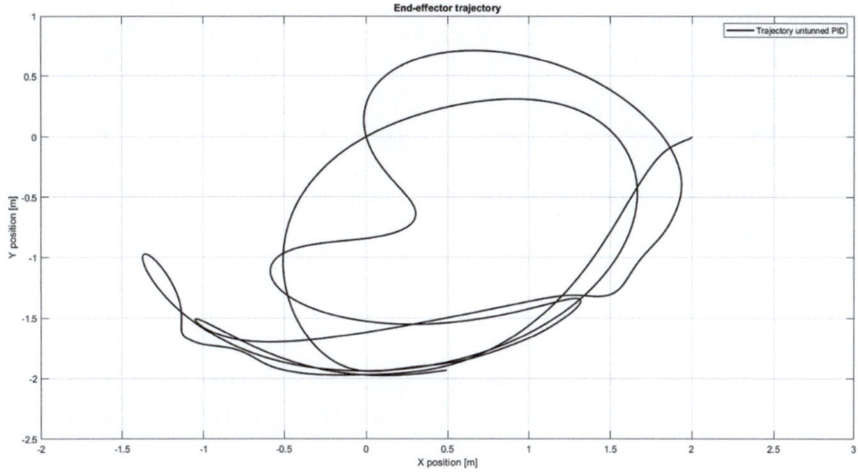

Fig. 12 Initial trajectory of the arm's end

Figure 13 represents how the controller is being unable to track the reference signal; it is clearly appreciable that the oscillations are increase with the time, not achieving the desired point at any case.

Reviewing the results of the first iteration, the need for parameter tuning is clear. After an iterative process of PID parameter tuning, two final configurations, one for each joint of the robotic arm, were defined, achieving a balance between stability, response speed, and tracking accuracy. Throughout the ten stages shown in the table, the process began with low generic values ($K_\mathrm{p} = K_\mathrm{i} = K_\mathrm{d} = 10$), which were progressively increased while observing the system's dynamic behaviour at each step. This approach has allowed for a reduction in both overshoot and steady-state error,

Fig. 13 θ_1 comparison with the reference value

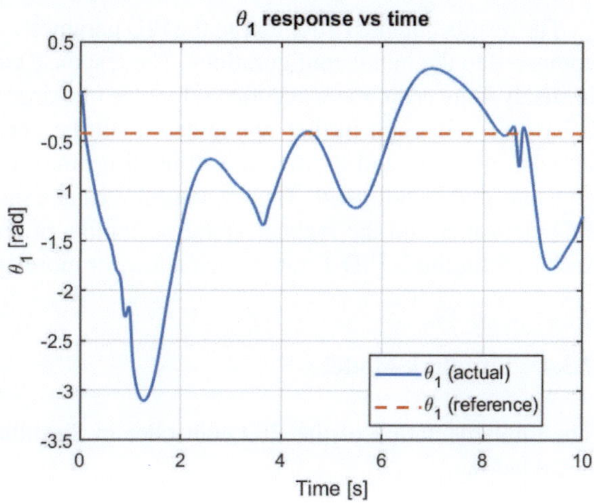

Table 10 PID controllers' parameters

	K_{p_1}	K_{i_1}	K_{d_1}	K_{p_2}	K_{i_2}	K_{d_2}
1	10	10	10	10	10	10
2	100	500	30	150	800	30
3	500	1200	50	800	2500	60
4	1000	2000	100	1600	5000	90
5	1500	2600	130	2200	8000	120
6	1800	3000	150	3000	11,000	130
7	2031	3359	161	3400	12,500	133
8	2031	3359	161	3700	13,500	134
9	2031	3359	161	3800	13,800	134.5
10	2031.44	3359.63	161.51	3876.78	13,852.27	135.12

while also improving the settling time. The intermediate combinations have been key to understanding the individual effect of each gain and making more informed adjustments in later stages.

After a few iterations with the PID tuner, the results improved significantly, allowing the controller to correct the error in a shorter time interval, enabling the arm to reach the desired position (Table 10).

With the last set of parameters, the results improved notably improved. As can be appreciated in Fig. 14, the end effector is reaching the desired point [0.5, 0.5] without problems, and reducing highly the amount and amplitude of the oscillations. The system has not only achieved the objective position, but it also has done it without taking much time.

- θ_1 takes 0.844 s in reaching stability for reference value.
- θ_2 takes 3.9 s in reaching stability for reference value. (Figs. 15 and 16).

The results obtained after tuning the PID parameters show a notable improvement compared to the initial configurations. The response curves exhibit a clear reduction in steady-state error, more accurate reference tracking, and more controlled dynamic behaviour. Although further tuning could still be performed to slightly reduce the initial oscillations and shorten the settling time, the current results have been considered acceptable and valid. They represent a clear example of the performance of a PID controller and the manual, iterative configuration process, supported by tools such as Simulink's PID Tuner, which allow for more efficient parameter refinement.

2.3.4 Simulink Model

The implementation of the PID controller in Simulink will be explained in more detail below.

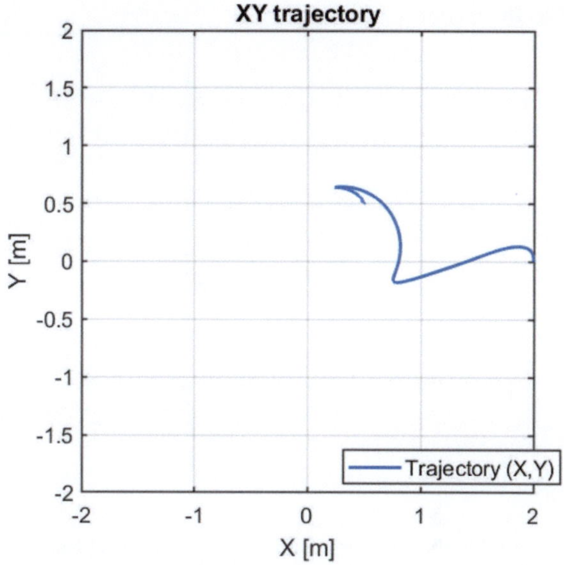

Fig. 14 Trajectory of arm's end

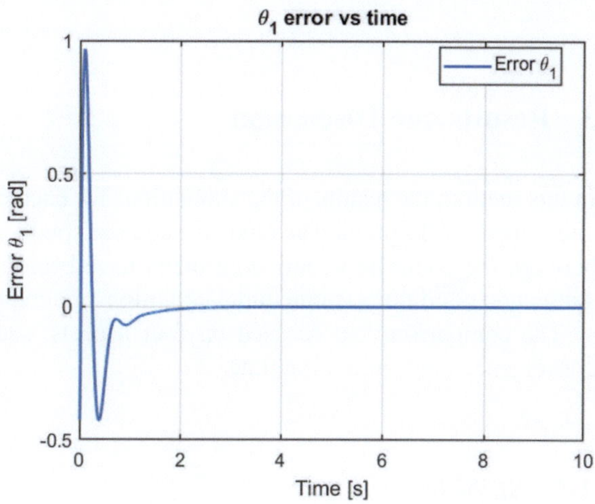

Fig. 15 θ_1 error

The model is composed by six blocks: first, the reference value, introduced by the user as a Cartesian vector as mentioned previously. This is connected to the inverse kinematics (in green). It contains the MATLAB code to transform the Cartesian vector into angular variables to be used as references by the controllers.

The controllers receive two inputs, the reference value from the inverse kinematics block, and the feedback angle variable coming from the dynamic model block. The torque, which is the input of the dynamic model block, corresponds also with the output of the controllers as it is the variable generated as the control action.

Fig. 16 θ_2 error

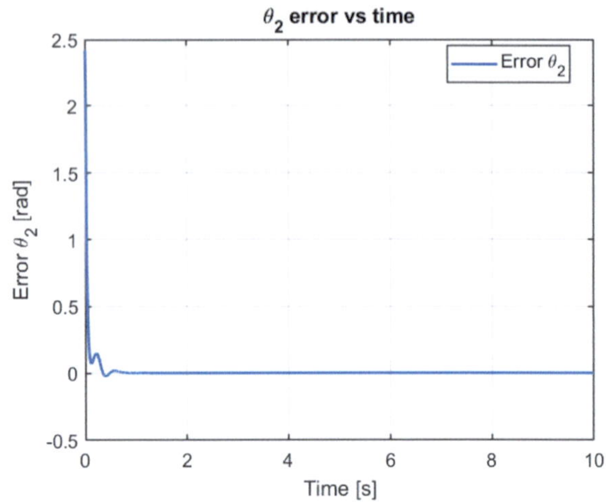

Finally, the forward kinematics subsystem block converts the angular variables θ_1 and θ_2 into Cartesian coordinates that represent the controlled position of the arm's end effector(Fig. 17).

3 Results and Discussion

In this section, the results of the simulations for each controller have been analysed and compared. To provide an objective comparison focused solely on the differences between the controllers, both controllers have been used with the same input and simulation conditions (same initial conditions, references, and limits).

The comparison has focused on four aspects, such as stability, settling time, convergence, and simulation time.

3.1 Stability

Figure 18 compares the θ_1 responses for both controllers. The PID line moves sharply at the start (a large dip followed by a brief rise), but the swings soon decrease, and the curve settles close to the reference within a few seconds. The fuzzy line begins more gently, yet the swings do not clearly decrease; it crosses the reference and then sits below it with noticeable ripples. In short, the PID stabilises the joint, while the fuzzy controller does not reach the target (Fig. 18).

Figure 19 compares the θ_2 responses. The PID line shows a brief overshoot and then the swings decrease quickly; the curve settles close to the reference within the first two seconds. The fuzzy line starts more gently but then dips well below the

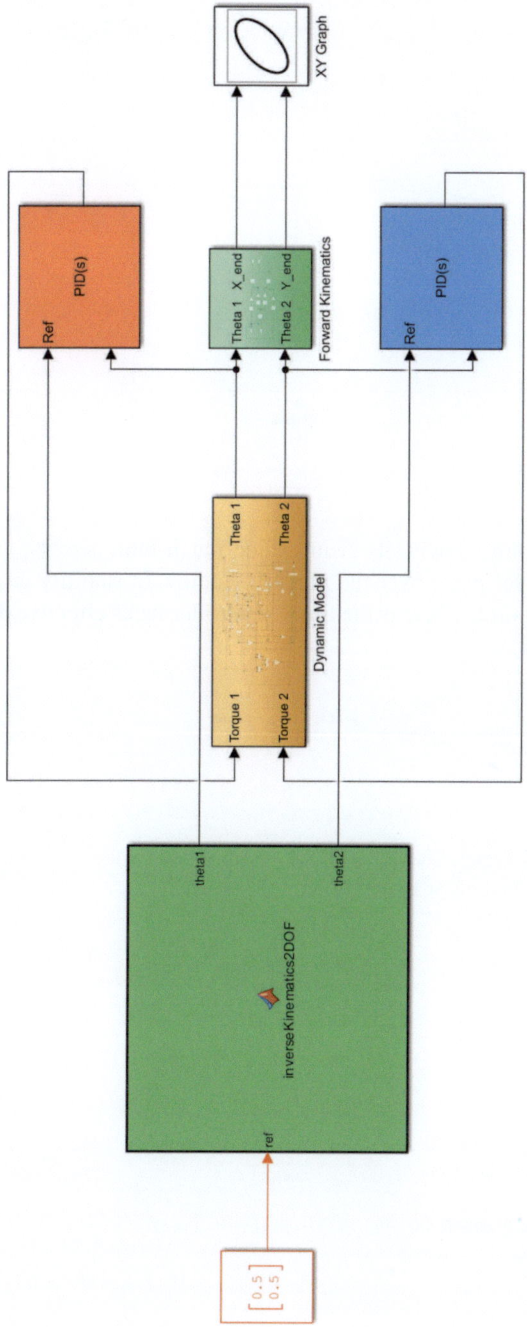

Fig. 17 Simulink model for PID control

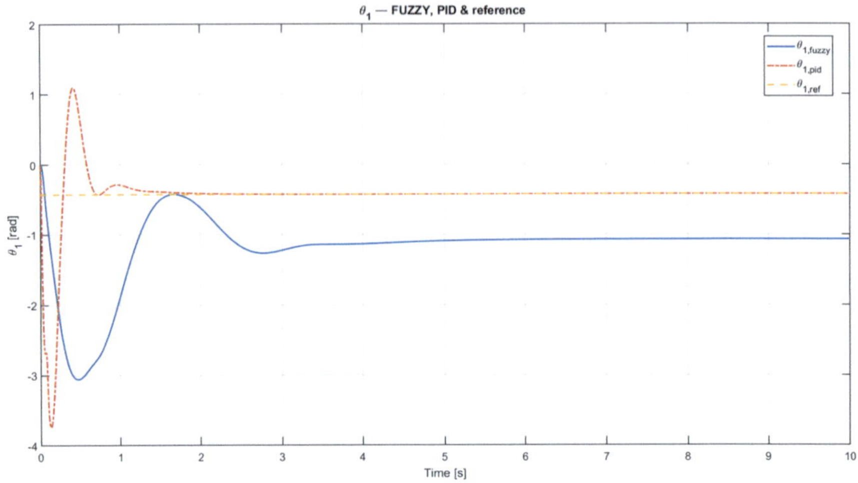

Fig. 18 θ_1 evolution comparison

target and recovers only slowly; its swings fade much more slowly, and the trace levels off below the set-point over the plotted window. In stability terms, the PID clearly stabilises the joint, whereas the fuzzy controller lacks effective damping and does not reach the target.

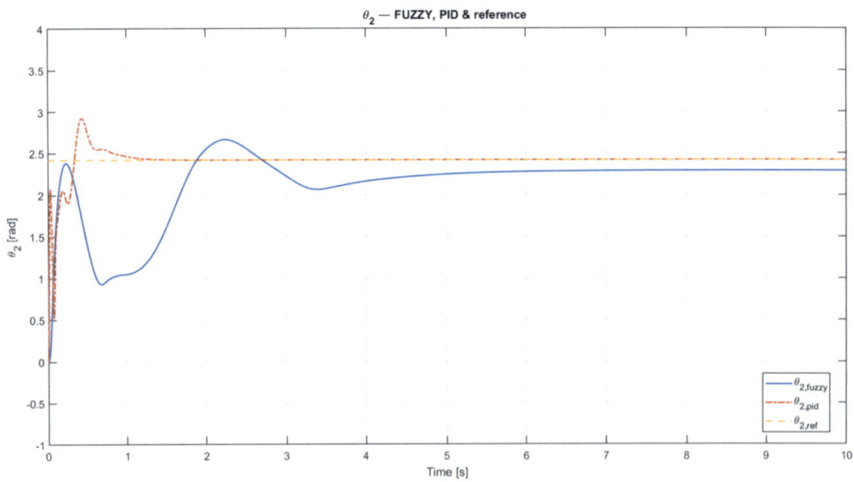

Fig. 19 θ_2 evolution comparison

3.2 Settling Time

For this section, the variables analysed could have been the controlled variables, as in the first case, however, because the reference is different from zero, the tracking error has been used here to provide a clearer explanation for the reader.

The θ_1 error, shown in Fig. 20 shows a clear difference in settling behaviour. The PID trace (dashed) enters a small tolerance band around zero within the first 1.5 s and then remains inside for the rest of the window, despite a tall initial overshoot. By contrast, the fuzzy trace (solid) does not achieve a sustained stay within such a band: after the early swings, it settles with a visible offset above zero, so a settling time cannot be assigned within the plotted horizon.

For θ_2, the pattern is similar. The PID error moves into the tolerance band quickly and stays there with only minor ripples, meeting the settling-time criterion. The fuzzy error, although it decreases after the initial transient, approaches the band only intermittently and ends slightly away from zero; therefore, it does not meet the settling-time requirement within the simulated window (Fig. 21).

Although the fuzzy controller does not reach the set-point and therefore has no settling time under the standard definition, one might adopt a more permissive view in which settling means stabilising around any constant level, even if it is biased. On that basis, the fuzzy response levels off much later than the PID: around 6 s for θ_1 and around 7 s for θ_2, compared roughly 1 s for the PID in both joints. This underlines that the PID not only attains the correct target but does so much sooner.

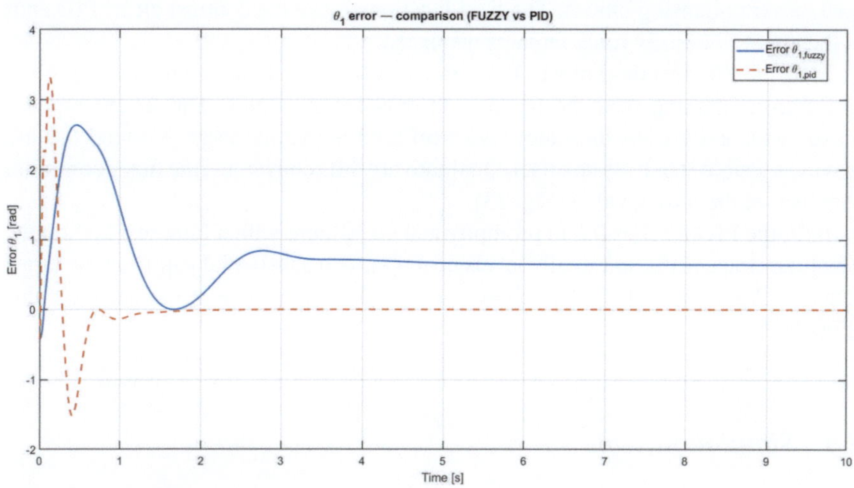

Fig. 20 θ_1 error comparison

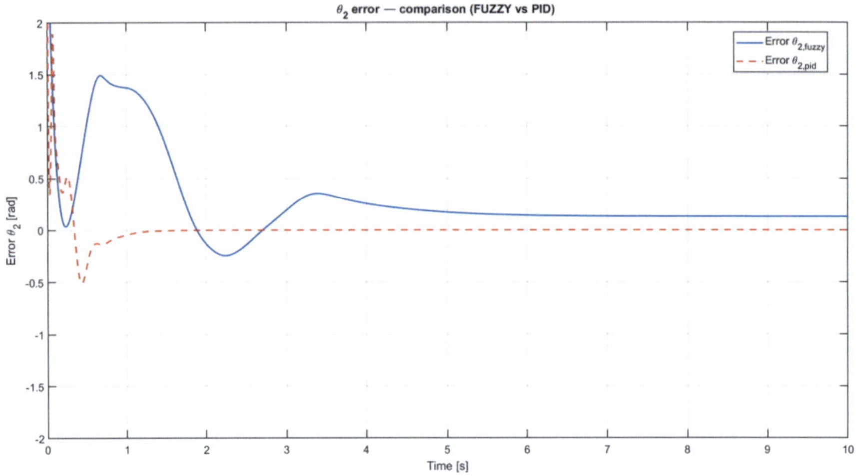

Fig. 21 θ_2 error comparison

3.3 Convergence and Steady State

As mentioned previously, only the PID reaches the desired point. In Fig. 22, the PID path heads towards the goal and finishes on the reference point, giving a clean closure of the trajectory. The fuzzy path loops widely and ultimately settles away from the goal, never collapsing onto it. The final locations make the contrast clear: PID ends at the target; the fuzzy track remains displaced.

In *X*, the PID response moves through an early undershoot and then settles close to 0.5 m, remaining near the target with only small ripples. The fuzzy response makes wide excursions and later levels off well above the target (around 0.8 m), leaving a visible steady-state offset. Qualitatively, PID converges; the fuzzy controller stabilises at the wrong value (Fig. 23).

In *Y*, the PID reaches 0.5 m promptly and stays there with a faint residual ripple. The fuzzy trace drifts and finally levels near 0 m, so a substantial gap from the target remains. Again, only the PID achieves a reliable steady state at the desired value (Fig. 24).

3.4 Simulation Time

Finally, the last aspect compared is the simulation time, which can be considered a good measure of the resources required by both controllers. With identical settings (ode4, fixed-step of 1 ms, simulation horizon of 10 s), the Simulink Profiler reports a wall-clock of 0.483 s for the PID and 27.048 s for the fuzzy controller. Normalised, this is 0.0483 s per simulated second for PID versus 2.7048 s/s for fuzzy, i.e. the

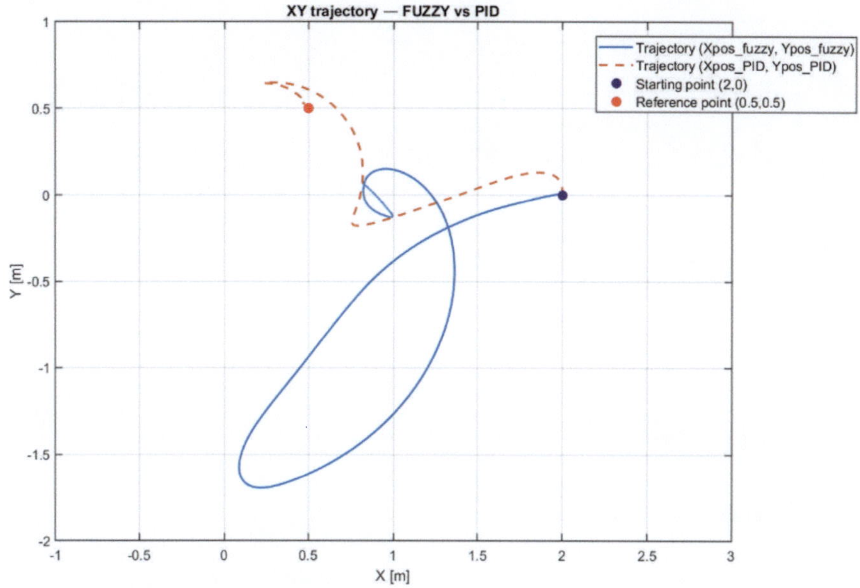

Fig. 22 *XY* trajectory comparison

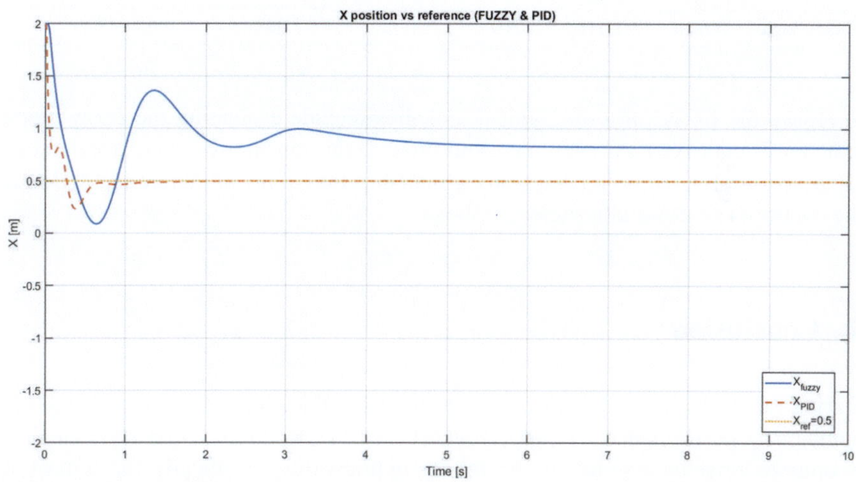

Fig. 23 *X* position comparison

fuzzy run is about 56× slower. Put differently, the PID executes at roughly 20.7× real time, while the fuzzy case runs at about 0.37× real time (Table 11).

This gap reflects the computational load of fuzzy inference and defuzzification at every 1 ms step, compared with the algebraically simpler PID. In practical terms, PID is the economical option for real-time feasibility and batch sweeps. If you need

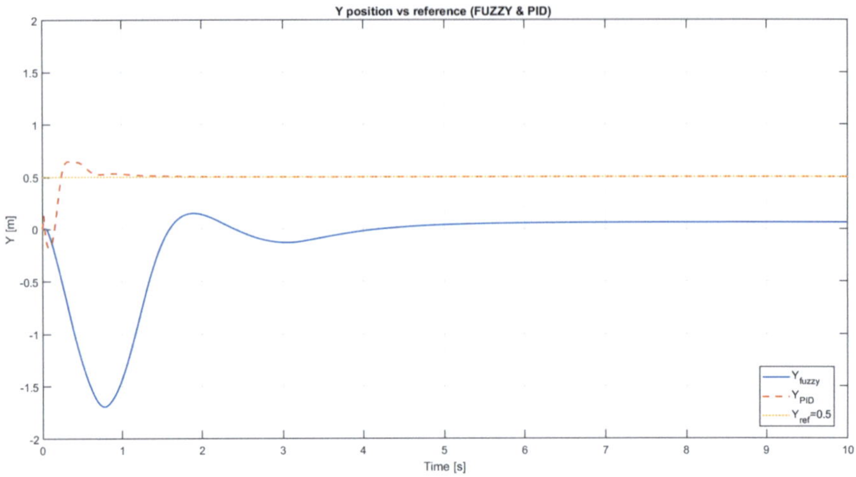

Fig. 24 *Y* position comparison

Table 11 Simulation time comparison

Controller	Solver	Step size	Simulated time [s]	Wall clock [s]	Time/sim time
PID	Ode4	0.001	10	0.483	0.0483
Fuzzy logic	Ode4	0.001	10	27.048	2.7048

to reduce the fuzzy run time, typical remedies include evaluating the fuzzy block at a slower controller sample time (with proper rate transitions), using accelerator/rapid accelerator, trimming rules or membership functions, or generating code for the controller to avoid interpreter overhead.

4 Conclusion

During the development of this research, the main objective has been to assess and compare the performance of two controllers based on very different control strategies, in order to judge their suitability for robotic applications, specifically, the control of a two-degree-of-freedom arm. Each controller brings its own advantages and drawbacks in terms of setup and tuning; however, when tested under identical conditions and evaluated through stability, settling behaviour, convergence/steady-state accuracy, and simulation time, a clear difference in performance emerges.

The PID controller is the only one that stabilises the arm and reaches the target. It does so rapidly (around 1 s to settle in both joints) and with a much lower computational cost, approximately 56× faster than the fuzzy controller under the same solver and step size. In Cartesian space, the PID also follows a shorter, cleaner path to the goal. Although there is a pronounced early overshoot, the oscillations reduce quickly, and the response remains close to the reference thereafter.

By contrast, the fuzzy logic controller, as tuned here, does not reach the set-point: it levels off at biased values (if "settling" is read as stabilising around its own constant level, this occurs around 6 s for θ_1 and 7 s for θ_2) and its final position remains displaced from the target. Its run time is also substantially higher, reflecting the additional effort of rule evaluation and defuzzification at every time step.

From a practical standpoint, these findings have clear implications. In applications where reliability, turnaround time, and computational economy are critical, as is common in industrial settings, the PID offers a safer and more efficient choice: it achieves the specified position, settles promptly, and enables quicker batch testing and on-line operation thanks to its modest computational load. While the fuzzy approach can produce a visually smoother early evolution, in this configuration it neither guarantees convergence to the desired pose nor offers competitive simulation times. Overall, for the plant and operating conditions considered, the PID is the more dependable and cost-effective controller.

Next steps could include testing the controllers on manipulators with more degrees of freedom; adding constraints explicitly, such as workspace limits, torque limits, and joint-angle limits; controlling complete trajectories, not only the final point; or even implementation on a real robotic arm to validate the conclusions in hardware.

References

1. Waseem S, Adnan M, Iqbal MS, Amin AA, Shah A, Tariq M (2025) From classical to intelligent control: evolving trends in robotic manipulator technology. Comput Electr Eng 127:110559. https://doi.org/10.1016/j.compeleceng.2025.110559
2. Tinoco V, Silva MF, Santos FN, Morais R, Magalhães SA, Oliveira PM (2025) A review of advanced controller methodologies for robotic manipulators. Int J Dyn Control 13:36. https://doi.org/10.1007/s40435-024-01533-1
3. Müftü S, Gökçe B (2024) Design and implementation of an optimized PID controller for two-limb robot arm control. Bitlis Eren Univ J Sci 13(1):192–204. https://doi.org/10.17798/bitlisfen.1370223
4. Ghaleb NM, Aly AA (2018) Modeling and control of 2-DOF robot arm. Int J Adv Comput Sci Appl 9(1):222–229
5. Li Noce A, Patanè L, Arena P (2025) A stable and safe method for two-leg balancing of a quadruped robot using a neural-network-based controller. Robot Auton Syst 186:104901. https://doi.org/10.1016/j.robot.2024.104901
6. Quiros ARF, Abad AC, Dadios EP (2015) Object locator and collector robotic arm using artificial neural networks. In: 8th IEEE International conference on humanoid, nanotechnology, information technology, communication and control, environment and management (HNICEM). IEEE. https://doi.org/10.1109/HNICEM.2015.7393270
7. Shneen SW, Gaber RK, Salih RS, Jiaad SM (2025) Artificial neural network (ANN) based proportional integral derivative (PID) for arm rehabilitation device. Kufa J Eng 16(1):80–103. https://doi.org/10.30572/2018/KJE/160106
8. Mishra P, Pandey MK, Mahia RN (2024) Model predictive control for a 2-DOF robotic arm: dynamics and control. In: International conference on modelling, simulation & intelligent computing (MoSICom). IEEE. https://doi.org/10.1109/MoSICOm63082.2024.10881967

9. Abdolkarimi A, Zahedi MH (2024) Controlling a 6DOF serial manipulator using parallel robot by ANN. In: 20th CSI International symposium on artificial intelligence and signal processing (AISP), Mazandaran University of Science and Technology. IEEE, Babol, Iran. https://doi.org/10.1109/AISP61396.2024.10475254
10. Navya M, Pai V, Muralidhara M, Rao M, Jain N, Rao A (2024) Estimation of inverse kinematics solutions of a 2D planar robotic manipulator using feed-forward neural network. In: International conference on IoT based control networks and intelligent systems (ICICNIS). IEEE. https://doi.org/10.1109/ICICNIS64247.2024.10823376
11. Gouda MM, Said FE-Z, Elmoushi ES, Abdelwahab SA (2022) Performance analysis and improvement of robot arm 5DOF using PID and fuzzy controllers: a comparative study. Port Said Eng Res J 26(3):110–133. https://doi.org/10.21608/PSERJ.2022.132466.1177
12. Aminiazar W, Najafi F, Nekoui MA (2013) Optimized intelligent control of a 2-degree of freedom robot for rehabilitation of lower limbs using neural network and genetic algorithm. J NeuroEngineering Rehabil 10(96). https://doi.org/10.1186/1743-0003-10-96
13. Han J, Shan X, Liu H, Xiao J, Huang T (2023) Fuzzy gain scheduling PID control of a hybrid robot based on dynamic characteristics. Mech Mach Theory 184:105283. https://doi.org/10.1016/j.mechmachtheory.2023.105283
14. Minh Nguyet NT, Ba DX (2023) A neural flexible PID controller for task-space control of robotic manipulators. Front Robot AI 9:975850. https://doi.org/10.3389/frobt.2022.975850
15. Tavoosi J, Jokandan AS, Daneshwar MA (2012) A new method for position control of a 2-DOF robot arm using neuro–fuzzy controller. Indian J Sci Technol 5(3):2253–2260
16. Krzz A (2020) Model reference controller approach for robot arm. Int J Robot Autom 35(2):122–131
17. Bi M (2020) Control of robot arm motion using trapezoid fuzzy two-degree-of-freedom PID algorithm. Symmetry 12(4):665. https://doi.org/10.3390/sym12040665
18. Gao T (2024) Optimizing robotic arm control using deep Q-learning and artificial neural networks through demonstration-based methodologies: a case study of dynamic and static conditions. Robot Auton Syst 181:104771. https://doi.org/10.1016/j.robot.2024.104771
19. Bahrami V, Kalhor A, Masouleh MT (2021) Dynamic modeling and design of controller for the 2-DoF serial chain actuated by a cable-driven robot based on feedback linearization. Proc Inst Mech Eng C J Mech Eng Sci 236(5):2546–2558. https://doi.org/10.1177/09544062211027922
20. Kabir U, Hamza MF, Haruna A, Shehu GS (2019) Performance analysis of PID, PD and fuzzy controllers for position control of 3-DOF robot manipulator. Zaria J Electr Eng Technol 8(1):18–25
21. D'Addato G, Carli R, Pedrosa E, Pereira A, Palopoli L, Fontanelli D (2024) Joint torques prediction of a robotic arm using neural networks. In: Proceedings of the Italian National Ph.D. programme in autonomous systems (DAuSy), Università di Trento
22. Safaei A, Koo YC, Mahyuddin MN (2017) Adaptive model-free control for robotic manipulators. In: International conference on intelligent robotics and industrial systems (IRIS). IEEE. https://doi.org/10.1109/IRIS.2017.8250090
23. Wazzan AN, Basil N, Raad M, Mohammed HK (2022) PID controller with robotic arm using optimization algorithm. Int J Mech Eng 7(2):3746–3751
24. Sharkawy AN (2022) Forward and inverse kinematics solution of a robotic manipulator using a multilayer feedforward neural network. J Mech Energy Eng 6(4):303–310. https://doi.org/10.30464/jmee.00300
25. Bouzid R, Narayan J, Gritli H (2024) Exploring neural networks for forward kinematics of the robotic arm with different length configurations: a comparative analysis. In: IEEE International conference on interdisciplinary approaches in technology and management for social innovation (IATMSI). IEEE. https://doi.org/10.1109/IATMSI60426.2024.10503452
26. Hasan AH, AlHamri M, Firoz O, Khan SR, Khan SG (2024) Design and control of 2DoF wearable robot arm for rehabilitation. In: Proceedings of the 6th international conference on robotics and automation in industry (ICRAI). IEEE. https://doi.org/10.1109/ICRAI62391.2024.10894684

27. Fitria S, Sariman, Nawawi Z, Kurnia RF, Dewi T (2024) Simulation of robot arm system control using fuzzy logic. Sriwijaya Electr Comput Eng J (SELCO) 1(1):20–29. https://doi.org/10.62420/selco.v1i1.2
28. Lin J, Liu X, Ren Z (2022) Fuzzy PID control for multi-joint robotic arm. In: 2022 IEEE International conference on industrial informatics (INDIN). IEEE. https://doi.org/10.1109/INDIN51773.2022.9976122
29. Bikova M, Latkoska VO, Hristov B, Stavrov D (2022) Path planning using fuzzy logic control of a 2-DOF robotic arm. In: 2022 IEEE 17th International conference on control & automation (ICCA), Naples, Italy. IEEE. https://doi.org/10.1109/ICCA54724.2022.9831903
30. Pham VC, Tran MH, Nguyen PA, Vu NS, Thi NN (2025) Robust adaptive fuzzy sliding mode control for trajectory tracking of cylindrical manipulators. arXiv preprint. https://doi.org/10.48550/arXiv.2508.05584
31. Kern J, Marrero D, Urrea C (2023) Fuzzy control strategies development for a 3-DoF robotic manipulator in trajectory tracking. Processes 11(12):3267. https://doi.org/10.3390/pr11123267
32. Kulkarni P, Kulkarni O, Sayyad JK (2024) Tuning of a robotic arm using PID controller for robotics and automation industry, 2024 6th International conference on energy, power and environment (ICEPE), ICEPE2024-10668952

Design and Implementation of a Neuro-Fuzzy Controller for a Robotic Leg

Milosz Chrobak

Abstract In this research, a neuro-fuzzy controller has been designed and implemented for a robotic leg system. The compound pendulum model uses all the required forces acting on the system, so it has been used to develop the basic model. The controller combines a neural network, which is used for the non-linear dynamic modelling, and a fuzzy logic system to model uncertainties with regards to the inputs. The simulations are done on the MATLAB/SIMULINK environment as it simplifies complex concepts through the intuitiveness of the program. A conventional PID controller has also been designed and the variable error of the output renders it unsuitable for modelling non-linear systems. The results show that the neuro-fuzzy system is the most suitable for the leg model, due to the immediate response time, accuracy, and stability. The controller successfully combines the strengths of these machine learning methods to accomplish high-level control. This reinforces the potential for further development of these control techniques for various robotic systems.

1 Introduction

Since the creation of the first robot by George Devol in the early 1950s, the possibilities and uses for these man-made machines have grown exponentially. In order to understand the inner workings of the human body, research was conducted to apply mathematical models that reflect the rules of nature. After seeing how many complex ideas could be broken down into simpler rules of physics, many of them were starting to be applied in machinery. Human abilities of walking, running, or jumping were being implemented, with these mathematical modelling techniques at the centre. The gait of leg movement is best described as continuous, based on a mathematical model that uses analytical mechanics [42]. Simultaneously, human understanding about the body and the central nervous system (CNS) further developed with various research. Artificial neural network (ANN) algorithms were a result of said research and they are now widely used within robotics. Alongside the research and development of

M. Chrobak (✉)
School of Engineering, Computing and Mathematics, Oxford Brookes University, Oxford, UK
e-mail: chrobakmilosz7@gmail.com

A. Azizi (ed.), *Intelligent Control and Autonomous Systems*, Emerging Trends in Mechatronics, https://doi.org/10.1007/978-981-95-6402-6_6

neuroscience, artificial neural networks are performing human-level behaviours such as language processing, motor learning and face recognition [20]. They allow the program to make connections between different parts of the robot, similarly to how the body communicates. Fuzzy logic is what simulates the decision-making capabilities of humans in robots. In this project, the aim is to utilise neural networks for the approximation of the basic mathematical model and apply fuzzy logic for the control of the output angle of non-linear robotic leg model. A PID controller is also designed, and its results are compared to the neuro-fuzzy controller. The system has been modelled in MATLAB/SIMULINK, as this environment provides tools that simplify mathematical concepts that usually requires a much greater knowledge and in-depth analysis.

2 Aim and Objectives

Aim:

The aim of this project is to design a neuro-fuzzy control system that can be used to control a robotic leg.

Objectives:

- Research different modelling techniques for robotics system. Explore how other methods can be used but why the chosen one is most suitable in the literature review.
- Derive the equation of motion and model the dynamic behaviour of the robotic leg, which will then be used on MATLAB. Apply any functions into the system that can make the model more realistic.
- Explore artificial neural networks in depth and how to use the MATLAB toolbox to further develop the model. Decide on the optimal parameters that will provide the most accurate results.
- Design the fuzzy logic controller that will be used to control the ANN plant previously developed.
- Build a PID controller and compare its functionality and results to that of the fuzzy logic controller. Consider whether both still provide satisfactory results or whether one of them is not suitable for the modelling of the robotic leg.

3 Project Justification

The use of legged robots has drastically increased in various career fields as the ability to control and expand the systems has developed. This research task focuses on the modelling of a single non-linear robotic leg, meaning that the mathematical principles that apply to this system can be repurposed in other projects. The same knowledge on dynamics used to control a single leg can be used as the starting point for the construction of quadruped robots [5]. Therefore, it could be possible to replicate the movements and behaviours of animals with more limbs and mimic these observations to understand their survival instincts. This could be useful in military applications, for example, where bipedal robots scout hard to reach areas or places with extreme weather conditions.

There is a large interest in robotics for medical applications, such as exoskeletons or prosthetic limbs. The control of a prosthetic leg is one of the many areas, where a pendulum system is necessary [48]. Since this project uses the compound pendulum model to derive the equation of motion, it can be a good reference point within this industry. As a result of this medical development, many people's quality of life and satisfaction can improve, making it a beneficial field of interest. Furthermore, the use of a neuro-fuzzy controller can be developed to control the systems. Vladareanu et al. [41] suggested that the neuro-fuzzy approach is also suitable for the control of rehabilitation robots.

4 Literature Review

4.1 Introduction

Every research task within the field of robotics requires a model that can be used for simulation. In order to best represent the forces applied onto a non-linear robotic leg, multiple methods of representations are explored and discussed to explain the suitability of the compound pendulum model. The mathematical analysis of the system has also been explored with the explanations behind its suitability. In order to monitor and control the system, neuro-fuzzy and PID controllers are both designed with the intention to analyse the results and examine the suitability of each one. All of these methods are further discussed in the following section.

4.2 Compound Pendulum Model

Many of the modern robots take inspiration from activities performed by humans. Due to the complex nature of everyday tasks performed, there is a need to consider any patterns and rhythms that can be used to create the foundations of a system. Walking, in terms of basic mechanical movements, is displacement from the original position due to periodic internal changes within the structure and the reaction forces from the environment [14]. These 'periodic' oscillations are what allows the robotic leg to be modelled as a pendulum. With this concept in place, the model can be developed from static positioning to motion analysis of the robotic system. As the humanoid robots are designed with multiple joints, their movement allows them to perform tasks that are similar to human behaviour, such as walking, jumping or avoiding obstacles [3]. Jain et al. [23] also suggest that the extended system of a pendulum can be used to model more complex systems like bipedal walking. This is further reinforced by Liao et al. [29] who agreed that walking movements of a robot are a non-linear periodic locomotion.

There are researchers who suggest the use of an inverse pendulum system instead, as it considers the centre of mass to be above the base, which may be deemed as more realistic. For example, the human posture and balancing abilities can be modelled using an inverse pendulum, as aspects of human nature that are fully dependent on leg and core strength can be mimicked [35]. All of the models (inverted pendulum and other variations) assume that the body mass is concentrated at a single point, representing the body's centre of mass [49]. These examples consider the string, which represents the robotic leg itself, to be massless. This can be deemed as unrealistic when it comes to modelling the system since the robotic leg will change where the centre of mass is located, which can have an effect on the amount of input torque required for certain angle movements. Simplifying the dynamics of the system to a point mass attached to a rod (inverted pendulum model) can be very limiting for the design with regards to the stabilisation as well as the physical limitations of the joints and motors [18]. However, through the use of the compound pendulum, the focus remains on the angular positioning of the robotic leg, whilst still considering the centre of mass and its effect on the whole system. Using a pendulum model can simulate the operations of a controller and the system's response time without the need for additional hardware [45]. It reduces mathematical complications from applying more forces onto the system, which do not affect the overall system in this project.

After considering the various modelling techniques discussed, the compound pendulum model has been selected and used for the development of the non-linear robotic leg system.

4.3 Neuro-Fuzzy Controller

Fuzzy logic is used as an attempt to introduce logical rules through artificial intelligence to improve a system. It works by executing certain 'fuzzy rules' that are used to control the inputs based on the desired output. Fuzzy systems and their hybrid derivations are used to computationally simulate the human ability of reasoning in an efficient way, which is why research is currently focused on the use of these systems for modelling, control and classification [9]. Aldair et al. [4] suggest that the main features of a fuzzy controller are its ability to deal with non-linear systems and adaptability for problem-solving, which is what makes them superior to other controllers. Dehyadegari and Khajehasani [11] also suggest that a fuzzy system allows for the modelling of any uncertainties of a dynamic model. The fact that this approach can be used for models that do not have a complex mathematical analysis further reinforces the suitability of the compound pendulum model for the control of the non-linear robotic leg.

As the leg model is based on human movement, the connections within the body that cause these actions must be considered. Just like the brain is a network based on neurons, the ANN is a system of nodes [25]. Learning the patterns and true data distribution is intractable, so a neural network can be used to develop a model that approximates the data distribution [21]. Neural networks are used to create a relationship between the input torque of the motor into the system and the angle of the leg, which can be used instead of the plant model for the implementation of fuzzy logic.

For many years, neural networks and fuzzy control are utilised in the control of non-linear systems that would be too complex to illustrate mathematically [37]. Abougarair [1] also suggests that when using fuzzy logic control, the mathematical model of the system is not required, meaning that it is directly applied to the neurological model. Using neuro-fuzzy systems is beneficial for both complex and non-linear problems, meaning that they can be used for a variety of applications, such as data classification or rule-based control [19]. This idea reinforces the suitability of the fuzzy neural network for the control of the non-linear robotic leg. A neuro-fuzzy controller combines the benefits from both neural networks and fuzzy systems as it provides smoothness and adaptability [32].

Another controller instead of the neuro-fuzzy that has been considered is the PID. A combination of the proportional, integral, and differential gains is used to monitor and control the output of the system. The benefits of using a conventional PID controller are its simplicity, efficiency, reliability, and robustness [30]. There are multiple variations of the standard controller which have a variety of applications in different engineering models. Some results show that using a neural PID controller over a classic PID is advantageous [46]. The most important task when using these controllers is finding the most suitable gains. In order for a PID controller to work, it must be tuned properly so that it suits the dynamics of the system [2]. In certain instances, the use of conventional PID [controller] methods of tuning could result in the system becoming unstable or inadequate control loop performance [8]. The problem with linear controllers is their inability to consider the systems non-linearity and any external disturbances [39]. PID controllers can be used effectively with the correct optimisation of the gains but the process of re-optimisation for individual scenarios can be a time-consuming and process, so they are not ideal for bipedal robots in constantly varying environments [34]. The two ideas will be further developed and compared in this project.

Considering the two types of controllers discussed, the combination of neural networks and fuzzy systems for the proposed controller provides a greater degree of accuracy and control in comparison with the PID controller. Therefore, the neuro-fuzzy controller has been selected to control the non-linear systems.

4.4 Kinematics Approach

Another approach that could have been taken for the control of the robotic leg instead of the derivation of the equation of motion is the use of kinematics which feeds onto the controller. Neuro-fuzzy systems can be used to solve inverse kinematics by creating a relationship between the motion and joint coordinates [7]. Gaed et al. [16]further reinforce this idea, as they explain that in robotic science, kinematics is considered a branch of mechanics where the relationship between position and orientation is investigated without considering the force. The use of kinematics requires a much more in-depth mathematical analysis of the model, which as previously mentioned, is not required for the use of neuro-fuzzy controllers. Also, this approach is more suited for the focus of stability of a model due to its ability to control the end effector through the use of angles, whereas the purpose of this project is to control the angle of the robotic leg provided an input at the motor. Wang et al. [43] suggest that a deep neural network can be employed so that the non-linear equations do not have to be solved, further reinforcing the suitability of the basic model using the equation of motion. This means that any additional calculations will create extra error in the main model that may cause unnecessary problems. Simplified models may have an imperfect representation of the system, but they are able to portray the robot's behaviour with a reasonable degree of accuracy in a way suitable for analysis, as well as providing a principled design for a controller [36]. A full, whole body dynamics model of a robot may provide more precise results but it comes at a computational expense that is platform dependent [24]. Therefore, potentially doing a more time-consuming, in-depth analysis of the system may not provide a much greater accuracy of results that could outweigh the cost of processing the data.

4.5 Conclusion

Based on the literature review, this project aims to design a neuro-fuzzy controller for a non-linear robotic leg through the use of a compound pendulum model for the derivation of the mathematical model. Many other researchers consider the use of an inverse pendulum instead as it allows for the consideration of the centre of mass above the pendulum and helps with stabilisation errors. The focus of this project is on the angle of the pendulum; therefore, the use of a compound pendulum allows for the disregarding of extra, unnecessary analysis that is more likely to increase error than improve accuracy. Due to the fact that both neural networks and fuzzy logic do not require complex mathematical solutions, the use of inverse kinematics is not considered. A PID controller will also be designed in order to compare the results of different types of controllers and discuss the suitability for the model.

Since the aims and objectives have been set and the relevant literature discussed, the following sections of the research are focused on the methodology, which includes the design and model implementation, results, and overall project discussion.

5 Methodology

An initial research period is required to gain a deeper understanding of each aspect of the task. Artificial neural network (ANN) and fuzzy logic have been explored separately, as they are interdependent topics that coincide into making the final neuro-fuzzy controller. If any part is not completely understood, more research was conducted to ensure that the required knowledge had been acquired. The relevant MATLAB Toolboxes have been comprehensively researched, so that the optimal conditions are recreated in the system. In order to create the initial plant model of the non-linear robotic leg, the equation of motion has been derived, and represented as a compound pendulum. This was then modelled on the MATLAB-SIMULINK environment. An ANN plant has been created to model the behaviour of the system. The previous research allowed for educated assumptions to made, which ensures that the system is functioning correctly. After this, a fuzzy system was designed, in order to control the output angle of the model through the input torque. If at any stage the results proved to be different from those theoretically expected, the models were reviewed before moving onto the next step. Finally, the results from the completed neuro-fuzzy controller have been compared to those of the designed conventional PID controller. The suitability of each model in multiple real-life applications was discussed, based on the results and conclusions of other researchers discussed in the literature review.

The flowchart in Fig. 1 summarises the key aspects of the methodology and the following sections discuss the step-by-step process to achieve the final design and controller models.

5.1 *Initial Plant Model*

The system used for modelling the non-linear robotic leg has been represented by the compound pendulum, as previously discussed in the literature review. Figure 2 is a diagram of the model used.

Table 1 displays the parameters that were used in the leg model to simulate the initial conditions.

There is a pivot point where the torque, T, is applied in order to control the angle of movement, θ. The centre of gravity is distance d away from the pivot and the length of the whole bar is L.

The angular equation of motion has been derived through the following equation:

$$T = I\ddot{\theta}, \tag{1}$$

where $\ddot{\theta}$ is the acceleration with regards to the angle.

When the pendulum is displaced from the centre (i.e. when $\theta \neq 0$), the perpendicular (to the bar) component of the weight became the restoring force, attempting

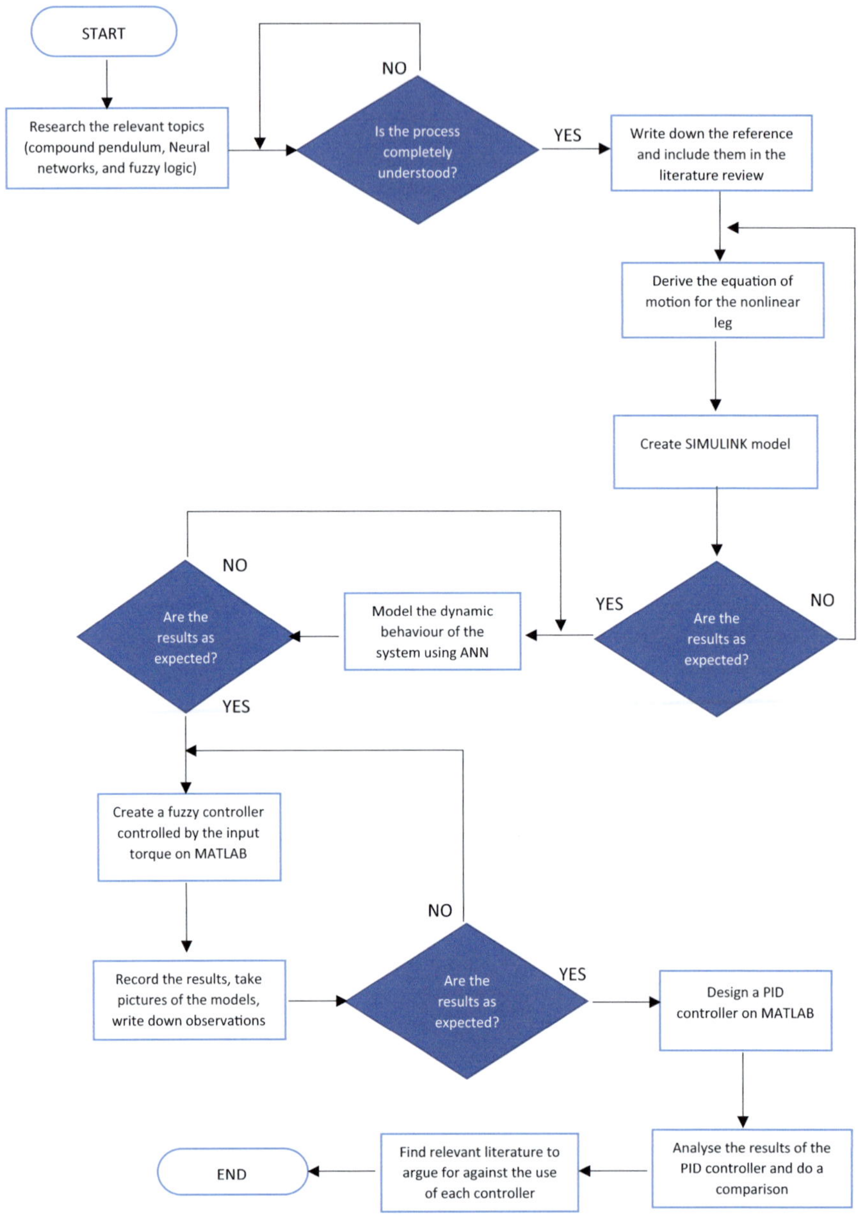

Fig. 1 Methodology flowchart

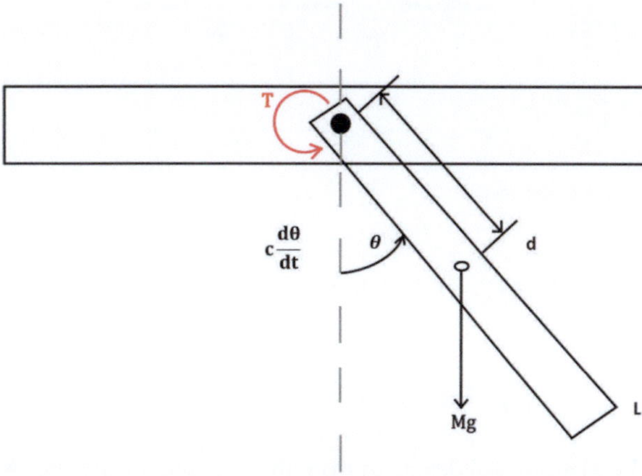

Fig. 2 Compound pendulum model for robotic leg system

Table 1 Parameters used for the leg model

Parameter	Symbol	Value
Torque	T	– (N)
Mass of pendulum	M	0.430 kg
Force of gravity	g	9.81 N
Angle	Θ	– (°)
Damping coefficient	c	0.00035
Bar length	L	0.495 m
Distance (from pivot to centre of gravity)	d	0.0230 m
Moment of Inertia	I	0.0090 kg m^2

to bring the system back to equilibrium. The damping rate also acts against the direction of motion, meaning that it adds to the total resistive force. Since the torque is equivalent to the moment, both of them are represented in a similar fashion, as shown in Eqs. 2 and 3.

$$M = F \times d \tag{2}$$

$$T = -mg\sin(\theta) \times d - c\dot{\theta}, \tag{3}$$

where $\dot{\theta}$ is the velocity with regards to angle.

To find the equation of motion, the two previous Eqs. (1 and 3) are combined and finally rearranged, so that the highest order derivative is the subject of the formula, shown in Eq. 4. This has been done to simplify the plant model design process on SIMULINK. The final equation is displayed through Eq. 5.

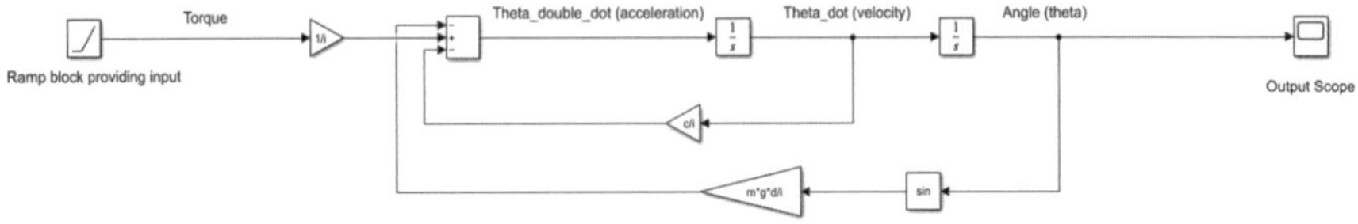

Fig. 3 Initial plant model on SIMULINK

$$T = I\ddot{\theta} + mgd\,\sin(\theta) + c\dot{\theta} \tag{4}$$

$$\ddot{\theta} = -\frac{c\dot{\theta}}{I} - \left(\frac{mgd}{I}\right)\sin(\theta) + \frac{T}{I} \tag{5}$$

Figure 3 displays the plant model designed from the equation of motion. A ramp block has been used to provide the input signal for the system. The first gain block represents torque, which as shown in equation X6, is divided by the inertia. This block is attached straight to the source as it is independent of the θ in the equation and since it is positive, it is attached to the positive part of the sum block. The acceleration (or $\ddot{\theta}$) is the subject of the formula, so it is the first value after the sum block and does not have anything attached to it. In order to represent the velocity, the previous acceleration has been integrated, which is why the integration block was used. This is where the damping coefficient is divided by the inertia, represented by the gain block. It is then attached to the negative terminal of the sum block. Similarly, the velocity had to be once again integrated to display the angle. The gain block represents restoring force and since in the equation this is multiplied by $\sin(\theta)$, a sin block is used. Table 1 displays the parameters that were used in the leg model to simulate the initial conditions.

Figure 4 displays the angle output of the initial plant model. Even though the desired output is achieved, it is not limited by any value. When a scope was attached, the graph reached values 17,000°. When considering the robotic leg, the maximum change in angle is 180°; otherwise, the leg moves far enough to enter back into the base. However, to ensure the best accuracy of the model in the later stages, a saturation block was not added to limit the angle at this point.

5.2 *ANN Plant*

Artificial neural networks are key machine learning tools used for data modelling, with a variety of implications. The first neural network model (single layer perceptron) was introduced by McCulloch and Pitts in 1944, following various research based on brain activity. It became a large topic of interest in neuroscience and computer science. In the 1960s it was shown that the artificial neurons from neural networks act similarly to those in a brain, due to their ability to recognise sophisticated patterns [27]. This was until the theory was disproved by Minsky and Papert

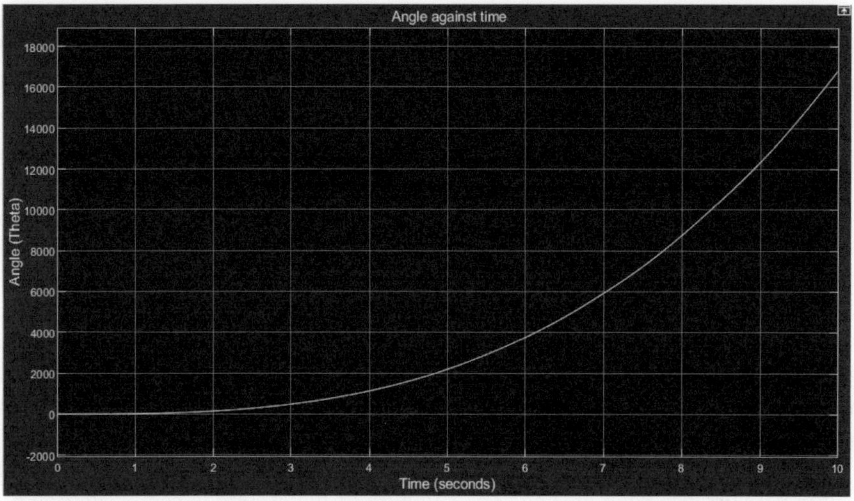

Fig. 4 Angle output from scope

in 1969, causing the research into neural networks to drastically slow down. With the development of the backpropagation learning algorithm and Hopfield's energy approach, interest in ANN renewed in the early 1980s [22]. The development of the multi-layer perceptron model solved the problems of its predecessor, and the neural network model continues to expand into other fields.

A neural network is a self-learning system, functioning similarly to the human brain and nervous system, which performs thousands of iterations to predict an output [12]. The input data provided is split between two datasets: training and testing. Each input is fed through multiple layers of interconnected neurons. Weights are applied between the layers and each neuron computes a weighted sum of the inputs, which is the sum of the products of the inputs and weights. Figure 5 displays a graphical representation of the processes within a neural network.

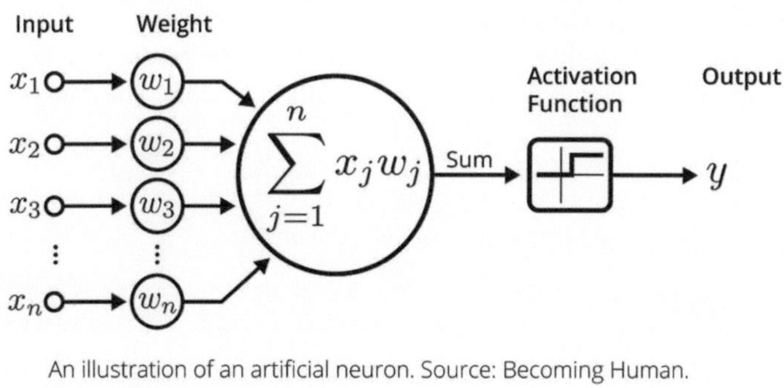

Fig. 5 Illustration of an artificial neutron [31]

After this, the values are run through an activation function, which displays the outputs. An activation function is used to decide whether a neuron will be activated [15]. A set of weights is initially assigned as the model is trained on the training dataset and adapted with each iteration until the optimum values are decided. This is done through a process of backpropagation, which constantly updates the weights based on the accuracy of the training data outputs. A bias is a constant added onto the weighted sum that is used to shift the result to either side of the activation function and avoid overfitting. The neural network has been modelled using the initial plant model on the SIMULINK environment.

Firstly, the input and output data from the plant were stored for the neural network to be able to analyse the model. The input data, provided by the ramp block, is the torque applied by the motor and the output data is the corresponding output angle of the leg. This data was then transferred from SIMULINK into the MATLAB workspace in two arrays. The extended plant model is shown in Fig. 6.

Utilising MATLAB's Neural Networks toolbox, a two-layered feed-forward neural network has been created. The structure of the system is best visualised by Fig. 7.

Since the regression line is important to show the fit of the data, the fitting tool was used, and the input and output data were imported. A balance between the computational time and the number of epochs (iterations) is necessary for the model to provide suitable results without taking an unrealistic amount of time to process the data. For the purpose of a trial run, 220 observations were used from the inputs and outputs in order to see how the neural network will operate.

With a 70/30 ratio for the data split of the training and testing data, respectively, and the use of the Levenberg–Marquardt training algorithm, a neural network was successfully trained. The Levenberg–Marquardt algorithm is a non-linear optimisation algorithm that combines the steepest descent and Gauss–Newton methods. An advantage of using this algorithm is its fast convergence [40]. Since the dataset was small and not too complex, two hidden layers were used. This can be seen in Fig. 8. The regression value, R is above 0.9 (90%), meaning that the neural network was reasonably tailored for the model. The testing value of R is marginally smaller, which is more than likely as this is where the neural network algorithm was tested to predict the later portion of the imported values.

The graphs displayed in Fig. 9 show that even though the gradient of the regression line is close to '1', the overall fit of the system is unacceptable. The data points in the

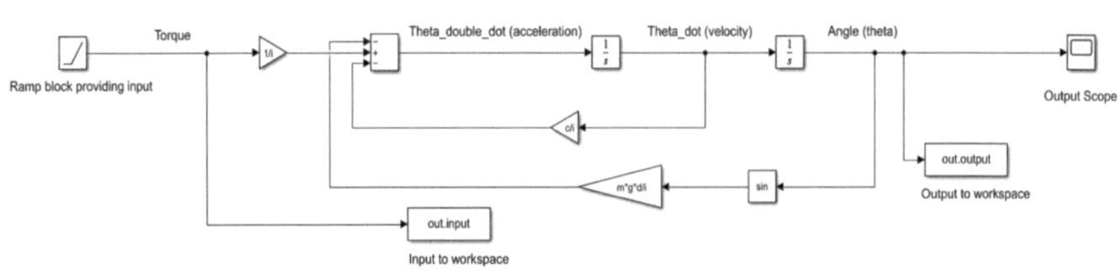

Fig. 6 Plant model with input and output data records

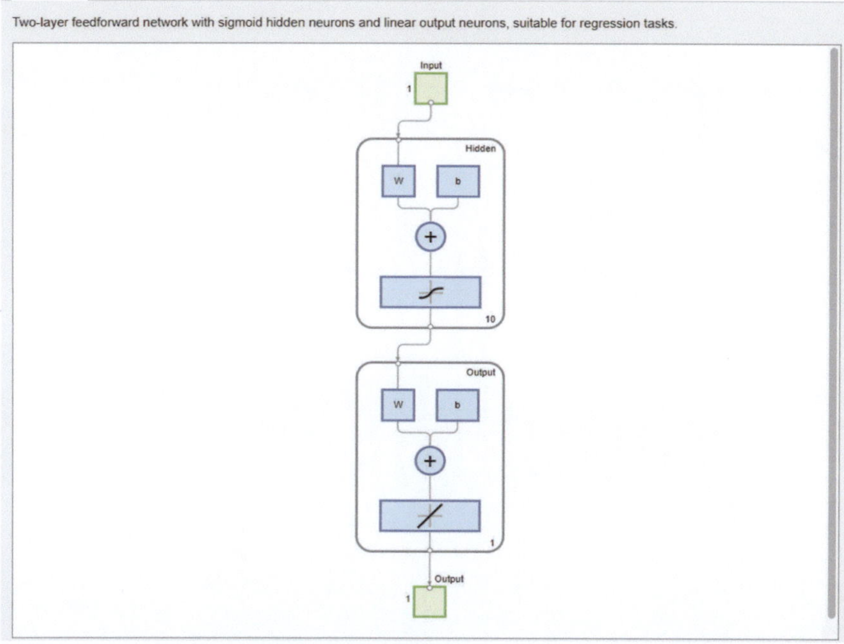

Fig. 7 Visualisation of the neural network model

training stage form a curve which the neural network does not reflect. The fact that in the other stages the data is accumulated at the top of the graph, suggests overfitting.

The neural network was remodelled using 4589 observations by increasing the number of data measurements per step. All the previously designated conditions remain the same. Figures 10 and 11 display the summary of model and the regression line graphs, respectively.

The regression value has now improved to a perfect '1'. In real life, this would be impossible as there will always be exterior factors that create error. However, since this is a mathematical model, this value can be achieved. Figure 11 clearly shows that the data points are aligned with the fit and are evenly distributed across the system. Increasing the number of observations drastically improved the accuracy and distribution of the results. This neural network has been used for the overall modelling of the robotic leg.

A SIMULINK block for the trained model can be created through the Neural Network application in MATLAB [38]. The torque applied has now been represented as a constant instead of a ramp block, and it is fed into the neural network block. This is used as a replacement of the entire plant model, with a scope connected to visualise the output. This is shown in Fig. 12.

The main system breaks down into the two main layers used within the neural network, as shown in Fig. 13. The first layer uses a tan sigmoid activation function

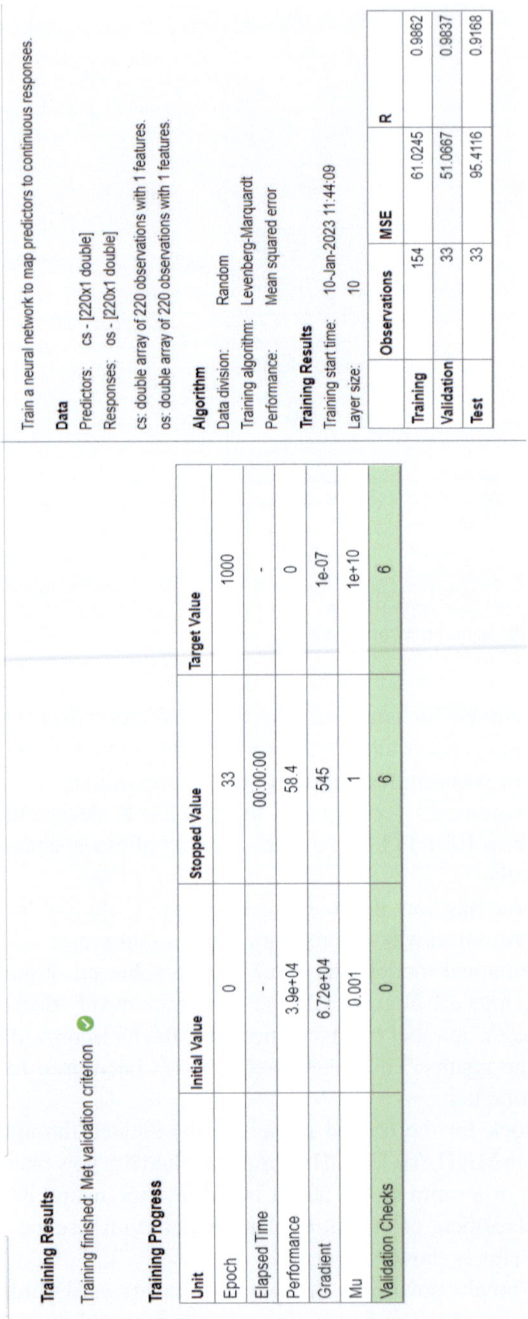

Training Results

Training finished: Met validation criterion

Training Progress

Unit	Initial Value	Stopped Value	Target Value
Epoch	0	33	1000
Elapsed Time	-	00:00:00	-
Performance	3.9e+04	58.4	0
Gradient	6.72e+04	545	1e-07
Mu	0.001	1	1e+10
Validation Checks	0	6	6

Train a neural network to map predictors to continuous responses.

Data

Predictors: cs - [220x1 double]

Responses: os - [220x1 double]

cs: double array of 220 observations with 1 features.

os: double array of 220 observations with 1 features.

Algorithm

Data division: Random

Training algorithm: Levenberg-Marquardt

Performance: Mean squared error

Training Results

Training start time: 10-Jan-2023 11:44:09

Layer size: 10

	Observations	MSE	R
Training	154	61.0245	0.9862
Validation	33	51.0667	0.9837
Test	33	95.4116	0.9168

Fig. 8 Summary of the neural network training

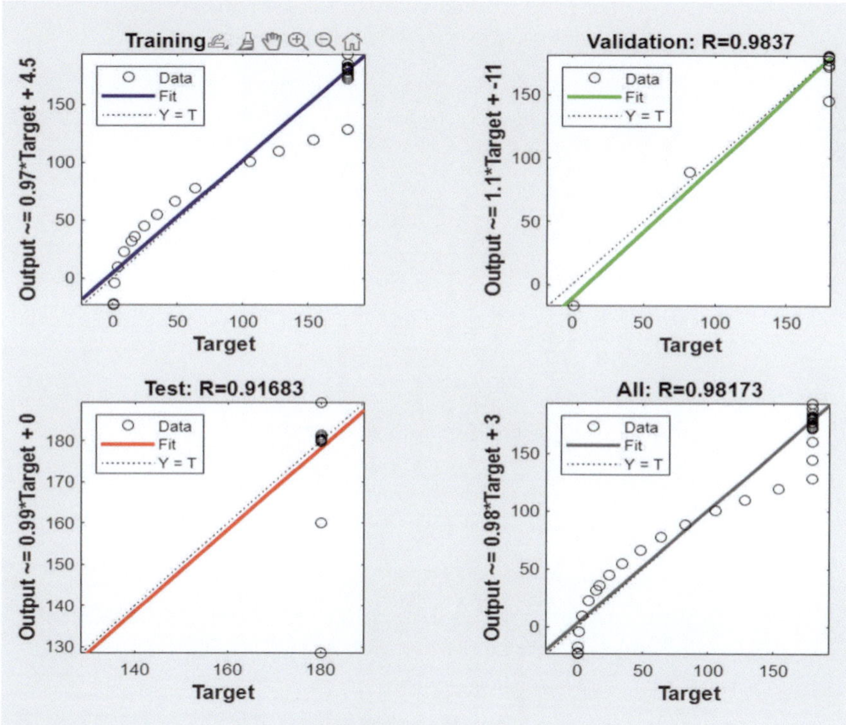

Fig. 9 Regression plots of the neural network

and the second uses a pure linear activation function for the prediction processes at each point.

5.3 *Fuzzy Logic Control*

Fuzzy logic is a way of portraying experience-based knowledge into a set of rules for the algorithm to be able to understand. It allows for logical reasoning where the answer is not always a binary output. In the 1920s, Polish logician Lukasiewicz first introduced a multivalued logic concept, where statements can hold fractional truths, not just ones and zeros. In 1937, Max Black used the multivalued logic and applied it to lists and sets of objects, creating the first fuzzy set curves and 'vague' sets [26]. Fuzzy sets were introduced in 1965 by Lotfi Zadeh, who built on the work of his predecessors. Zadeh [47] explained that classes of objects in the real world do not always have clear criteria with regards to membership definition, yet they play an important role in pattern recognition, thinking and communication. This became the foundation to further development of the fuzzy logic system.

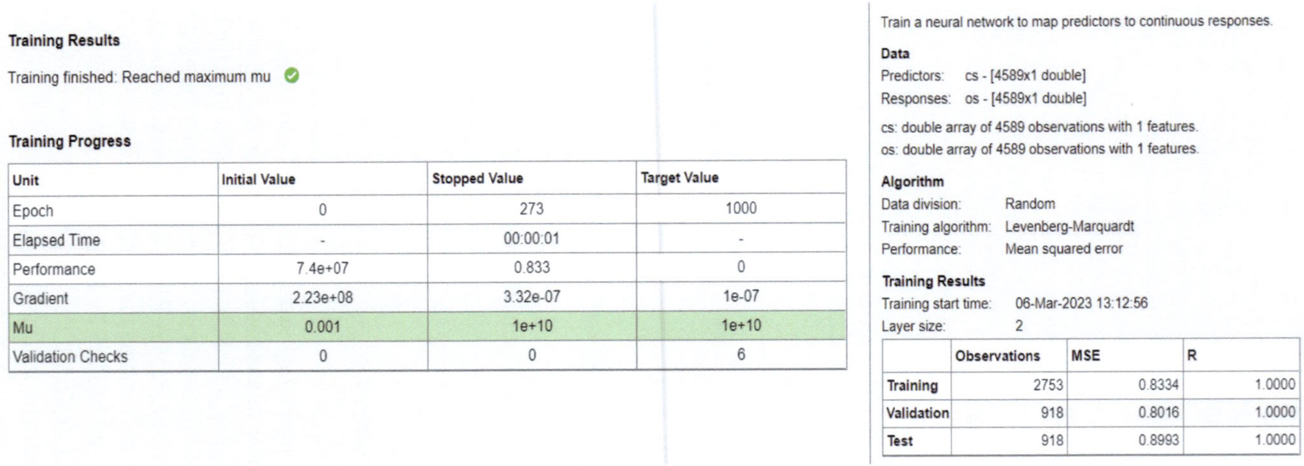

Training Results

Training finished: Reached maximum mu

Training Progress

Unit	Initial Value	Stopped Value	Target Value
Epoch	0	273	1000
Elapsed Time	-	00:00:01	-
Performance	7.4e+07	0.833	0
Gradient	2.23e+08	3.32e-07	1e-07
Mu	0.001	1e+10	1e+10
Validation Checks	0	0	6

Train a neural network to map predictors to continuous responses.

Data

Predictors: cs - [4589x1 double]

Responses: os - [4589x1 double]

cs: double array of 4589 observations with 1 features.

os: double array of 4589 observations with 1 features.

Algorithm

Data division: Random

Training algorithm: Levenberg-Marquardt

Performance: Mean squared error

Training Results

Training start time: 06-Mar-2023 13:12:56

Layer size: 2

	Observations	MSE	R
Training	2753	0.8334	1.0000
Validation	918	0.8016	1.0000
Test	918	0.8993	1.0000

Fig. 10 Summary of retrained network

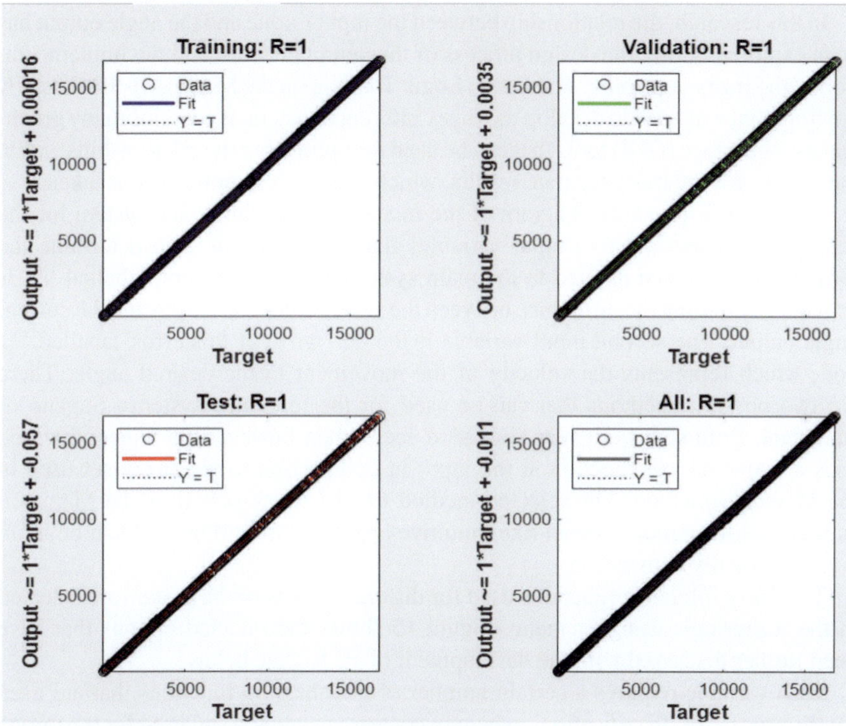

Fig. 11 Regression plots of retrained neural network

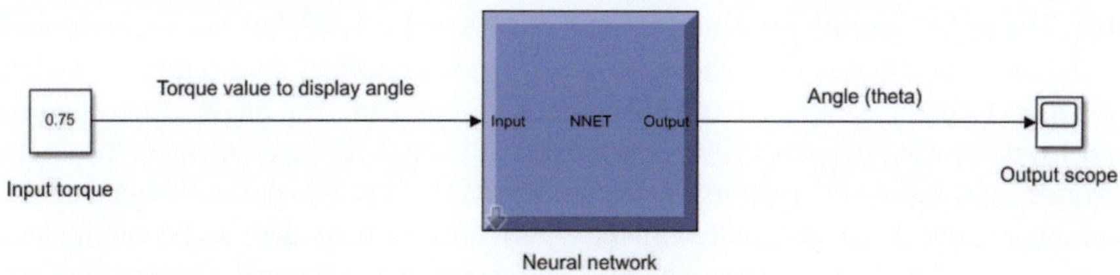

Fig. 12 Neural network block model

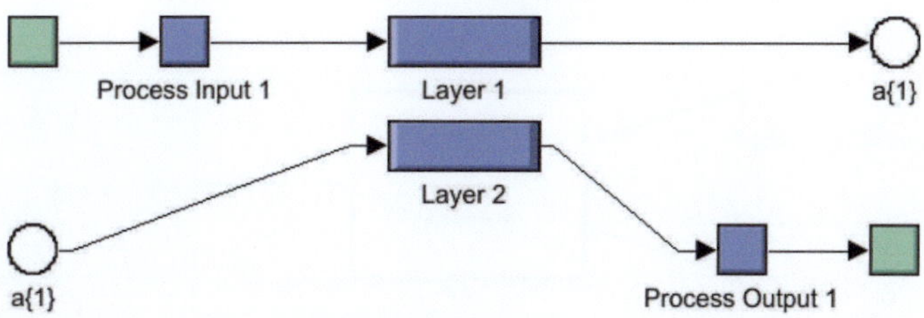

Fig. 13 Layer breakdown of the neural network block

In this research, the relationship between the input torque and the angle output has been explored during the design process of the neural network and the implementation of the fuzzy controller. The Fuzzy Logic Toolbox on the MATLAB-SIMULINK environment was used to develop the fuzzy inference system. It contains many graphical user interface (GUI) tools that can be used for optimising the relationships within the system for the most accurate results, which is why this approach was taken.

Figure 14 displays the diagram of the modelled fuzzy inference system for the robotic leg. There are two input variables that have been used to determine the torque, which is then applied to the main system. Firstly, the error, labelled '*e*', is provided by finding the difference between the selected desired output and the current angle output. The second input variable is the derivative of this error, labelled 'e_ dot', which represents the velocity of the movement to the desired angle. There are two possible methods that can be used for the inference system—Sugeno or Mamdani. Both styles are very similar to each other, however the Sugeno method uses a mathematical function at the input in comparison to the fuzzy set used in the Mamdani method. The selected method for the fuzzy system is the Mamdani as it allows for a more human-like, intuitive approach, even though it can be more computationally demanding.

The fuzzy inference functions used for different stages of the system are selected in the fuzzy logic designer menu. Figure 15 shows the selected options that have been further discussed with the development of each stage below.

Each variable requires a certain number of membership functions that are used for the process of 'fuzzification', where fuzzy sets are created and used for the torque estimations.

Figure 16 displays the membership functions used for the angle error input variable. The initial starting position, where the angle value is 0°, has been represented by the membership function labelled 'Zero'. On the plot, the y-axis portrays to what extent each membership function has an impact at the specific angle error value (0 and 1 representing 0% and 100%, respectively). 'Low_1' is the area where the angle error reaches the $-45°$ region. The closer the value is to this point, the greater the weight of 'Low_1' on the result, with the 'Zero' membership function becoming less important leading towards zero weight. There are five major areas within which the

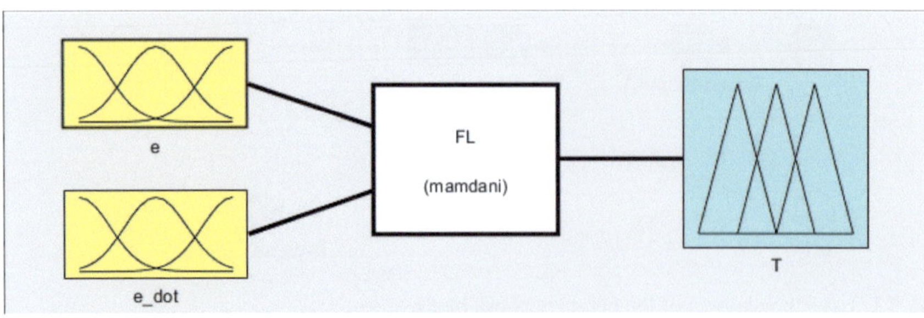

Fig. 14 Diagram display the fuzzy inference system

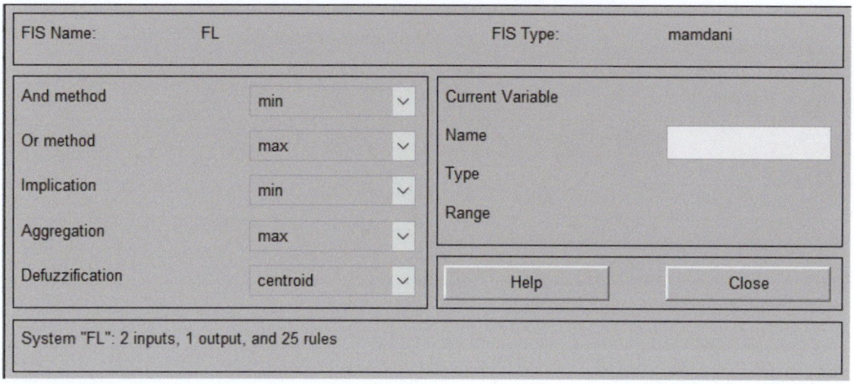

Fig. 15 Inference function definitions summary

robotic leg can be located. The direction of movement is represented not only by the angle of the error having opposite signs, but also by the numbers '1' and '2' in the names of the membership functions. The relationships between each membership function follows the same pattern, therefore, both the velocity input variable and the torque output variable have the same ranges and values. The membership function plots for the remaining variables are shown in Figs. 17 and 18.

Even though the angle range of each variable needs to be within ± 90°, an extra 10° is added on either end in order to make room for any additional errors that may be imported by the fuzzy inference system itself.

After the fuzzy variables are formed, fuzzy rules need to be applied in order to define the complex system. The application of fuzzy rules to the fuzzified input in order to calculate the fuzzy output is the basis of a fuzzy inference system [6]. Normal linguistic rules cannot be defined for a computer system due to the vague nature of descriptive words that cannot be easily transported into the environment. The speed and accuracy of the fuzzy controller when identifying the inputs and outputs is dependent on the fuzzy rules, which adopt the 'if…then' form [28]. Therefore, by

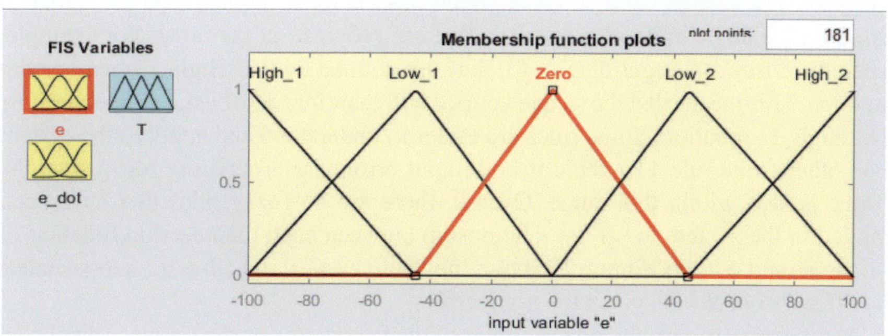

Fig. 16 Membership function summary for angle error

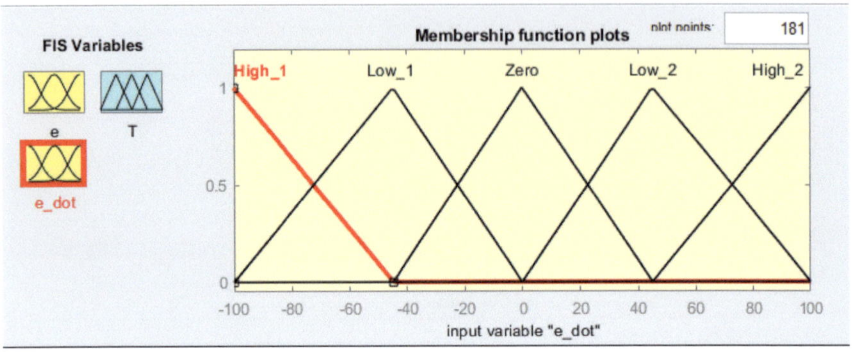

Fig. 17 Membership function summary for velocity error

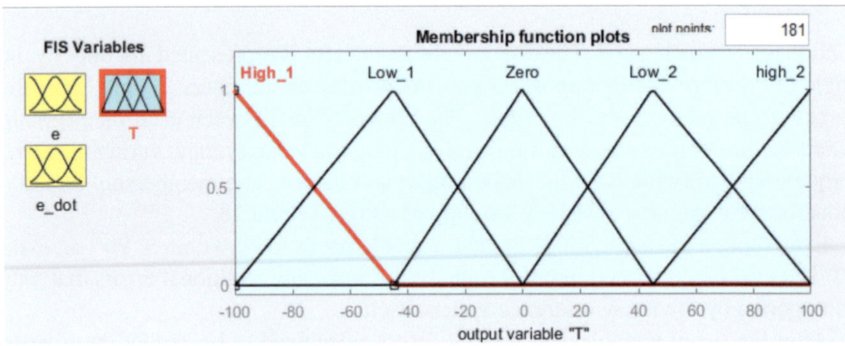

Fig. 18 Membership function summary for torque

taking rules from the real world and transforming them into fuzzy rules, a relationship between the membership functions of each input variable has been formed to define its impact on the desired torque output.

Table 2 displays the fuzzy rules applied to the system in order for it to function correctly. The membership function which has the most influence on the torque is based on the angle and velocity errors that are provided at the input. For example, when the errors are larger than -45, they are defined by the 'High_1' membership function. This means that the torque applied will therefore also be most influenced by the 'High_1' function. Some rules are easier to understand and apply to the system than others, like rule 11, where if both input errors are around the zero value, the torque is also within this range. Overall, there are 25 fuzzy rules that have been applied to the system to set the relationship between each membership function of the inputs and outputs. Figure 19 shows the 'Rule viewer', which is used to visualise the effects of each rule given the input errors.

Another way of visualising the effects of the fuzzy rules on the torque is displayed in Fig. 20. Using this interface, the total output surface can be visualised based on the

Table 2 Fuzzy rules

	Angle ('e')					
Angular Velocity ('e_dot')		High_1	Low_1	Zero	Low_2	High_2
	High_1	High_1	High_1	High_1	Low_1	Zero
	Low_1	High_1	High_1	Low_1	Zero	Low_2
	Zero	High_1	Low_1	Zero	Low_2	High_2
	Low_2	Low_1	Zero	Low_2	High_2	High_2
	High_2	Zero	Low_2	High_2	High_2	High_2

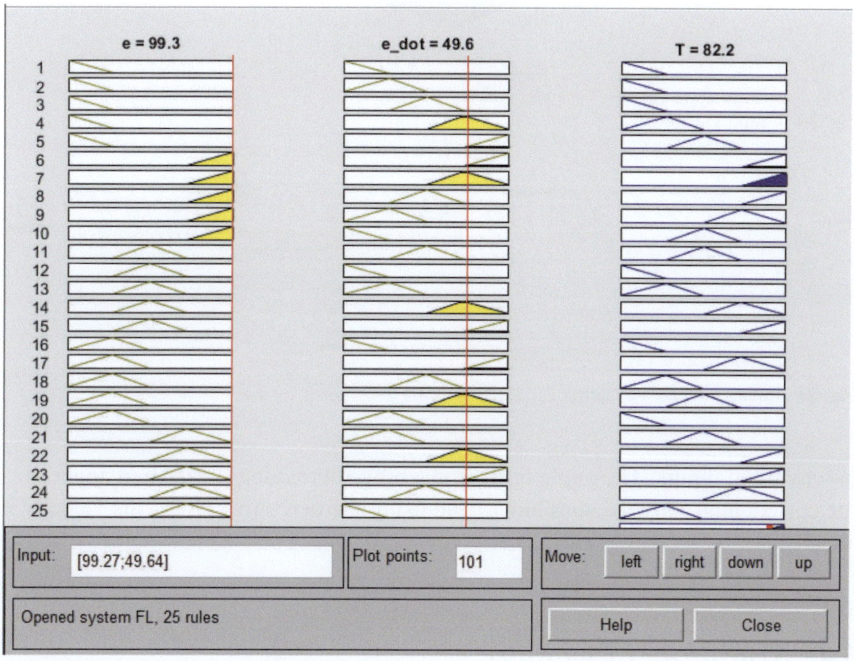

Fig. 19 Rule viewer for the applied fuzzy rules

selected input variable, unlike the rule viewer which only shows single calculations at a time.

The final step for the completion of the fuzzy logic system and the determining the output torque is the process of 'defuzzification'. The fuzzy variables, after the fuzzy rules have been applied, are converted back into the original format. There are multiple methods that can be used; however, the selected one is the centroid. This is the most common method to determine the point that represents the centre of gravity (COG) for the given range.

Figure 21 displays the SIMULINK model with the fuzzy logic block implemented. This block is attached to the main plant model, and it has been provided with the

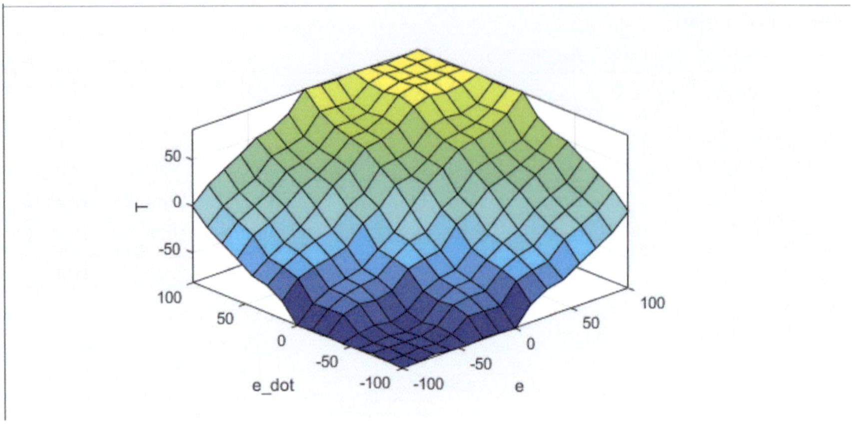

Fig. 20 Graph viewer

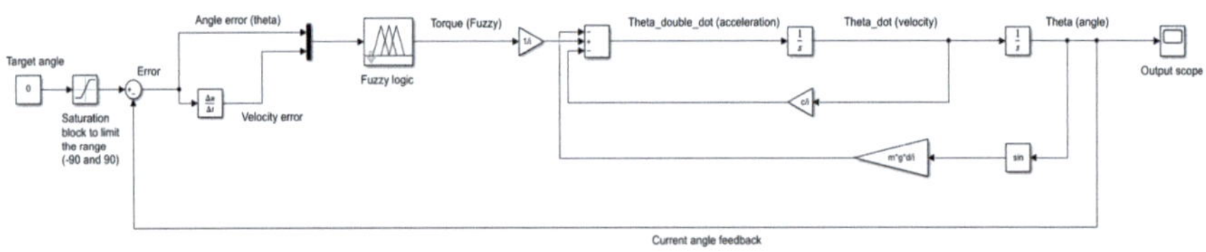

Fig. 21 Fuzzy inference system

two required inputs. The angle error is given by subtracting the desired angle from the current angle (by the sum block) and using the derivative block, the velocity is attached as the second input.

5.4 *Neuro-Fuzzy Controller*

Once both the neural network and fuzzy logic inference system have been designed, both of the SIMULINK blocks were combined to produce the final system. The neural network block replaces the initial plant model since it reflects the relationship between the input and output. The fuzzy logic block has been attached in the same way as to the plant model in Fig. 21, with a feedback loop for the error connected from the output back to the input. The complete neuro-fuzzy controller is displayed in Fig. 22.

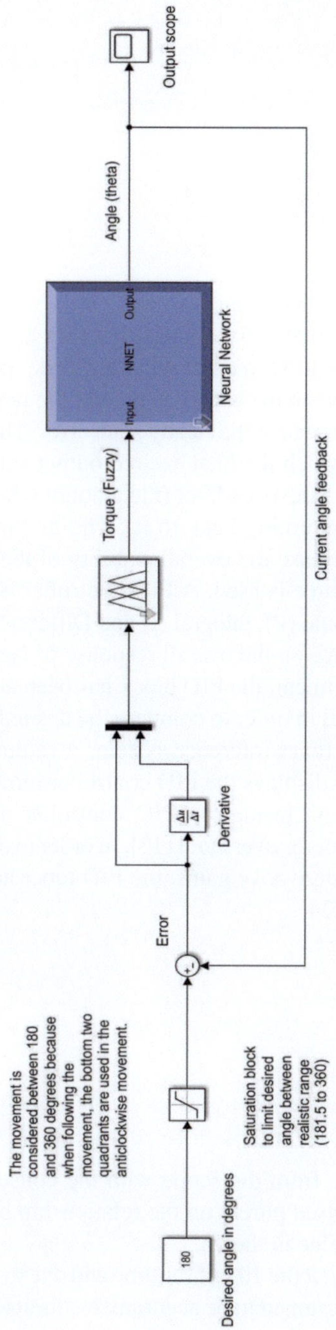

Fig. 22 Complete neuro-fuzzy controller model

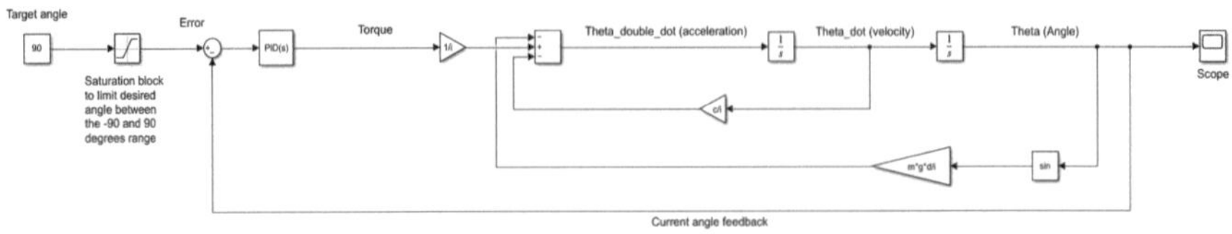

Fig. 23 PID controller model

5.5 PID Controller

The other type of controller that has been designed for the control of the robotic leg is the PID. There are three modifiable parameters that each have a different effect on the output. Proportional gain is used to control the response time of the system as it determines the difference between the actual angle and the target one. Integral gain is applied with the intention to reduce the steady state error. This parameter controls the speed and accuracy with which the final desired output value is reached. As the integral gain increases, so does the overshoot (the amount which the desired output value is exceeded by before returning back to it). The differential gain is used to dampen the overshoot and increase the overall stability of the system. For the best results, a combination of all three is used. A PID controller is strongly affected by its three parameters; proportional (P), integral (I) and Differential (D), where two of them can have a larger influence on the overall response of the system [44].

On the SIMULINK environment, the PID block has been attached onto the plant model. A feedback loop is used in order to compare the desired target angle with the current angle, similarly to the fuzzy inference system. A saturation block is used on desired angle input. Figure 23 displays the PID controller model.

The automatic tune function inside the PID controller allows for the fastest response without any oscillations or overshoot [13]. In order to obtain the best balance of proportional, integral, and derivative gains, the PID tuner has been used, and the final values are shown in Fig. 24.

6 Results

6.1 ANN Block

Figure 25 displays the results from the scope with the constant input being 0.75. Since the neural network is based purely on the relationship between the input and the output value, a constant value is shown.

The result is stable throughout the 10 s of runtime and due to the perfect regression value, the angle displayed is assumed to be accurate. No limits were set on the initial model and so they were also not considered for the neural network system. Both variables increased in value at different rates, meaning that an angle could not be

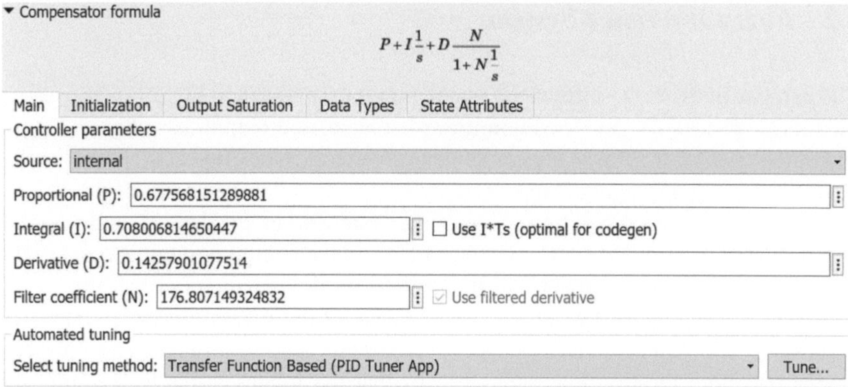

Fig. 24 PID controller block parameters

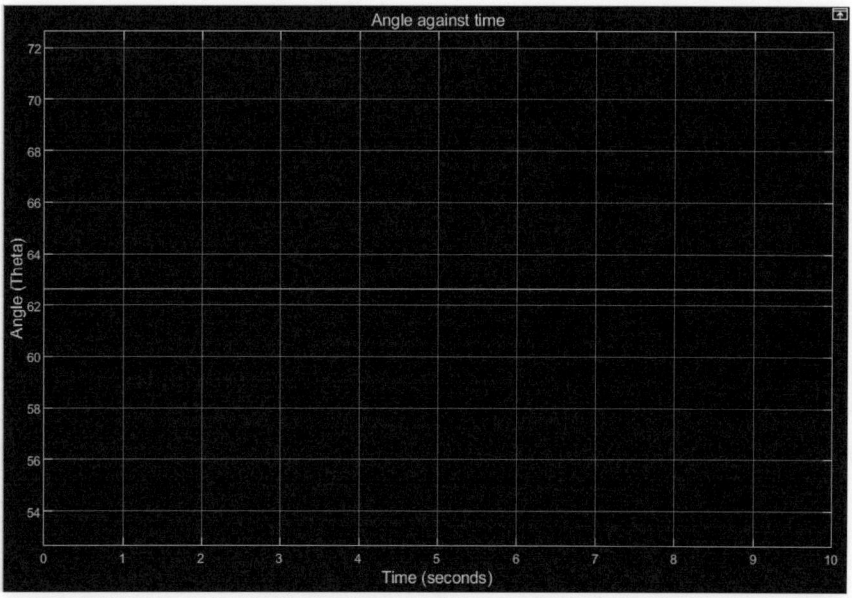

Fig. 25 Angle output of the neural network model

estimated based on a manually selected input at this stage. Therefore, the completed neuro-fuzzy controller has been used to further reinforce the correct functioning and suitability of the neural network model.

6.2 *Fuzzy Inference System*

The purpose of the fuzzy inference system, previously shown in Fig. 21, was to move the robot leg from its 'rest' position towards the desired output, based on the angle and velocity error feedback. Since the neural network block does not allow for this visualisation, the plant model has been used.

Figure 26 displays the output scope of the system when the desired angle was set to 90°. It shows a smooth curve, which starts from the central point, that reaches the target value after 5 s and maintains the position for the remainder of the time.

Figure 27 reinforces the correct functionality of the fuzzy inference system, as it shows a smooth curve reaching the – 90° target angle. Being able to correctly navigate to both the limiting values proves that the model works well within the given range and allows for a full range of motion for the robotic leg.

It can be noticed on the two previous figures that an output angle error was present in the system. When the time, t is 0, the corresponding angle value was not at 0°

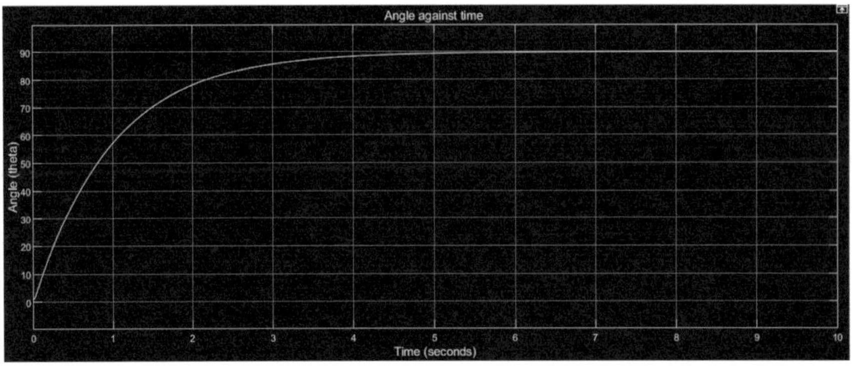

Fig. 26 Output scope for the fuzzy system

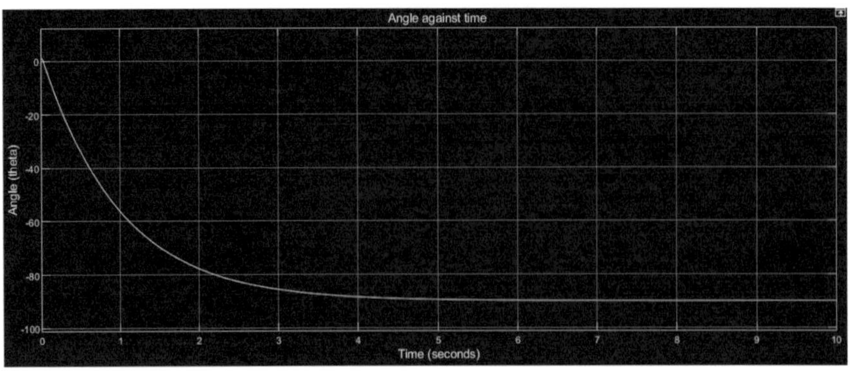

Fig. 27 Output scope for – 90° angle

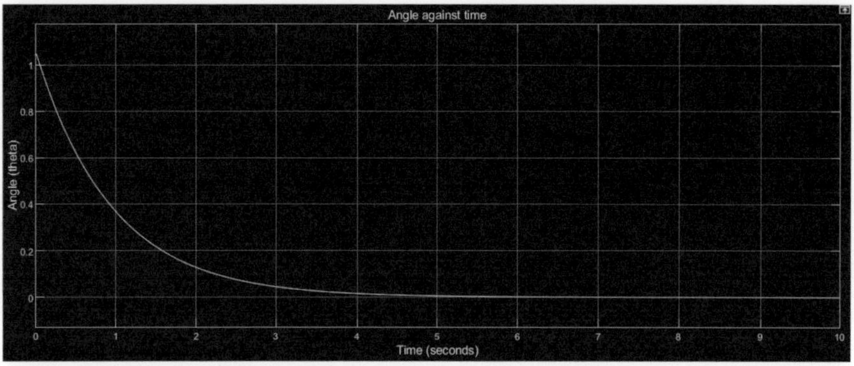

Fig. 28 Output scope displaying the output error of the fuzzy system

where it should be. This was caused by the fuzzy logic during the fuzzification and defuzzification processes. It was expected from the model since the fuzzy system itself is used to solve unclear and imprecise data. In order to get a clearer view of the error, the target value is set to zero and Fig. 28 displays the output scope below.

It can be seen that the overall angle error of the fuzzy inference system was just over 1°. This is a small angle that did not make a difference on the effectiveness on the model, since it would not be noticeable in real life. Initially, the model would fail when the desired angle was set to – 90°, as with the addition of the error, the actual angle was – 91.2°, which was beyond the scope. This is why the previously mentioned extra range was added onto the membership functions. That way, every necessary angle for the model was reachable. Therefore, even with this error, the system was classed as reliable.

A problem occurred when the desired angle was set to a value outside of its range. The system did not run as it was unable to go through fuzzification with a value that did not follow any of the modelled relations. This always resulted with the output being unchanged and so nothing would be displayed on the scope. In order to solve this, a saturation block has been placed after the desired angle input, so that all values beyond the range are set to the maximum limit value (either – 90° or 90°).

6.3 Neuro-Fuzzy Controller

Figure 29 displays the output result from the neuro-fuzzy controller when the desired angle is set to 360°.

The movement is not displayed due to the nature of the neural network, as previously mentioned. The output scope above clearly shows that the target angle was achieved and maintained through the use of the controller. An angle error is visible on scope. Figure 30 shows how far the result was from the target value.

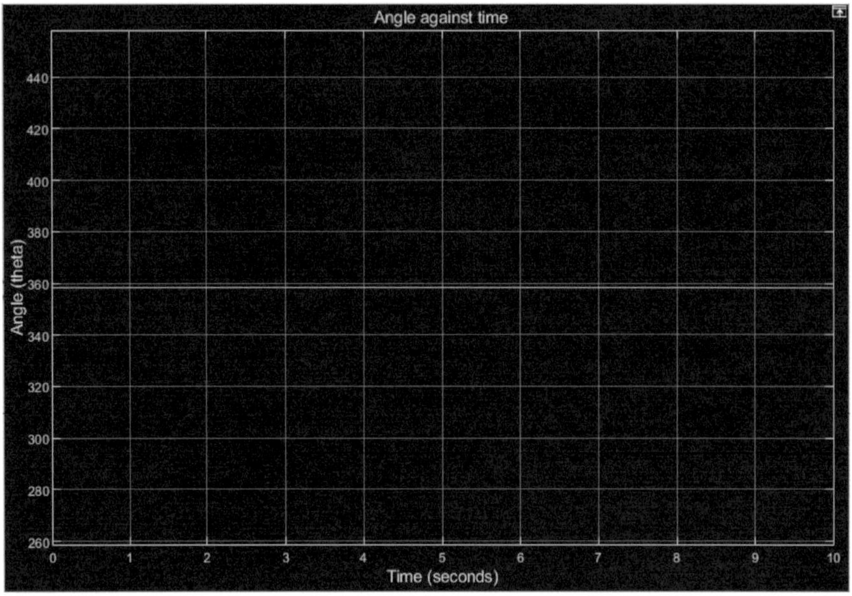

Fig. 29 Output scope for 360° target angle from neuro-fuzzy controller

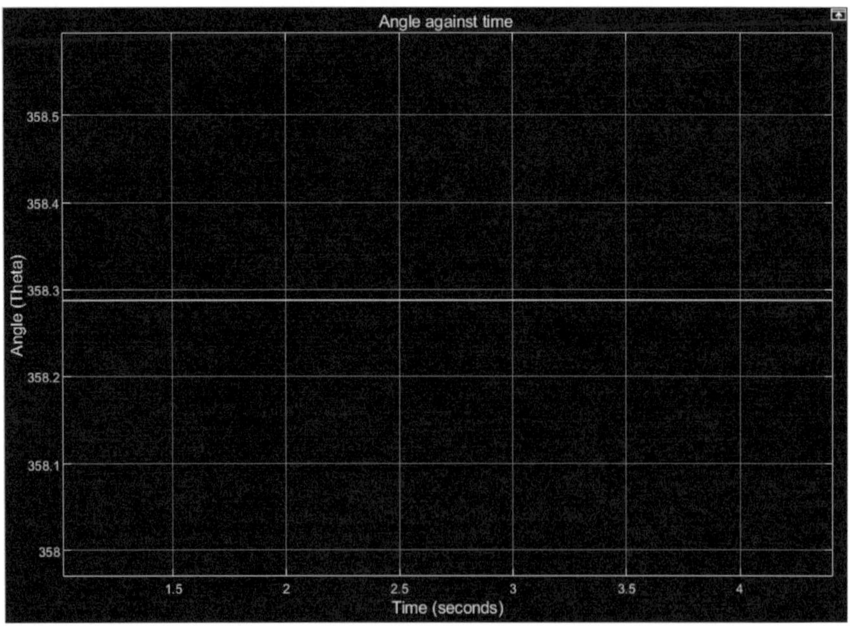

Fig. 30 Output scope displaying the system error

The completed controller has less than 2° error, which means that only the fuzzy inference system contributes to it. Therefore, the neural network functions accurately reflect the relationship between the torque and angle output. Figure 31 displays the unit circle with four main quadrants. The robotic leg model uses the bottom two quadrants, which is why the range has been now set between 180° and 360° for the neuro-fuzzy controller.

The saturation block ensures that the desired angles are kept within this range. The lower range limit has been increased to 181.5° to consider the error of the fuzzy system. Figures 32 displays the output angle when the target was set to 180°.

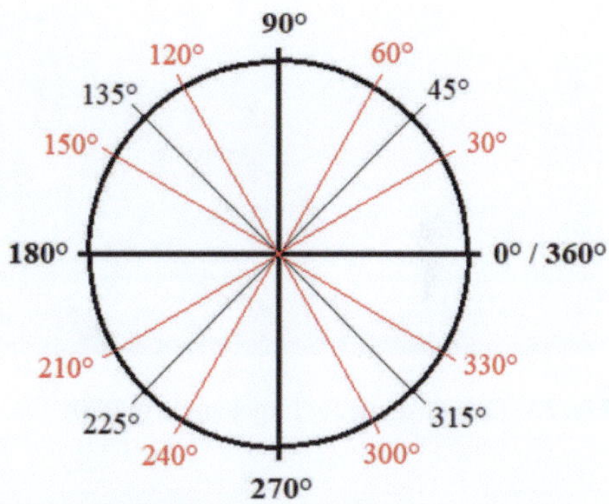

Fig. 31 Angle quadrants [33]

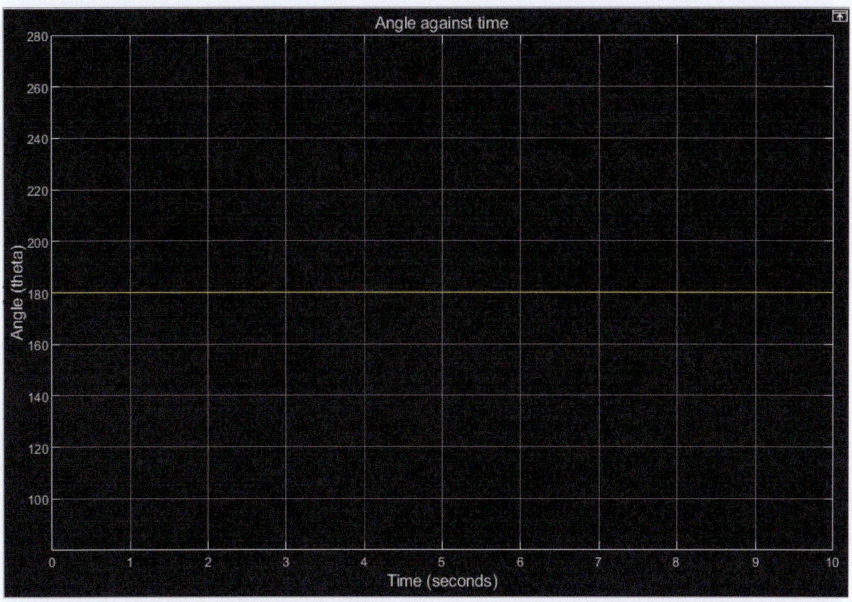

Fig. 32 Output scope for target of 180°

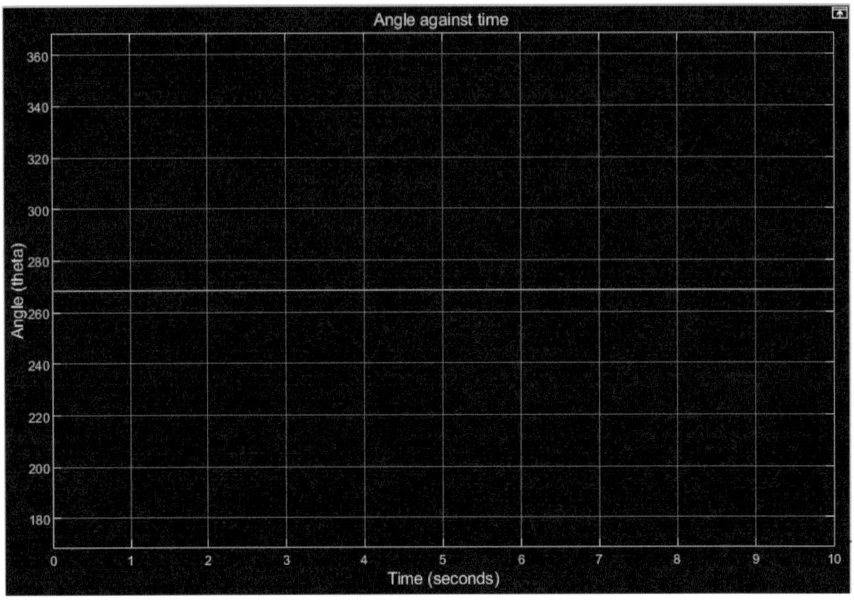

Fig. 33 Output scope for target angle of 270°

The graph clearly showed that the 180° target was reached. This is the only value at which there was no overall angle error in the system. In Fig. 33, the desired angle was set to 270°, which is the neutral position. The output angle reached the target value with a less than 2° error, showing that the error remains the same throughout the whole range of motion whilst using the neuro-fuzzy controller.

6.4 PID Controller

The proportional and integral gains have the largest influence on the tuned output. Increasing the derivative gain does dampen the overshoot, but it has a large effect on the response time and stability, which are the more important factors with regards to the robot leg. Figure 34 displays the output for a desired angle of 10°.

The image clearly shows that the PID controller can effectively provide the correct torque, which ultimately reaches the set target. Even after tuning there is a small overshoot of around 2° in the first 0.5 s, but this is quickly stabilised with no steady state error present. Figure 35 displays the output based on a desired angle input of 90°.

The system is able to stabilise with no steady state error in the same amount of time. This is a reasonable settling time, meaning that the feedback loop is functioning correctly. However, the overshoot size now increased to over 10°. Therefore, the greater the target angle, the larger the overshoot. The added saturation block limited

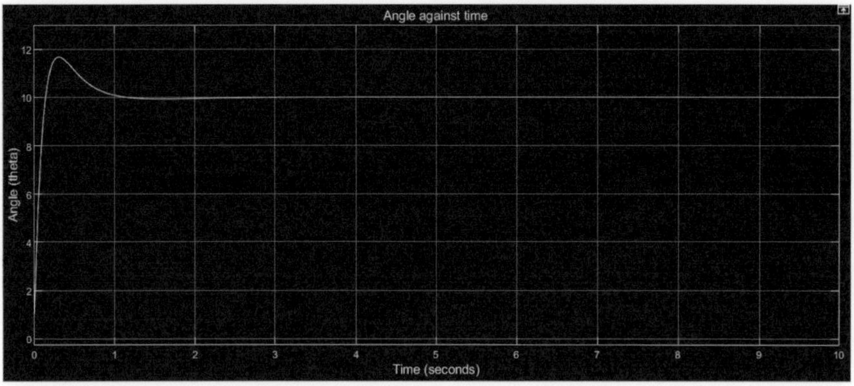

Fig. 34 Output target angle of 10° using PID controller

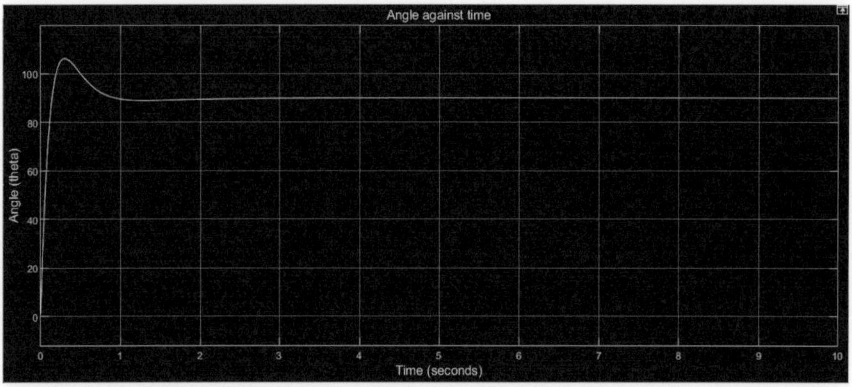

Fig. 35 Output scope of tuned PID response

the angle input between – 90° and 90°. Figure 36 proves that the system functioned in the same way in the opposite direction.

6.5 Fuzzy System Against PID Controller

The neural network did not contribute to the errors of the neuro-fuzzy controller as displayed in Fig. 29 of the results, so its overall output angle was the same as the fuzzy system. For the best comparison, the output of the fuzzy inference system and PID controller model have been used, since they both display the movement of the robotic leg with time. This is shown in Fig. 37.

The fuzzy system reaches the target angle without any overshoot in about 6 s and stays there for the remainder of the time. The PID controller is able to reach this

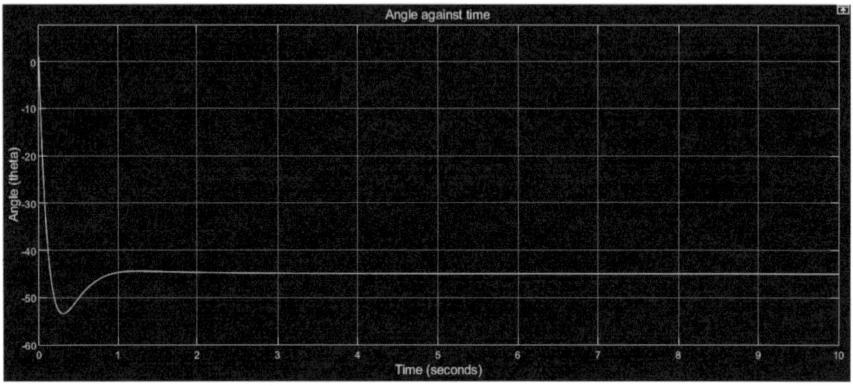

Fig. 36 Output for – 45° desired angle using PID controller

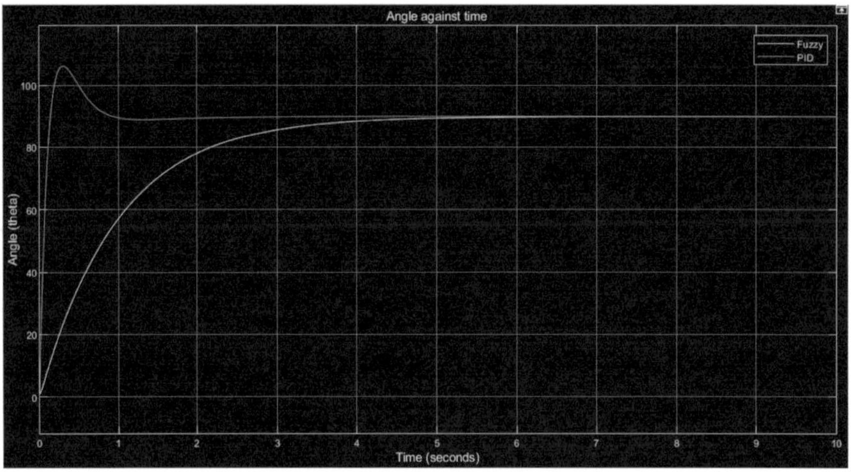

Fig. 37 Output scope of both Fuzzy and PID controllers with target angle of 90°

same stability in around 2.5 s but has a large overshoot of about 12°, as previously mentioned.

7 Discussion

7.1 General Observation

Both controllers have been successful with regards to the control of the robotic leg. However, there are multiple factors that prove the neuro-fuzzy controller is the most suitable one. Even though there are constant developments in control methods, the PID is still widely used in industry due to the simplicity of 3 tuneable parameters and simple structure [17]. It is simple and easy to use, whilst still producing reasonable results. This was to be expected, as gathered from the research of Mahdi and Raafat [30], who confirm that conventional PID controller is simple, efficient, and reliable. Figures 24 and 35 reinforce this idea, as the MATLAB PID toolbox has been used to tune the model for optimal results without any extra inputs. Although not a lot of experience is required since software can be used for automatic tuning, if it is not done correctly the system can become unstable.

A much more in-depth knowledge about the system is required for the correct design of a neuro-fuzzy controller. In order to determine the fuzzy set of rules shown in Table 2 and the correct membership functions parameters, a lot more research has to be done into the topic for the correct assigning of each variable. Applying changes to the controller proved to be more difficult than with the PID controller, due to the larger possibility for error in both the neural network and fuzzy logic design. This meant, however, that the output results have been much smoother and contain less errors, as shown in Fig. 26. The controller itself was unable to display the stability of the system due to the nature of the neural network model used, which made it difficult to spot any errors that may occur during the processes. Also, due to the large amount of background processes that take place, the computing time required was longer than with the PID controller.

7.2 Real-Life Applications

When considering real-life applications, such as a prosthetic leg, the shorter time taken to reach an angle would be beneficial, but not crucial. Faster response time combined with an overshoot can cause major stability issues for the patient. This would be the same for any use in bipedal robots, as it can cause the system to fall and damage internal parts. Even with PID angle input saturated, there is a potential for motor damage when reaching values close to the limit. The results from Sect. 6.4 show that with a target angle of 90°, an angle of 112° can be initially reached. This risk would be eliminated by reducing the maximum angle to 75°, but at the cost of limited movement.

The greatest drawback with regards to the PID controller is its incompatibility with non-linear systems. Results obtained with the use of the controller further reinforced this idea, as the overshoot, steady state error and time taken to reach stability varied with the desired angle input. This means that the PID gains have to be retuned every time the target changes, which makes it very impractical and time-consuming. This worked as expected from the research of [39], who stated that the PID controller cannot consider non-linearity and the resulting need for re-optimisation is a time-consuming process, marking them unideal for bipedal robots.

The smooth curve of the fuzzy system ensures that the output angle will always stay within the set range. This has been done by using the saturation block in the same way as in the PID controller. The fact that the model could hold the angle position without any steady state error made it reliable and well suited for robotic leg modelling. Adding limits has made it more realistic as the robotic leg will still move in the desired direction even if it is outside of the range, whilst still keeping the components safe. The fuzzy rule set can be used for any desired output angle within the range, making it the more suitable controller for non-linear robot leg model. Once the system is initially designed, no adjustments are necessary, unlike the PID controller. This reinforces the research of [4], who believed that the fuzzy controller's main features are the ability to deal with non-linear systems and problem-solving adaptability. There is still an error in the output result of the fuzzy system, but its impact is much less significant. For a prosthetic leg or bipedal robot, a difference of less than 2° will not affect the balance or stability of the system. However, the fact that it takes around 6 s to move to the desired angle is a drawback, as a system implementing it would have to walk slowly. This may not be a problem for military applications as the speed would ensure safety, but for prosthetic legs, this can be an inconvenience for the user.

The neuro-fuzzy controller is the most suitable for the control of the robotic leg as it resolves the issues of both the PID controller and fuzzy system. The combination of neural networks and fuzzy logic provided superior control over the non-linear system in comparison to the other controllers, reinforcing the findings of [37]. The system's quick response time and lack of overshoot allow for swift output reaction and ease of manoeuvrability for the user, whether it is for a prosthetic leg or a robot in a hostile environment. It still has the 2° angle error due to the mathematical representations of vague language in the fuzzy system, but the constant output value eliminates any overshoot complications of the PID controller. The saturation block within the system protected it from any motor damage, like it did with the other models. The simulation allows for a regression line value of '1' for the neural network, but this value is unattainable in real life. As a result, the overall error of the system would increase. However, due to the simplicity of the mathematical model, as well as the large number of observations, the regression value is unlikely to drop low enough for the error to substantially increase.

The compound pendulum model used for the robotic leg system was enough to represent the robotic leg movements and develop a functioning controller. By modelling the leg as a whole and not just the centre of mass attached by a string, the focus remained on the immediate forces acting on the system and reduced unnecessary mathematical complications. This model was justified through the use of fuzzy logic, which can be used for modelling uncertainties of dynamic models [11]. The fact that the neuro-fuzzy controller successfully displayed the correct output angle based on the initial plant model reinforces the suitability of the compound pendulum model for the robot leg. Although the kinematic approach allows for a more in-depth mathematical analysis, using a fuzzy logic control system meant that it was not a necessity, making it compatible with the neural network. Therefore, the modelled controlled aligned with the expectations of [1].

7.3 Future Improvements

The neuro-fuzzy controller is the most suitable with regards to non-linear systems, which is why it would still be selected for future projects and modifications. A possible improvement to the model would be the implementation of a double compound pendulum. Although it would not have an effect on the efficiency or reliability of the controller, the realism of the model itself would improve. This system would consider the knee joint angles, which would allow a greater manoeuvrability. Each placement of the robot leg would have to correspond with a correct knee placement that would be trained through the neural network, just like in the current model. The height of the whole system could be controlled by the knee joint, meaning that modelling actions such as jumping would be an easier task. This is supported by Ahirwar et al. [3], who suggested that humanoid robots are modelled with multiple joints to perform behaviours such as jumping or avoiding obstacles. The key relationship between the torque and output angle would remain the same, further reinforcing the suitability of a neural network. With the gained experience of fuzzy systems, the fuzzy rules can be easily optimised towards the double compound pendulum system. For the purpose of basic control of a non-linear robotic leg, the single compound pendulum model encompasses all the required features. Another possible improvement would be the addition of a mass spring into the model. Che et al. [10] suggest that spring mass model is able to mimic human motion well. This spring is used to mimic the behaviour of muscles within human and animal legs, which means that the realism of the model would improve. However, this adds an extra complication to the model that can contribute the overall errors of the output angle.

8 Conclusion

The overall aim of the research task has been to develop a fully functioning neuro-fuzzy controller for a non-linear robotic leg, based on the input torque. All the outcomes outlined in the introduction have been fully achieved. Multiple modelling techniques for robotic legs were explored alongside current literature. Since using a neuro-fuzzy controller does not require a complex mathematical model and it is suitable for modelling non-linear systems, the compound pendulum model was used for the initial plant. The equation of motion was derived, rearranged, and the plant has been modelled in the SIMULINK environment. A neural network was designed to reflect the relationship between the input torque from the motor and the angle output of the robot leg. Multiple uses of artificial neural networks were researched in depth, which ensured the correct implementation within the system. A large number of iterations and observations ensured a regression value of 1, which can only be achieved through simulation. The fuzzy logic system was designed through the development of fuzzy variables and rules, used to convert the vague language of 'low' and 'high' movements of the leg into actual numerical values. This was applied to determine the torque needed for a specific angle output. The final controller was then completed by attaching the fuzzy logic block to the neural network, along with the angle and velocity error feedback. A PID controller has also designed in order to compare its suitability for the non-linear robotic leg model with the neuro-fuzzy system.

The neuro-fuzzy controller was able to instantaneously display the desired output angle within the range that was set through the use of saturation blocks. There was an overall response error, which was not large enough to have an effect on the overall performance of the system. The lack of overshoot ensured that if the controller was to be made in real life for a prosthetic leg or bipedal robot, there would be no risk of any components getting damaged. Comparing the fuzzy inference system with the PID controller, since they both displayed the actual movement of the angle, showed that the PID system cannot be used for non-linear models due to inconsistency in the output. The greater the desired angle, the larger the overshoot and steady state error of the system, which would be problematic if used in a prosthetic limb. For the PID controller to be functional, it would need to be constantly retuned, making it impractical for dynamic systems. The addition of the neural network to the fuzzy inference system fixed its major drawback, which was the slow response time of the system. Therefore, it can be concluded that the neuro-fuzzy controller is the superior choice, with regards to the control of a non-linear robotic leg.

References

1. Abougarair AJ (2020) Neural networks identification and control of mobile robot using adaptive neuro fuzzy inference system. In: Proceedings of the 6th international conference on engineering & MIS 2020, pp 1–9
2. Adarsh SR, Selvakumar S (2020) Model identification and control of prosthetic leg. J Phys: Conf Ser 1706(1):012109
3. Ahirwar D, Purohit J, Semwal VB, Gawre S, Rajpurohit M (2022) The recent advancements in humanoid robot technology. In: 2022 IEEE international students' conference on electrical, electronics and computer science (SCEECS). IEEE, pp 1–6
4. Aldair AA, Rashid AT, Rashid MT, Alsaedee EB (2019) Adaptive fuzzy control applied to seven-link biped robot using ant colony optimization algorithm. Iran J Sci Technol, Trans Electr Eng 43(4):797–811
5. Bakırcıoğlu V, Şen MA, Kalyoncu M (2022) Numerical investigation and experimental verification of the proposed robot leg virtual model. Proc Inst Mech Eng, Part C: J Mech Eng Sci 09544062221076769

6. Ben Jabeur C, Seddik H (2021) Design of a PID optimized neural networks and PD fuzzy logic controllers for a two-wheeled mobile robot. Asian J Control 23(1):23–41
7. Bodo G, Di Bello P, Tessari F, Buccelli S, Boccardo N, De Michieli L, Laffranchi M (2022) Comparative analysis of inverse kinematics methodologies to improve the controllability of rehabilitative robotic devices. In: 2022 International conference on rehabilitation robotics (ICORR). IEEE, pp 1–6
8. Burzynski P (2022) Discrete fractional order PID controller in case of the FLHex robot leg position control system. In: 2022 26th International conference on methods and models in automation and robotics (MMAR). IEEE, pp 264–269
9. de Campos Souza PV (2020) Fuzzy neural networks and neuro-fuzzy networks: a review the main techniques and applications used in the literature. Appl Soft Comput 92:106275
10. Che J, Pan Y, Yan W, Yu J (2022) Leg configuration analysis and prototype design of biped robot based on spring mass model. In: Actuators, vol 11, no 3. MDPI, p 75
11. Dehyadegari L, Khajehasani S (2021) Robot control by using intelligent systems considering complete constraints. Int J Ind Eng Prod Res 32(1):79–91
12. Desai M, Shah M (2021) An anatomization on breast cancer detection and diagnosis employing multi-layer perceptron neural network (MLP) and convolutional neural network (CNN). Clinical eHealth 4:1–11
13. Eldirdiry O, Zaier R, Al-Yahmedi A, Bahadur I, Alnajjar F (2020) Modeling of a biped robot for investigating foot drop using MATLAB/Simulink. Simul Model Pract Theory 98:101972
14. Farid Y, Siciliano B, Ruggiero F (2022) Review and descriptive investigation of the connection between bipedal locomotion and non-prehensile manipulation. Annu Rev Control
15. Feng J, Lu S (2019) Performance analysis of various activation functions in artificial neural networks. J Phys: Conf Ser 1237(2):022030
16. Gaeid KS, Nashee AF, Ahmed IA, Dekheel MH (2021) Robot control and kinematic analysis with 6DoF manipulator using direct kinematic method. Bull Electr Eng Inform 10(1):70–78
17. Ghith ES, Tolba FAA (2022) Real-time implementation of tuning PID controller based on whale optimization algorithm for micro-robotics system. In: 2022 14th International conference on computer and automation engineering (ICCAE). IEEE, pp 103–109
18. Gong Y, Hartley R, Da X, Hereid A, Harib O, Huang JK, Grizzle J (2019) Feedback control of a cassie bipedal robot: walking, standing, and riding a segway. In: 2019 American control conference (ACC). IEEE, pp 4559–4566
19. Harifi S, Khalilian M, Mohammadzadeh J, Ebrahimnejad S (2020) Optimizing a neuro-fuzzy system based on nature-inspired emperor penguins colony optimization algorithm. IEEE Trans Fuzzy Syst 28(6):1110–1124
20. Hasson U, Nastase SA, Goldstein A (2020) Direct fit to nature: an evolutionary perspective on biological and artificial neural networks. Neuron 105(3):416–434
21. Hu B, Simon AM, Hargrove L (2019) Deep generative models with data augmentation to learn robust representations of movement intention for powered leg prostheses. IEEE Trans Med Robot Bionics 1(4):267–278
22. Jain AK, Mao J, Mohiuddin KM (1996) Artificial neural networks: a tutorial. Computer 29(3):31–44
23. Jain A et al (2021) Real-time swing-up control of non-linear inverted pendulum using lyapunov based optimized fuzzy logic control. IEEE Access 9.https://doi.org/10.1109/ACCESS.2021.3058645
24. Kasaei M, Lau N, Pereira A (2019) Comparison study of well-known inverted pendulum models for balance recovery in humanoid robot. arXiv preprint arXiv:1906.01936
25. Kim P (2017) Matlab deep learning : with machine learning, neural networks and artificial intelligence. Apress, New York, NY. https://doi.org/10.1007/978-1-4842-2845-6
26. Kosko B, Isaka S (1993) Fuzzy logic. Sci Am 269(1):76–81
27. Krogh A (2008) What are artificial neural networks? Nat Biotechnol 26(2):195–197
28. Kunyou H, Lumin C (2021) Research of fuzzy PID control for lower limb wearable exoskeleton robot. In: 2021 4th International conference on intelligent autonomous systems (ICoIAS). IEEE, pp 385–389

29. Liao F, Zhou Y, Zhang Q (2021) Stability analysis of a flexible biped robot in periodic walking. In: 2021 40th Chinese control conference (CCC). IEEE, pp 583–589
30. Mahdi SM, Raafat SM (2022) Robust Interactive PID controller design for medical robot system
31. McCullum N (2021) Deep learning neural networks explained in plain English, freeCodeCamp.org. freeCodeCamp.org. Available at: https://www.freecodecamp.org/news/deep-learning-neural-networks-explained-in-plain-english/. Accessed: 19 Apr 2023
32. Mukhter M, Khuder D, Kalganova T (2021) A control structure for ambidextrous robot arm based on multiple adaptive neuro-fuzzy inference system
33. Muniz H (2018) 3 expert tips for using the unit circle, 3 expert tips for using the unit circle. Available at: https://blog.prepscholar.com/unit-circle-chart-radians-degrees. Accessed: 19 Apr 2023
34. Napoli M (2022) Robust balancing for bipedal robot via model predictive control. Doctoral dissertation, The Ohio State University
35. Nise NS (2019) Nise's control systems engineering. Wiley, Hoboken, NJ, p 123
36. Paredes VC, Hereid A (2022) Resolved motion control for 3d underactuated bipedal walking using linear inverted pendulum dynamics and neural adaptation. In: 2022 IEEE/RSJ international conference on intelligent robots and systems (IROS). IEEE, pp 6761–6767
37. Raheema MN, Kadhim DA, Hussein JS (2021) Design an intelligent hybrid position/force control for above knee prosthesis based on adaptive neuro-fuzzy inference system. Indones J Electr Eng Comput Sci 23:675–685
38. Schabron B, Alashqar Z, Fuhrman N, Jibbe K, Desai J (2019) Artificial neural network to detect human hand gestures for a robotic arm control. In: 2019 41st annual international conference of the IEEE engineering in medicine and biology society (EMBC). IEEE, pp 1662–1665
39. Sherwani KI, Kumar N, Chemori A, Khan M, Mohammed S (2020) RISE-based adaptive control for EICoSI exoskeleton to assist knee joint mobility. Robot Auton Syst 124:103354
40. Tchórzewski J, Wielgo A (2021) Neural model of human gait and its implementation in MATLAB and Simulink environment using deep learning toolbox. Stud Inform: Syst Inf Technol 1
41. Vladareanu L, Vladareanu V, Wang H, Feng Y, Radulescu M, Ciocîrlan AC (2019) Versatile intelligent portable robot platform developed through adaptive neuro-fuzzy control. Int J Model Optim 9(5)
42. Vukobratovic M, Juricic D (1969) Contribution to the synthesis of biped gait. IEEE Trans Biomed Eng 1:1–6
43. Wang R, Hudson SJ, Li Y, Wu H, Zhou C (2019) Normalized neural network for energy efficient bipedal walking using nonlinear inverted pendulum model. In: 2019 IEEE international conference on robotics and biomimetics (ROBIO). IEEE, pp 1400–1406
44. Widhiada W, Nindhia TGT, Widiyarta IM, Budiarsa IN (2020) Control of bionic robot leg performance with proportional integral and derivative controller. Int J Mech Eng Robot Res 9(10)
45. Wiebe F, Babel J, Kumar S, Vyas S, Harnack D, Boukheddimi M, Popescu M, Kirchner F (2022) Torque-limited simple pendulum: a toolkit for getting familiar with control algorithms in underactuated robotics. J Open Source Softw 7(74):3884
46. Yu W (2018) Pid control with intelligent compensation for exoskeleton robots. Academic Press, London, an imprint of Elsvier. Accessed: 14 Oct 2022
47. Zadeh LA (1965) Fuzzy sets. Inf Control 8(3):338–353
48. Özdemir C, Öztürk, S, Şengül Ö, Kuncan F (2022) Position control of the suspended pendulum system with particle swarm optimization algorithm. El-Cezeri 9(2):669–679
49. Živanović S, Lin B, Dang HV, Zhang S, Ćosić M, Caprani C, Zhang Q (2022) Evaluation of inverted-pendulum-with-rigid-legs walking locomotion models for civil engineering applications. Buildings 12(8):1216

Multi-modal Supply Chains: A Generative AI Framework for Intelligent Logistics Optimization, Freight Exchange, and Dynamic Routing

Morteza Pourjahangiri Pilehroud and Jens Wollenweber

Abstract In the era of globalization and escalating sustainability demands, multi-modal supply chains, integrating road, rail, air, and sea transport offer significant efficiency potential but face challenges in coordination, freight exchange, and adaptive routing. This research introduces a generative AI framework based on a hybrid Variational Autoencoder-Generative Adversarial Network (VAE-GAN) architecture to enhance intelligent logistics optimization, enabling scenario generation for cost-effective, resilient, and eco-friendly operations. The framework is developed using Python, leveraging synthetic datasets generated to simulate realistic logistics scenarios, including emissions, disruptions, and multi-leg routes. A mixed-methods approach incorporated model training with PyTorch, discrete-event simulations for validation, and benchmarking against Mixed-Integer Linear Programming (MILP) and heuristic baselines. Evaluations focused on quantitative metrics such as total cost, transit time, CO_2 emissions, and constraint satisfaction, alongside visual analyses of latent spaces and distributions. Results demonstrate that the VAE-GAN achieves near-optimal logistics costs, VAE-GAN, with superior computational efficiency—reducing execution times significantly compared to MILP—while maintaining scenario diversity and realism across baseline, disruption, high-demand, and sustainability conditions. The model exhibits zero constraint violations, stable prediction errors in feedback loops, and realistic trade-offs between cost and emissions. This research bridges a gap in applying generative AI to multi-modal logistics, improving flexibility, delivery efficiency, and ecological impact. Future extensions include real-time API integration and reinforcement learning for adaptive deployment.

Keywords Multi-modal supply chains · Generative AI · VAE-GAN · Logistics optimization · Freight exchange · Dynamic routing

M. P. Pilehroud (✉) · J. Wollenweber
Technical University of Applied Sciences Wildau, Wildau, Germany
e-mail: mopo1940@th-wildau.de

A. Azizi (ed.), *Intelligent Control and Autonomous Systems*, Emerging Trends in Mechatronics, https://doi.org/10.1007/978-981-95-6402-6_7

1 Introduction

1.1 Context of Multi-modal Supply Chains

The complexity and globalization of supply chains in recent years have driven the need for efficient, reliable, and durable logistics systems. Multi-modal supply chains, employing two or more modes of transport (such as air, sea, rail, and road), are considered to be more economically efficient, environmentally sustainable and provide an optimum level of service. SteadieSeifi et al. [29] argues that multi-modal transport schemes are a key factor to combat congestion, emissions, and asset utilization, mostly on long distance and international trading transport context.

But the realization of multi-modal supply chains is accompanied by large difficulties. Fragmented decision-making, no data integration among various modes of transport, and asynchronism in information flows are the causes for not achieving the maximum potential expected from multi-modal logistics [24]. In addition, disruptions, including traffic, infrastructure downtime, and demand irregularities add extra layers of complexity to the management of the supply chain.

Those challenges have created an increasing momentum for digitization and smart decision-making systems. The European Green Deal and the Sustainable and Smart Mobility Strategy of the European Union aim to support a modal shift to rail and inland waterways, with the aim of decreasing the carbon footprint of freight transport by 90% by 2050 [12]. Inventions and advances in information technologies are able to provide data-driven solutions to control (near) real-time-based and end-to-end coordination in multi-modal supply chain.

In the meantime, new research work has been focusing on improving the flexibility of logistics systems using advanced Artificial Intelligence methodologies. For example, Blessing [3] presented a simulation GAN for capacity release management and GAN simulation may potentially be used to enhance operational efficiency. These trends illustrate the pertinence of AI-based methodologies to meet the complexity of multi-modal logistics environments.

1.2 Significance of Generative AI in Logistics

The rapid advancement of Artificial Intelligence has significantly disrupted decision-making environment in the logistics and supply chain domain. Among different types of AI, generative models, especially GANs and VAEs, have become the workhorses for solving sophisticated, data-rich logistics problems. These models naturally have the ability to learn complex data distributions and then sample new high-quality solutions making them particularly beneficial for problems such as multi-modal logistics where the environment is uncertain and dynamic.

Goodfellow et al. [14, 15] observed that GANs are basically to generate the synthetic data through the adversarial training of two neural networks, denoted by the generator and the discriminator. This natural functionality enables the creation of better logistics scenarios that take into account dynamic freight assignments and agile routing algorithms based on complex analysis of historical and real-time data patterns. Similarly, VAEs use probabilistic encoding and decoding processes to infer latent data structures and to generate realistic synthetic outputs, which can produce more reliable demand forecasting and more nuanced disruption simulation Wang et al. [32].

Recent academic study results have proven the utility of generative AI in the logistics-sensitive area. For instance, Khlie et al. [18] VAE-RNN hybrid models were successfully used for demand prediction in supply chain, and significant accuracies increase and operational cost savings were reported. At the same time Naji et al. [23] applied GANs for taxi demand prediction and demonstrated promise and ability of generative models in real-time transportation forecasting. Altogether, these results show that generative AI models increase the flexibility of logistics systems and help in meeting sustainability goals, they reduce the amount of freight unnecessarily going through the systems and cut down on wastage.

The search for hybrid architectures that merge VAEs and GANs has shown potential to improve the strength and interpret ability of generative models [26]. These combined approaches provide a path to the development of advanced logistics optimization that smoothly combine freight exchange mechanisms, dynamic routing, and sustainability.

Thoughtful use of generative AI in multimodal logistics solutions squarely fits a rising demand for autonomous and real-time agility. This allows logistics planners to be proactive in reacting to disruptions, to use the resources more efficiently with higher precision, and to align the operational decisions with the prime objective of sustainability.

1.3 Research Problem and Gap Identification

Given the potential theoretical and practical benefits of multi-modal logistics networks and the fast advance of AI methods, the combination of state-of-the-art generative AI models in operative supply chain systems is still not explored . Existing multimodal logistics methodology tends to use deterministic optimization models and rule-based systems that are not very good at handling real-time disruptions, dynamic fluctuations in demand, and intermodal coordination complexity [29]. To this end, logistics networks have often inefficiencies, higher costs, inadequate freight exchange matching and greater environmental burdens.

Known freight exchange platforms are mostly static and influenced by transactions and do not have the intelligence to dynamically pair shippers and carriers according to latest operational conditions. Route choice, critical for the performance of multi-modal supply chains, is made without full regard to the possibility of using alternative transport modes, the observance of sustainability criteria as well as the resilience of the system.

Generative AI approaches including GANs and VAEs provide powerful tools to overcome these challenges by learning to model data patterns and to generate better logistics settings. Recent studies, e.g., Blessing [3], Chen et al. [6], have shown that GAN-based models can be used for logistics optimization and predicting vehicle trajectories. In addition, Khlie et al. [18] applied VAE-based models to demand forecasting in supply chains, emphasizing their versatility in data-driven decision-making settings.

Survey through the literature shows there is still a huge and important gap. To the best of author's knowledge, very few studies have formulated a unified modular framework that integrates generative AI models into simulation-based validation for the efficient allocation of logistics resource, freight exchange efficiency, and dynamic routing under sustainability aims. Second, even though the hybridization of VAE-GAN models has been only sporadically studied in neighboring fields [26], it is largely neglected in the context of multi-modal logistics optimization.

This research aims to fill this gap by introducing an extensive generative AI framework for multi-modal supply chains. This framework and philosophy has the objective to improve the decision-making in the logistics optimization, freight exchange, and dynamic routing enabling the ongoing principles of sustainability and resilience supply chain management.

1.4 Research Objectives and Questions

The research aims to form a complete framework for designing, developing, and validating the proposed generative AI approaches. This architecture is designed to dramatically improve the operational performance, agility, flexibility, and environmental sustainability of large-scale, multi-modal supply chain networks. This project addresses the gap between the current situation, state-of-the-art knowledge and technology by making a novel use of cutting-edge hybrid Generative Adversarial Networks (GANs) and Variational Autoencoders (VAEs) for operational improvement to the very heart of the logistics of resource allocation, efficient exchange of freight, and dynamic routing in a world of massive uncertainty and system disruption. To meet this overarching goal, the study is driven by the following inter-related and operational objectives:

1. **Framework Conceptualization and Development**: To design and implement a modular Python-based generative AI framework (based on a VAE-GAN hybrid model) embedded into a simulation-based analytics logistics setting for learning, optimizing complex logistic patterns.
2. **Empirical Performance Evaluation**: To empirically evaluate the operational effectiveness of the proposed framework, through strong empirically grounded simulation testing, by quantifying its improvements in key performance indicators (KPIs) such as cost efficiency, delivery performance, environmental

impact (e.g., carbon emissions), and systemic resilience, relative to conventional optimization approaches.

3. **Sustainability and Practical Implication Analysis**: To examine the sustainability implications and practical viability of implementing generative AI solutions in multi-modal logistics systems, thereby contributing to the broader academic and industry discourse on digital transformation, green supply chains, and strong decision-making in dynamic environments.

In direct alignment with these stated objectives, the study is structured around three core research questions, which will be investigated:

1. **Research Question 1 (RQ1)**: What would be the optimal design, training schedule, and integration strategy for a VAE-GAN hybrid generative AI model in a simulation-based environment for learning and optimizing complex logistics decisions on resource assignment, cargo exchange, and dynamic routing in the context of a multi-modal supply chain?
2. **Research Question 2 (RQ2)**: How can the adoption of this generative AI framework be quantified, compared to traditional optimization methods (such as Mixed-Integer Linear Programming (MILP) and reinforcement learning (RL)-based methods) in a simulated multi-modal logistics context, improve performance metrics, total logistic cost, transit time, carbon emissions, and system resilience?
3. **Research Question 3 (RQ3)**: What are the critical sustainability implications and practical operational considerations associated with adopting an AI-driven decision-making framework, particularly emphasizing its alignment with environmental objectives and its capacity to enhance resilience against disruptions, within the contemporary logistics sector?

These clearly defined objectives and precise research questions establish a strong foundation for an empirical investigation into the practical relevance and value of generative AI in improving logistics optimization under sustainability goals.

1.5 Scope, Assumptions, and Methodological Boundaries

This research aims to articulate its scope of research to conceptualization, design, development, and simulation-driven empirical validation of a new generative AI framework for multi-modal logistics optimization. Although the general architecture involves conceptual functions for freight exchange and dynamic routing, their empirical validation, due to the required depth and live API-based data access (the place holder exists in the framework and ready to be connected via API), is not within the scope of this research, and is therefore included as a main research area. The main focus of the empirical work in this paper is to illustrate the effectiveness of the proposed framework for managing logistics resource allocation and increasing logistics key performance indicators in a controlled simulated environment.

Rationale for Placeholder Modules:

- Freight exchange and dynamic routing are conceptually critical to multi-modal logistics.
- Their preliminary design within the framework validates interoperability, modularity, and extensibility.
- Full-scale empirical testing requires real-time APIs, live carrier data, and advanced traffic prediction engines, which exceed the resource and scope limitations of this research.

The precise scope of this study is judiciously delimited by several key considerations:

1. **Specific Generative AI Models**: The two principal generative AI models used and investigated in this work are the GAN and the VAE, as well as the special focus on a unified hybrid combination of both models. Other AI categories, while relevant to the general area of logistics, have been deliberately omitted from the direct focus of this study.
2. **Simulation-Based Validation**: The empirical validation is carried out in simulation, not on a live operational basis, under a controlled custom programmed Python environment in Pycharm with relevant libraries. Although this methodological decision makes it possible to validate scenario-based and structured investigations in a controlled environment, it also limits the extent to which the performance of the framework can incorporate many of the complex and emergent behaviors and unanticipated disruptions typical of actual logistics operations.
3. **Reliance on Synthetic Data**: Priming data that is used both in model training and simulation is mostly synthetic and designed to mirror realistic multi-modal logistics setting. While substantial effort was made to provide accurate representation of the real world and preserve data integrity through the life cycle, purely synthetic data does have limitations in representing the surprise anomalies, emergent complexity, and subtle stochasticity present in live operational data.

Despite the validity of the methodology, this study has some intrinsic limitations that should be explicitly addressed:

1. **Generalizability**: Due to the purely synthetic nature of our experimental setup, the results from this study have potential limitations concerning the straightforward extrapolation and generalizability to a wide range of real-world heterogeneous multi-modal logistics operations.
2. **Model Complexity and Interpretability**: Advanced generative AI models, and VAE-GAN hybrids in particular, are complex by their nature, and interpreting these models is inherently difficult and computationally intensive. Although the framework develops coping mechanisms for some of these problems, they are recognized features of this methodological strategy.

3. **Real-Time Data Processing Simulation Constraints**: Even if the framework is fundamentally conceived to be adaptable to nonstationary conditions, the practical ability of processing real-time data simulation in the simulated environment is only limited by the adopted methodology for synthetic data generation and the computational performances of the simulation engine. This speed itself may have an effect on how large and fast highly dynamic, continuous, real-time flows of data can be processed.

Despite these restrictive focuses and limitations, this study deliberately results in a strong conceptual and empirically grounded base for how generative AI can be used strategically in multi-modal logistics. It provides deep theoretical contributions and practical implications, and thus suggests clear directions for further advances and applications into the new phenomenon of intelligent supply chains.

2 Literature Review

This chapter aims to provide a comprehensive literature survey on the relevant academic literature on multi-modal supply chain systems, the application of Artificial Intelligence in logistics problems, and the emerging use of generative models, including but not limited to GANs and VAEs, in the area of supply chain optimization. Its main focus is to detail theoretical and empirical motivation for the research that follows, critically establishing where the present body of knowledge falls short, and in so doing unequivocally legitimizing the case for the generative AI framework created. The review shows in a systematic manner in thirteen main sub-sections, in order to enable an organized and comprehensive approach to the relevant scholarly discussion. The development and operation of multi-modal supply chains is explained in the first part, in Sect. 2.1, which emphasizes its importance and challenges as well. Then, Sect. 2.2 presents a brief overview of the classical and state-of-the-art optimization models for supply chain, including mathematical models and data-driven models. Section 2.3 subsequently explores the concepts and technical means of core generative AI models. Section 2.4 offers us an explicit taxonomy on Generative Adversarial Networks, classifying the plethora of architectures and applications into categories. Then, Sect. 2.5 describes the Generative Adversarial Networks, and they are followed by Variational Autoencoders in Sect. 2.6. Section 2.7 further clarifies the synergistic unification of the VAE-GAN hybrid architectures. Section 2.8 provides a comprehensive survey of how AI, and sustainable generative models specifically, are being applied in a variety of logistics and supply chain management settings. In the final part of the review, to summarize the findings from the literature and critically establish the main research gaps that will be focused on in this study, Sect. 2.9 concludes the review and Sect. 2.10 considers the inter-related facets of sustainability in multi-modal supply chains followed by application of AI in logistics, identification of research gaps and sustainability dimensions in multi-modal supply chains in 2.11, 2.12, and 2.13.

By critically engaging the existing scholarly discourse across different scholarly fields, this section thus positions the present study squarely at the intersection of the supply chain management and AI, and then provides a solid conceptual and empirical basis for the ensuing design and construction of our proposed generative AI framework.

2.1 Evolution of Multi-modal Supply Chains

The multimodal supply chain has dramatically evolved over the last century, influenced by the growing complexity of the global economy, environmental concerns, and the development in logistics infrastructure. Intermodal: Supply chains that use two or more modes, road, rail, sea, and air, to get your goods from point A to point B as quickly as possible in the most efficient manner. This method presents clear merits such as cost decrease, better service reliability, and environmental enhancement [29].

The emergence of multi-modal logistics systems at an early stage was predominantly attributed to the need to harmonize varied transport systems in order to overcome physical and infrastructural barriers. With the push for global trade, multimodal transportation played an increasing role in providing routes for international trade and long distance freight passages. By the time of writing, advancements in intermodal terminals, containerization, and digital tracking systems have produced more and more efficiency in the operation of multi-modal logistics systems [24].

However, nevertheless, in spite of these developments, multi-modal supply chains still experience a number of operational issues. Decentralized decision-making, little integration of real-time data in transport modes, and capacity restrictions on intermodal transfer locations typically lead to lack of effectiveness and higher logistics costs. In addition to these disruptions, multi-modal logistics systems are vulnerable to the risks of traffic jams, bad weather, and infrastructure failures, which can greatly impact the reliability and resilience of the systems.

Recent policy measures put forward in the European Union have also manifested the need to transfer freight transport from road to environmentally more friendly modes, rail, and inland waterways. The European Commission's on Sustainable and Smart Mobility of the European Commission (EC) pursues to reduce the environmental impacts of freight transport through digitization, interoperability, and multi-modal integration [12]. The above described policy signals a clear demand for smart, data-driven solutions that can improve the agility and the sustainability of multi-modal supply chains.

The attention on the theoretical tractable optimization models and decision-support systems to the multi-modal logistics has been incrementing in the academic literature. Traditional methodologies have used deterministic optimization methods, like Mixed-Integer Linear Programming (MILP) to solve routing and resource allocation decisions [9]. However, these models are insufficient for capturing the dynamic and uncertain nature of actual logistic systems, which lead to data-driven and AI-based techniques for solving logistic systems.

Recently, some studies start to investigate the application of generative AI models to further strengthen the functions of multi-modal supply chain. As an example, Blessing [3] presented a simulation-based GAN approach for supply chain optimization, showing that AI-driven optimization is possible in complex logistics. The incorporation of hybrid generative models, such as VAE-GAN architectures, has also been suggested as a way of improving decision-making in logistics and transportation networks [26].

The multi-modal supply chains evolution is a continuous movement toward digitization, sustainability, and intelligent decision support. The work presented here extends these advances by suggesting an AI-based framework to improve logistics operations that involve several modes of transport using generative models.

2.2 Theoretical Foundations of Supply Chain Optimization

The optimization of the supply chain has been the subject of attention for both academia and industry over many decades with an aim of improving efficiency, reducing cost, and service quality delivery. Supply chain optimization is rooted in the legacy of operations research and mathematical programming and decision-support systems are rooted in the operations research and mathematical programming. Classical optimization solutions have relied heavily on deterministic and stochastic models to deal with diverse logistic problems ranging from routing, inventory, and capacity management.

Mixed-Integer Linear Programming (MILP) is one of the most used optimization methodologies on logistics. The MILP models are constructed to minimize or maximize an objective function such as cost or service level/utility, while considering a linear constraint set and decision variables that may be continuous or integer valued [9]. From the perspective of multi-modal supply chain, MILP has been widely applied to seek a set of the optimal routing scheme, to distribute the freight volume among the transport modes, and to schedule the shipments with respect to the constraints of capacity and time.

Although traditional optimization models (e.g., MILP) are powerful in static and well-defined settings, they have several limitations. These models generally work under full and deterministic information which is not well-suited to dynamic, uncertain, and large-scale logistics operators. It is also the case that the computational complexity of MILP is exponential with respect to the size of the problem and does not scale easily or lend itself to real-time usage in operative contexts [29].

To overcome these limitations, there have been studies about heuristic and metaheuristic methods such as Genetic Algorithm, Tabu Search, and Ant Colony Optimization. However, learning-based approaches are more scalable and flexible than traditional registration approaches, while they are also for better performance but have relatively high demand on parameter tuning and need not be adapted to real-time changes.

More recently, with the advances in reinforcement learning (RL) methodologies, the interest in using RL techniques for modeling supply chain behaviors has also increased. RL algorithms, such as Q-learning and deep reinforcement learning, train agents by interacting with environments through iterations to learn their optimal policies [17]. These techniques have potential for solving the uncertain routing and resource allocation problem. However, the RL methods suffer from issues such as convergence stability, and computational complexity that require huge training examples.

Meanwhile, the rise of data-driven and AI-driven optimization methods has provided new opportunities for making supply chain decisions better. With generative models (e.g., GANs, and VAEs) capable of learning complex data distributions, more realistic logistics scenarios can be generated. There is also work such as [3, 6] showing that GAN-based models can be useful in supply chain simulations and vehicle trajectory prediction, respectively. Hybrid VAE/GAN models have also been proposed to improve resilience and ability to interpret in generative optimization frameworks [26].

Supply chain optimization has many theoretical underpinnings, with pros and cons for each. We extend these studies by embedding generative AI models into a simulation-based environment for multi-modal logistics optimization.

2.3 *Generative AI: Concepts and Techniques*

Generative Artificial Intelligence (AI) is a class of advanced machine learning models that can learn and internalize the underlying probability distributions of training data, and then can generate novel, synthetic data instances that are similar to the input data. While discriminative AI models tend to concentrate on transforming inputs to their corresponding outputs (like for example, in classification or regression), generative models are in turn designed to encode an understanding of the structure and range of variability present in a dataset. This basic difference gives generative AI the ability to do more than copy existing patterns, but to interpolate and create new and very different points of data that are plausible.

The attractiveness of generative AI is the fact that it allows to tackle multifaceted issues in data-rich and uncertain contexts. Outside of synthesis, generative models can be used in a variety of tasks: data augmentation in the case of a sparse dataset, anomaly detection in the form of detecting deviations from learned distributions, imputation of data values for missing data, and advanced scenario planning via the generation of realistic future states. These features are more transformative in the fields such as multi-modal logistic where the decision-making is commonly affected by information scarcity, dynamical changes, and an improbably solution space exploration.

Generative Adversarial Networks (GANs) and Variational Autoencoding (VAE) are two of the most popular and influential technology in generative AI recently. The different ways in which they both learn data distributions and generate synthetic outputs make them particularly well-suited to the hard optimization and scenario generation problems at the core of this research. The conceptual motivations, architectural details, and generative dynamics of these two key types of generative models, and how to integrate the two models, will be discussed in the following subsections.

2.4 Taxonomy of Generative Adversarial Networks (GANs)

In the last decade, there has been a accelerated divergence of Generative Adversarial Networks (GANs), leading to a wide and varied range of architectures and methodological improvements. In order to better understand the complexity of this space and, more importantly, to provide some concrete context for where exactly in the broader GAN space, the particular GAN-based models used in this research sit, below is the systematic view of GANs. It is based on the classification by seminal and comprehensive reviews [10, 16, 31], and hence provides a solid framework for understanding their diverse capabilities and mechanisms.

2.5 Generative Adversarial Networks (GANs) Detailed Mechanisms

There exists a remarkable class of generative models relying on the adversarial learning mechanisms, that is Generative Adversarial Networks (GANs) invented by Goodfellow et al. [13]. GAN consists of two main neural networks: a **generator** (G) and a **discriminator** (D), trained simultaneously against each other in a zero-sum, minimax game. The role of the generator is to approximate the true data distribution of real samples ($p_{data}(x)$) and generate synthetic data ($G(z)$ from a latent noise vector z) that looks identical to real data. On the other hand, the role of the discriminator is to reliably distinguish between real-world data samples and the generated (i.e., synthetic) ones. (2) Adversarial Learning: The competition between the two networks forces them to progressively boost their performance; one of them progressively learns to generate increasingly realistic data, while the other improves its ability to distinguish between produced fakes and real. Such a dynamic mechanism effectively leads the generatorâŁ,to fit the underlying complex manifold of the real data distribution.

Mathematically, the training of a Standard GAN is articulated through a value function $V(D, G)$ that the discriminator aims to maximize and the generator aims to minimize: $\min_G \max_D V(D, G) = \mathbb{E}_{x\sim p_{\text{data}}(x)}[\log D(x)] + \mathbb{E}_{z\sim p_z(z)}[\log(1 - D(G(z)))]$. This formulation implicitly minimizes the Jensen-Shannon divergence between the real and generated data distributions, though practical challenges like vanishing gradients and mode collapse can impede stable convergence.

Today, in logistics and transportation, GANs have shown an impressive potential for generating realistic and situationally relevant scenarios for simulation-based analyses and for enabling various complex decisions. For example, Blessing [3] used GANs for simulating realistic supply chains to optimize their management, as plausibility of logistic scenarios' parameters, such as demand patterns or fleet movements, could be well provided by those models. Goodfellow et al. [23] also successfully applied GANs for fine-grained taxi demand forecasting, which demonstrates that they can be useful in real-time transportation systems. In addition, a variety of GAN architecture improvements (including hybrid versions) have been investigated to enhance model resiliency and prediction accuracy. Chen et al. [6] proposed a special convolutional autoencoder-GAN (CAE-GAN) for vehicle trajectory prediction, which achieved better prediction accuracy in dynamic routing tasks by learning complex spatial-temporal dependencies.

However, despite such strengths, instability of training process, mode collapse (few varieties of the generated outputs), and hyperparameter sensitivity are major issues in implementing GAN models.

2.6 Variational Autoencoders (VAEs) Detailed Mechanisms

Variational Autoencoders (VAEs) by Kingma and Welling [19] are a different, powerful probabilistic generative model. Unlike the competitive structure of GANs, VAEs use an encoder-decoder structure to model latent data representation and to reconstruct input data by sampling from a trained probabilistic distribution. The **encoder** network takes an input data point (x) and maps this to the parameters (mean and variance) of a latent distribution ($q_\phi(z|x)$), which is usually assumed to be a Gaussian. The **decoder** network (also called the generator) samples from learned latent distribution (z) and decodes to reconstruct the input ($\hat{x}$ from $p_\theta(x|z)$). The training objective of a VAE is to optimize a variational lower bound on the data likelihood, and in doing so, to encourage tight data reconstruction, while also constraining the latent space to lie in some (usually pre-specified) prior distribution (such as a standard normal). Such dual task leads to a well-defined, smooth, and interpretable latent space. The basic objective function of a VAE comprises two main parts:

(a) **Reconstruction Loss**: This term measures how faithfully the decoder reproduces the input data. For continuous (binary) vectors, this could be the Mean Squared Error (MSE) (Binary Cross-Entropy (BCE)).
(b) **Kullback-Leibler (KL) Divergence Loss**: This regularization term quantifies the distance between the learnt latent distribution ($q_\phi(z|x)$) and the assumed prior distribution ($p(z)$). It enforces a well-behaved latent space and encourages the model to learn a variety of representations, reducing mode-collapse problems.

Such optimization-based approach can produce more stable training dynamics than GANs do, and the structured latent space enables more smooth interpolations between the generated samples as well as better interpretability of the learned features.

VAEs have received growing interest in the supply chain area, by their capacity to model complex data distributions and learn interpretable latent variables. Wang et al. [32] even designed a VAE-based model to achieve multimodal integration of transportation data with the feature of better fusing different sources of data to support decision-making. The only related work we have found a detailed review on the application of AI for the forecasting of fashion sales by Khlie et al. [18], which applied a VAE-RNN network to demand forecasting in supply chain management with remarkable improvements in predictive accuracy, and an increased understanding of the demand driving factors via analysis of the latent space. In tasks where probabilistic modeling, uncertainty quantification, and the generation of diverse, but still plausible, data samples are important, VAEs are especially useful.

2.7 Hybrid VAE-GAN Architectures and Synergies

With the development of generative models in recent years, it has become more and more popular to combine VAE to GAN. It is the optimal trade-off between two different architecture's advantages, which help to alleviate their drawbacks and contribute to more solid model, better generation quality, and better interpretability. Yet, in spite of their well-posed learning and controlled latent space, VAEs can produce rather blurry or less sharp samples than GANs. In contrast, GANs are found to produce high-rez and sharp images, but suffer from issues such as training instability and mode collapse.

The VAE-GAN combination architecture uses the best of both networks. A common integration strategy is to include the VAE decoder as the generator in the GAN (or have the GAN discriminator furnish an additional adversarial loss term that punishes the VAE reconstructions to push them toward realistic outputs). This joint-objective optimization on the one hand results in a more semantically meaningful and adversarial strong representation learned by the VAE model and on the other hand forces the GAN to yield better attention maps and more diverse generations. The first work is to integrate such approach [20], showing its potential to improve image synthesis. A related work [26] also demonstrated that VAE-GAN models can generate complex topological magnetic states and learn nontrivial data features. Similarly, [4] provided a comprehensive study on the performance, stability, and representational power improvements obtained, by (different) techniques/methods of introducing GAN modules into a VAE in different domains.

Though being empirically successful in related areas such as computer vision and materials science, the adaption of hybrid VAE-GAN models as solutions to the logistics and transportation problems is significantly lagging behind. This work specifically intends to fill this gap and strategically combine the advantages of both architectures to create an intelligent and sustainable multimanual logistics optimization framework. The hybrid VAE-GAN model proposed in this research (see `src/models/optimization/vae_gan.py` in the provided codebase) is designed to address domain-level challenges pertaining to logistics domain such as heterogeneity, dynamics in temporal sense, realistic scenario creation, and interpretability in decision-making. By leveraging the VAE to characterize a structured, probabilistic latent space of logistics features (e.g., demand deviations, disruptions) and the GAN to sharpen this into high-realism and actionable logistics scenarios (e.g., route optimizations, freight assignments), this hybrid model offers a promising solution for adaptability and resilience in complex multi-modal supply chains.

Although there have been impressive successes by the standalone GANs and the VAEs, mode collapse for the GANs and blurry reconstructions for the VAEs can be limitations for generating the high-fidelity scenarios. These limitations are addressed by hybrid VAE-GAN models that combine the structured latent representation of VAEs with the realism-inducing training dynamics of GANs. In logistics settings, such synergy makes it possible to construct multimodal supply chain situations that both respect operational constraints (e.g., capacity, transit time) and capture emergent patterns of disruption in an authentic fashion [8, 14].

2.7.1 Based on Objective Function

The adversarial loss function is the principle fundamental characteristic of GANs that directly determine the training, convergence, and the final quality and realism of their downloaded outputs. These objectives specify the adversarial game that the generator and discriminator play against each other:

(a) **Standard GAN (SG-GAN)**: The seminal Standard GAN proposed by [15] is based on a minimax game. Generator's purpose is to generate samples that are indistinguishable from the real data that the discriminator tries to accurately classify between the real and fake samples. This iterative process is implicitly trying to minimize the Jensen-Shannon (JS) divergence between the true data distribution and the generated data distribution.

(b) **Wasserstein GAN (WGAN)**: To solve the training instabilities and mode-collapse problems of Standard GANs, [1] introduced WGAN. It uses Earth Mover's Distance (Wasserstein-1) as the loss instead of the JS divergence. This measure has a smooth gradient and confines higher quality generator outputs to less homogenous regions in the output space.

(c) **Least Squares GAN (LSGAN)**: Introduced by [21], LSGAN replaces the binary cross-entropy loss of the Standard GAN's discriminator with a least squares loss. This perturbation is shown to alleviate the vanishing gradients issue especially when generated samples are far away the data manifold, leading to more stable training and superior generation.

2.7.2 Based on Architecture

Although the architectural design principles of GANs are highly diversified to accommodate different strategies of boosting generative capability, incorporating auxiliary information, or improving the training properties:

(a) **Conditional GAN (cGAN)**: As proposed by [22], cGANs generalize the standard GAN architecture to account for auxiliary information (class labels, text sentences, or other data modalities) provided as additional inputs to the generator and discriminator networks. Such conditioning enables data generation that is conditioned on particular input characteristics.
(b) **Auxiliary Classifier GAN (AC-GAN)**: Based on cGAN, Odena et al. created AC-GAN [25]. In this architecture, the discriminator is required to recognize not only real from fake samples, but also the associated auxiliary information (e.g., class label) for real samples. This co-training objective incentivizes the generator to generate more coherent and class-specific samples.
(c) **VAE-GAN Hybrid**: This architecture is an innovative fusion between VAEs and GANs. As originally proposed in [20], VAE-GAN hybrids combine the stable probabilistic encoding and ability to learn good representations in the latent space with GANs' ability to learn complex data distributions via adversarial training. This connection usually means making the VAE's decoder the GAN's generator, or using the GAN's discriminator to supply an extra loss term to the VAE, in the hopes of generating better samples and interpretable latent spaces.

2.7.3 Based on Application Domain

GANs are extremely versatile and have been extensively used in many domains, thus they have been specifically classified in the following categories according to the intended application objectives:

(a) **Data Augmentation GANs**: These GANs are designed to enlarge and enhance the training data, especially in situations with little data or class imbalance. They improve the strenth and generalization of subsequent machine learning models by creating new realistic samples.
(b) **Simulation-Based GANs**: In specific applications such as the logistics and supply chain management, simulation-based GANs are used to: (a) generate complex and require realistic scenarios on-the-fly, and (b) optimize complex operational processes. For example, they were used in Blessing [3] to model supply chain behaviors to find best strategies.
(c) **Domain Transfer GANs**: Such models, often referred to as image-to-image translation GANs [36] are designed to learn the mapping between two different data domains. They can be used for the task of converting photographs to paintings, or satellite images to maps, and can be applied to cross-domain data synthesis more generally.

This organized taxonomy provides a clear understanding of the diverse range of GANs and their associated capabilities. Importantly, the VAE-GAN hybrid model introduced in this research, providing performance within a simulation-based logistics scenario, strongly relates to both the advanced architectural space (2.4.2c) and the application-driven area of simulation-based GANs (2.4.3b). This alignment emphasizes its designed purpose for demanding logistics optimization under high degree of uncertainty.

2.8 Applications of AI in Logistics and Supply Chain Management

Artificial Intelligence (AI) technologies penetration in the complex context of logistics and supply chain management has recently received both scholarly and practical interests substantially. AI methods have started to be used for a wide range of operational problems, including demand forecasting, inventory management, freight exchange optimization, and routing. Among many AI methods, advanced generative models, Generative Adversarial Networks (GANs) and especially, Variational Autoencoders (VAEs), have clearly shown promise as a class of powerful techniques to fundamentally improve logistics decision-making in highly complex, changing environments.

Demand forecasting is one of the key applications of AI, in supply chain. For the companies, highly accurate and responsive demand forecasting is core to achieving the product availability, fast lead times, and customer satisfaction of their wildest dreams. Conventional forecasting approaches such as statistical models have often failed to identify latent nonlinear relationships and adjust to the fastly changes of the market at the same speed. In contrast, AI-based models, and generative models in particular, are more accurate and can adapt to new inputs. For example, Khlie et al. [18] convincing confirmed the superiority of a VAE-RNN model on demand planning in supply chain management, as they observed remarkable gains over traditional model in regression performance. In this respect, generative models can generate a large number of realistic demand scenarios in order to perform risk analysis and proactive planning.

In the important field of traffic and routing, generative AI strategies have been carefully used to great effect for real-time routing choices and to deal with operational inefficiencies. Naji et al. [23] effectively used the GAN-based method for fine-grained taxi demand forecasting to demonstrate that generative models have very strong potential for practical use in real-time transportation systems. Inspired by this Chen et al. [6] proposed a convolutional autoencoder-GAN (CAE-GAN)-based vehicle trajectory prediction framework, which outperforms the compared methods in dynamically routed scenarios by capturing complex spatial-temporal movement distributions. These are not only limited to static route optimization, but also provide adaptive data-driven solutions for real-time traffic conditions and anomalies.

Freight Exchanges is another interesting domain where AI can present a good use case. Traditional freight matching system are static-based on rigid rule-based mechanisms which restricts their capability of responding to real-time variations in freight supply and demand, also, in operational constraints differences. More recent academic studies have investigated the novel application of AI-based methodologies to freight exchange processes, to support a more agile and immediate allocation of resources. For instance, [35] recently proposed a GAN methodology to model credit risk in supply chains, which reveals the broader utility of generative models in improving trustworthiness and transactional efficiency in freight logistics as they come to understand underlying patterns. The generating aspect here is the optimal matches that may not be noticeable at first glance.

In addition, simulation-based AI applications have been widely utilized for objective assessment and performance optimization of supply chains over numerous intricate operational settings. Blessing [3] indeed first applied a GAN-based simulation in the fieldâŁ,of supply chain management, providing evidence of its capacity to depict realistic logistics scenarios and enhance strategic decision-making. These applications collectively and convincingly highlight the strong potential value of generative AI'sâŁ,tremendous importance in overcoming the intrinsic complexity, high uncertainty, and critical sustainability requirements that underpin today's supply chain networks.

While substantial progress has been undeniably achieved in applying diverse AI techniques to logistics and supply chain management, extant studies frequently remain confined to addressing isolated tasks such as the aforementioned demand forecasting or singular routing optimization. Critically, there remains a conspicuous absence of comprehensive, truly modular frameworks that comprehensively integrate advanced generative AI models across multiple interdependent logistics functions, particularly within the multifaceted context of multi-modal supply chains. Although hybrid VAE-GAN models have consistently demonstrated very good performance in adjacent fields like image generation, topological structure learning, and synthetic data augmentation [4, 26], their systematic application within the intricate domain of logistics and supply chain optimization is still notably limited. This underutilization primarily stems from several acute domain-specific challenges: (1) logistics datasets are often inherently sparse, imbalanced, and profoundly heterogeneous, comprising complex temporal, spatial, and categorical variables, which significantly complicates the training of sophisticated hybrid generative models compared to more structured domains like computer vision [7, 32]; (2) GANs, despite their generative power, frequently demand substantial computational resources and meticulous hyperparameter tuning, which can impede their practical adoption in operational supply chain contexts where interpretability and strict real-time constraints are paramount; and (3) a notable dearth of standardized benchmarks and universally accepted evaluation metrics specifically tailored to the unique demands of logistics optimization further inhibits a comprehensive assessment of hybrid generative models in this domain.

2.9 Synthesis and Identification of Research Gaps

Previous sections have provided a review of multi-modal supply chains and their evolution over time, the theoretical insights into logistics optimization and the scientific fundamentals of generative AI, and the present uses of AI in the field of logistics. This extensive academic discussion shows a number of critical issues that require original approaches and thus marks key unexplored avenues for research which this research will seek to address.

Firstly, classical procedures for optimization such as MILP and heuristics (as described in: Sect. 2.2 -classical methods), are powerful techniques for static and deterministic cases but essentially fail to address the real-time adoption, the high level of uncertainty, and the scale when it comes to multi-modal logistics of today's need. Their drawbacks in terms of real-time resiliency and elasticity restrict their applicability for responding to unpredictable disruptions, demand variation, and volatile capacity alteration. Although more flexible learning methods, like RL, look promising, they suffer from the problem of convergence stability, computational demand, and the necessity for large, curated training datasets to work properly [17].

Secondly, despite the growing capabilities of individual generative AI models (as discussed in Sects. 2.5 and 2.6), their standalone application presents distinct limitations in the context of multi-modal logistics optimization. Generative Adversarial Networks (GANs), while proficient in synthesizing high-fidelity data, are often plagued by training instabilities and mode collapse, which can lead to a lack of diversity in generated scenarios. Conversely, Variational Autoencoders (VAEs), though offering second, while the capabilities of any single generative AI model are increasing as shown in Sects. 2.5 and 2.6, its independence has clear limitation for multi-modal logistics optimization. Although Generative Adversarial Networks (GANs) are capable of generating high-fidelity data, such models generally suffer from training instabilities and mode collapse that results in poor diversity of generated scenarios. On the other end, Variational Autoencoders (VAEs), which have stable optimization and interpretable latent spaces, may occasionally yield outputs without the sharp realism required for more complex logistical simulations.

Importantly, there is a general and meaningful missing link involving the creation of embedded, modular generative AI frameworks designed for the optimization in multi-mode supply chains. Existing uses of generative AI in logistics (refer to Sect. 2.8) tend to emphasize individual tasks, e.g., combining demand forecasting with routing in a local area. A lack of such generic frameworks can be observed, which are designed to flawlessly integrate the advantageous features offered by the various generative modes, while being thoroughly tested and validated in a dynamic simulation ground, to meet the multi-dimensional challenges in logistics optimization, freight exchange, and dynamic routing, all at the same time. The feature-wise hybridization of VAE-GAN models, however, is substantially less explored than in other complex domains, where impressive success has been documented [4, 26], due to the peculiar challenges of the logistics domain, such as the natural sparsity, imbalance and heterogeneity of logistics data [7, 32], and the high demands from the computational and interpretability point of view.

This research tackles the two identified gaps by introducing and empirically substantiating a new, hybrid, integrated VAE-GAN framework. Specifically, training and interpretable latent spaces, may sometimes produce outputs that lack the sharp realism important for complex logistical simulations.

A pervasive and significant research gap exists concerning the development of integrated, modular generative AI frameworks specifically designed for comprehensive optimization within multi-modal supply chains. Current applications of generative AI in logistics (Sect. 2.8) often focus on isolated tasks, such as specific demand forecasting or localized routing. There is a noticeably absence of comprehensive frameworks that smoothly combine the strengths of different generative paradigms and are strictly validated within a dynamic simulation environment to address the multifaceted challenges across logistics optimization, freight exchange, and dynamic routing simultaneously. The hybridization of VAE-GAN models, despite its proven efficacy in other complex domains [4, 26], remains conspicuously underutilized in multi-modal logistics due to unique domain-specific challenges, including the inherent sparsity, imbalance, and heterogeneity of logistics datasets [7, 32], alongside the demanding computational and interpretability requirements.

This research directly addresses these identified gaps by proposing and empirically validating a novel, integrated VAE-GAN hybrid framework. Specifically:

(a) Seeks to design and train a VAE-GAN hybrid model capable of effectively learning from and optimizing complex logistics decisions in multi-modal environments (addressing **RQ1**). This directly tackles the limitations of standalone generative models and the need for a whole approach.
(b) Evaluates the framework's performance against traditional optimization methods in a simulated environment to quantify improvements in key performance metrics (addressing **RQ2**). This fills the gap in comprehensive, simulation-based validation of generative AI in this domain.
(c) Examines the critical sustainability and practical implications of adopting such an AI-driven framework, with an emphasis on environmental objectives and resilience against disruptions (addressing **RQ3**). This contributes to the broader discourse on ethical and practical AI implementation in logistics, addressing the real-world constraints and opportunities.

This work addresses the theoretical gap between generative AI technology advancements and experimental practice of a comprehensive sustainable and resilient multimodal supply chain operations. It does this by offering new insights on a unified VAE-GAN model and the resilience of the proposed network trained on an changing simulation.

2.10 *Sustainability Dimensions in Multi-modal Supply Chains*

The emerging global concern for climate change, resource efficiency, and corporate social responsibility has radically redefined the dialogue around supply chain management. As a result, sustainability has become an unavoidable issue in today's logistics, where it was no longer sufficient to consider purely environmental aspects, but to consider a variety of economic, environmental, and social dimensions in the round [5]. Multi-modal transportation networks inherently raise benefits to sustainable logistics, but complete sustainability does not occur on its own but the result of a coordinated and integrated effort.

One well-known structure for the conceptualization of corporate sustainability is the **TBL** model provided by [11]. The TBL concept supposes that real sustainability is insensible without an ideal balance of traditional financial profitability and equally critical ecological and social performance. These pillars of integration work together:

(a) **Environmental (Planet) Dimension**: The main focus of his pillar is reduction of the ecological load of the logistic operations. Key factors are minimization of greenhouse gas (GHG) emissions, specifically CO_2, energy consumption, waste generation, and pollution. Multi-modal transport inherently provides relative advantages in this respect, because moving freight from carbon-intensive road to lower-emitting modes like rail and sea can dramatically lower the overall environmental impact [12, 29]. The evolution of transport routes, mode structure, and transport capacity utilization is a critical challenge toward becoming environmentally sustainable in multi-modal networks.

(b) **Social (People) Dimension**: This dimension is about the effects of the supply chain on human welfare within the company and outside in its environment. Its scope includes issues like labor rights, worker safety, "good corporate citizens, fair access to logistics services, and the humane treatment of all parties." Less frequently as an explicit goal in the traditional optimization models of logistics but ensuring fair working conditions for drivers, reducing noise in residential areas and makes sure safe way of operations has not been established are social sustainability concerns.

(c) **Economic (Prosperity) Dimension**: This dimension involves the economic performance aspects of the supply chain, such as sustained profitability, cost-effectiveness, and attractiveness. In terms of sustainability, economic sustainability is considered not an end in itself, but the ability to create sustainable value that enables the other two pillars. This dimension is centering on the scope of reducing operating costs, maximizing resource utilization, increasing services levels, and promoting economic sustainability.

The balance between the three TBL dimensions is not always straightforward as multi-modal supply networks tend to contain trade-offs. For example, the most environmentally friendly mode of transportation (e.g., rail) could be associated with longer lead times (or more costly infrastructure up front), which can affect economic efficiency or service performance. Likewise, favoring societal welfare (e.g., no night deliveries in residential areas) might result in higher operation costs or longer lead times.

This combinatorial nature implies that we need sophisticated decision tools to work our way through these complex trade-offs. Generative AI models, such as the VAE-GAN hybrid model explored in this research, are particularly well-suited to capture the interconnected relationship between operational variables, environmental response, and financial implications. By creating optimal logistics scenarios accounting for such interdependencies, the framework is intended to enable better data-backed decisions pursuing a balanced and comprehensive sustainability in multi-modal logistics, as opposed to single optimization objectives and toward multi-objective, sustainable value provision.

3 Theoretical Framework

3.1 Foundations of Generative Modeling in Logistics

Sustainability-aware planning increasingly impacts today's logistics systems that are confronted with uncertainty, multimodal routing complexity. Conventional optimization techniques, such as MILP, does not work well in quickly developing operation contexts. Generative models offer a new solution to the problem by learning the distribution of complex data and generating new instances of scenarios while maintaining structural soundness [33].

3.2 Variational Autoencoders in Supply Chain Scenario Generation

Variational Autoencoders (VAEs) are latent-variable models that approximate the true data distribution by mapping high-dimensional logistics data to a probabilistic latent space. This enables the model to learn compressed yet meaningful representations of complex logistics scenarios—such as emissions, cost, volume, and time—and generate plausible variations thereof.

The VAE loss function consists of a reconstruction term and a regularization term:

$$\mathcal{L}_{\text{VAE}} = \mathbb{E}_{q_\phi(z|x)}[\log p_\theta(x|z)] - D_{\text{KL}}(q_\phi(z|x) \| p(z)) \tag{1}$$

where $q_\phi(z|x)$ is the encoder, $p_\theta(x|z)$ is the decoder, and D_{KL} represents the Kullback–Leibler divergence between the approximate and true latent distributions.

3.3 Generative Adversarial Networks for Structural Realism

Generative Adversarial Networks (GANs) comprise a generator $G(z)$ and discriminator $D(x)$ in a two-player minimax game. The generator attempts to produce samples indistinguishable from real logistics data, while the discriminator learns to separate real from generated samples. The GAN loss is:

$$\mathcal{L}_{\text{GAN}} = \mathbb{E}_{x \sim p_{\text{data}}}[\log D(x)] + \mathbb{E}_{z \sim p(z)}[\log(1 - D(G(z)))]. \tag{2}$$

GANs are particularly suited for enforcing realism in synthetic logistics scenarios by learning high-dimensional interactions across emissions, time, delays, and multimodal complexity [14].

3.4 Hybrid VAE-GAN Integration

To benefit the best properties of both models, in this work, we use a hybrid VAE-GAN model, where the decoder of the VAE acts as the GAN's generator. The overall loss is a weighted sum:

$$\mathcal{L}_{\text{total}} = \mathcal{L}_{\text{VAE}} + \alpha \cdot \mathcal{L}_{\text{GAN}}. \tag{3}$$

Here, α is a controllable coefficient to balance the fidelity of reconstruction and adversarial realism. This hybrid of deterministic and probabilistic approaches lets the system maintain the feasibility of scenarios (e.g., plausible volume/cost trade-offs) while learning to produce context-sensitive disruptions and routing patterns [2].

3.5 VAE-GAN Hybrid Engine Architecture

The model consists of several Python-written programs PyCharm built on this theory. The engine is comprised of:

- **Synthetic Generator**: Creates scenario data with emissions, cost, multimodal legs, disruptions
- **Encoder–Decoder**: Compresses and reconstructs features in the latent space
- **Discriminator**: Enforces realism through adversarial training
- **Feedback Loop**: Enables dynamic scenario learning via simulation integration.

The architecture has been implemented using modular files such as `vae.py`, `gan.py`, `vae_gan.py`, and `train_optimization.py`, ensuring reproducibility and scalability for future extensions like freight exchange and dynamic routing.

4 Research Methodology

This chapter offers complete research methodology of this research which includes both theoretical and technical base. The format is based on an academic discipline which is logistics, SCM, and AI. Philosophical paradigms, data strategies, model architecture, simulation setup, and evaluation mechanisms are kept in one, which help to justify the VAE-GAN generative framework for multi-modal logistics optimization.

4.1 Research Philosophy and Design

4.1.1 Research Model

The underlying research model in this study is **positivist**. According to Saunders [27], "Positivism holds that a unified objective, measurable, empirical reality exists and seeks that to be described as accurately as possible using methodical observation," which is also consistent with our purpose to develop, implement, and test an AI-driven optimization system that is grounded on tangible logistics data. This framework can be used for bringing formal modeling, statistical testing, and simulation-based analysis to assess the proposed system in a replicable way [27].

4.1.2 Research Approach

The criteria used is **deductive** and **explorative**. It derives from well-defined concepts in supply chain optimization, such as MILP and AI-based modeling, and experiments with novel VAE-GANs applications for freight exchange and dynamic routing. The deductive side aims to benchmark plan synthesis tools with respect to MILP; in contrast, the inductive side looks at how generative AI can automatically create realistic logistics scenarios in a multi-modal fashion.

4.1.3 Methodological Framework

The research design is **quantitative** and **simulation-based**, incorporating both numerical modeling and experimental validation. The core phases include:

(a) Data collection and synthetic generation
(b) VAE-GAN model design and training
(c) Simulation of logistics scenarios
(d) Benchmarking against MILP and RL baselines
(e) Evaluation using technical and sustainability KPIs.

4.2 Data Sources and Collection Methods

4.2.1 Open-Source Logistics Datasets

Public logistics data was examined and partially and very limited used to benchmark outputs from model, e.g., shipping patterns, routing paths, emission factors. These sources include the European Freight Transport Database.

4.2.2 Synthetic Data Generation

The synthetic data and associated scripts are included under `syn` folder. The main training data was created using the `synthetic_generator.py`script. It models logistics transports, by road, rail, and sea, including metadata like:

- Emissions and carbon penalties
- Multi-modal route duration and costs
- Disruption and delay patterns.

These scenarios were essential for simulating complex, rare, or sustainability-driven events not found in public datasets [3, 32].

4.2.3 Preprocessing

To prepare the data for model training:

- Missing values were imputed; categorical variables encoded
- Continuous variables normalized
- Feature vectors formatted into PyTorch tensors.

4.3 Generative AI Model Development

4.3.1 Architecture Overview

The model combines a VAE and a GAN, reusing the VAE decoder as the GAN generator. It is used in the following scripts:

- `vae.py`: Encoder/decoder logic
- `gan.py`: Generator and discriminator
- `vae_gan.py`: Hybrid loss and training pipeline.

4.3.2 Mathematical Formulation

VAE Loss:

$$\mathcal{L}_{\text{VAE}} = \mathbb{E}_{q_\phi(z|x)}[\log p_\theta(x|z)] - D_{\text{KL}}(q_\phi(z|x)\|p(z)). \tag{4}$$

GAN Loss:

$$\mathcal{L}_{\text{GAN}} = \mathbb{E}_{x\sim p_{\text{data}}}[\log D(x)] + \mathbb{E}_{z\sim p(z)}[\log(1 - D(G(z)))]. \tag{5}$$

Total Loss:

$$\mathcal{L}_{\text{total}} = \mathcal{L}_{\text{VAE}} + \alpha \cdot \mathcal{L}_{\text{GAN}}. \tag{6}$$

4.3.3 Training Workflow

The training process includes:

(a) VAE pretraining (reconstruction and KL loss)
(b) Adversarial training using GAN loss
(c) Joint tuning of combined loss using `train.py`.

4.4 Simulation Environment and Scenario Design

A discrete-event and agent-based simulation environment was implemented using PyCharm with modular Python files such as `simulation_engine.py`, `resilience_testing.py`, and `feedback_loop.py`. Scenarios include:

- **Baseline**: Traditional logistics
- **Disruption**: Road closure, port congestion
- **Sustainability**: Prioritize green transport
- **High-demand**: Stress test during peak conditions.

4.5 Evaluation and Benchmarking

4.5.1 Quantitative Metrics

- Total logistics cost (€)
- Transit time (hours/days)
- Emissions (kg CO_2)
- Empty returns penalty
- Feasibility and realism scores.

4.5.2 Generative Model Metrics

- FID (Frechet Inception Distance)
- MSE (VAE reconstruction)
- Discriminator accuracy
- Mode Coverage Rate (MCR).

4.5.3 Baseline Comparison

Performance was benchmarked against:

- MILP using deterministic constraints
- Q-learning-based reinforcement learning for routing.

4.6 Ethical and Sustainability Considerations

The study complies with academic ethical standards:

- **Data Privacy**: No PII used; synthetic data ensures GDPR compliance
- **Reproducibility**: All source code, model weights, and notebooks available
- **Sustainability**: GPU resource optimization and emissions monitoring were implemented.

It adheres to the Triple Bottom Line (3P)—*People, Planet, Prosperity* [11].

4.7 Real-World Data Integration and API Infrastructure

Though the current design mainly focuses on synthetic datasets produced by the `generate_synthetic.py` script, it has been designed that in future versions of the framework to include modular tools to consume data from the real-world. This design decision allows for generalization and relevance in different logistics environments.

To make this possible, an extensible thin API layer has been developed with Flask (located under `api/endpoints.py`). This is a real-time module, enabling the capture and transference of the operational data from logistical platforms, TMS (transportation management systems) public API's for freight. Although not physically used in the current context, placeholders for e.g., `/predict` and `/scenario_upload`endpoints are present.

Real-life datasets, for example truck-level emissions, shipment risk information, and delivery delays can greatly enhance the context of the generated scenarios. Public repos and freight data services such as Freightos, Convoy, or EU Open Transport Data are other sources to be added later. Encoder-Decoder-Based Model: The encoder-decoder architecture of the model allows dynamic batch ingestion and so is flexible in coping with different formats and schema.

The integration of real-world data is also in accordance with advising literature to validate at operational conditions from current literature [28, 34]. We also demonstrate the (strategic) enactment of ESG-compliance with and in digital-based supply chain ecosystems.

4.8 Summary

This chapter has presented a very strong and replicable approach that combines logistics domain knowledge with generative AI and simulation-based analysis. It rationalizes the choice of synthetic data, describes the architecture of the VAE-GAN, validates through key performance indicators, and complies with ethical guidelines. Chapter 6 will present how the system was implemented and what result came out of it.

5 Overview of the Proposed Generative AI Framework

This section aims to provide deeper understanding of the proposed Multi-modal Logistics Generative AI Framework by presenting the relationship of the core modules and the conceptual architecture. Developed to support intelligent decision-making in dynamic and complex supply chain settings, the model smoothly connects advanced generative models with a simulation-based validation system. The framework accounts for the full cycle from data creation and ingestion to model training, simulating, and assessing performance. Its modular design in Fig. 1 allows to scale up, maintain, and extend it. It is convenient for academic research and also to be possibly deployed in industry in the future.

5.1 Overall Conceptual Structure and Interaction Flow

The Generative AI Logistics Framework is designed as a flexible multi-modular system that consumes changing logistics data, trains on complex patterns using a hybrid VAE-GAN model, producing optimal scenarios, and validates from simulations. High-level organization is organized in the fashion of a pipeline, i.e., the output of one stage feeds the input of the next. The core components, as shown in the framework's directory structure (`VAE_GAN_Logistics_Framework/`, as per Module Documentation , support a closed-loop learning and optimization process. Key interactions include:

(a) **Data Flow Initialization**: It starts with the Synthetic Data Generation (Module 1) where realistic, time-varying logistics datasets are generated to specific use case scenarios. This data is subsequently processed by the Data Loading & Preprocessing (Module 2), deployed to clean, preprocess, and partition data.
(b) **Model Training Integration**: The preprocessed data can then be used to train the models in Model Definitions (Module 4), where the main VAE-GAN hybrid architecture resides. The Training Pipelines (Module 6) schedule the multi-stage training for the VAE-GAN model in which the model starts to generate logistics scenarios that are optimized and also to learn more complex distributions of the data.
(c) **Simulation-Based Validation**: The trained generative AI model produces outputs to the Simulation Environment (Module 7). This environment runs specified logistics cases (i.e., baseline, disruption, high-demand, and sustainability) to enable the realistic validation of the framework's control performance.
(d) **Feedback and Iteration**: Importantly, the framework includes a feedback mechanism (within Module 7, and supported by Utilities, Module 10), where simulation results and performance statistics can guide modifications to the model and scenario design. This online learning strategy makes the framework more adaptive.
(e) **Evaluation and Benchmarking**: The Evaluation and Benchmarking (Module 8) module calculates KPIs and compares to standard optimization solutions so that quantification of the framework's strength can be provided.
(f) **Configurability and Maintainability**: The other modules (Global Configuration, Module 3) and (Utilities, Module 10) provide a centralized control on the parameters, the logging, and on the state of the models, providing reproducibility and extensibility on the whole workflow.

Such an interlinked architecture allows the framework to be an intelligent decision-support system, to be adaptable to diverse logistical constraints and to refine its optimization capabilities as the scenario unfolds.

5.2 Logistics Optimization Module

The Logistics Optimization Module constitutes a central smarts in the system, which is activated by the **VAE-GAN hybrid model** (`src/models/optimization/vae_gan.py`). It aims at discovering complex patterns in multi-modal logistics data and providing users with optimal operational scenarios for effective resource allocation. Unlike most of the classical deterministic optimizers, this module takes advantage of the VAE-GAN's generating properties to:

- **Generate Novel Logistics Scenarios**: Generate variety and realistic logistic configurations like best modal split, fleet allocation, and routing options, and novel challenging transportation scenarios including extreme events. This is done by sampling from the learned latent space and generating the respective logistics parameters.
- **Capture Complex Interdependencies**: By the deep learning structure, the VAE-GAN could recognize and model the nonlinear format among business logistics variables (e.g., how the weather affects the transit time and CO2 emissions at the same time), leading to more insightful optimization solutions.
- **Support Multi-objective Optimization**: By recognizing these trade-offs present in the data, the model can guide decisions that optimize between competing interests, for example, minimizing cost while minimizing carbon.

The output of the module translates directly in controlled simulation environment for validation and actionable knowledge for logistics planners. It is adaptive, meaning that it could learn from nonstationary inputs, which is especially advantageous when real-time changes were being made.

5.3 Freight Exchange Module

Freight Exchange Module(`src/models/placeholders/freight_exchange.py`) in this architecture is considered to be a key ingredient in advancing the effectiveness, and speed of multi-modal freight matching. As existing freight platforms are generally passive in action, we consider that the module ought to have an intelligent AI capability for active matching.

Current Implementation and Scope: As discussed in the scope of the research (Sect. 1.5) and in the module documentation, the implementation of the module is largely a **placeholder**. Its proximate role in the framework is showing the *integrating point* to mature, future AI-assisted freight matching algorithms. Although it potentially has reduced, rule-based logic for testing framework compatibility, its empirical refinement with state-of-the-art AI in optimization for real-time freight exchange is beyond the scope of the work outlined in this research. With the addition of this module, the completeness and openness of the framework are highlighted.

Future Development: In the future, more sophisticated AI models (e.g., reinforcement learning, graph neural networks, or higher-order generative models) for real-time capacity, demand, sustainability criterion, network condition depend on dynamic matching of shipper-carrier pairs that will be available in the market which will change the way today's logistics and supply chain is.

5.4 Dynamic Routing Module

The Dynamic Routing Module (`src/models/placeholders/dynamic_routing.py`) is a critical step toward efficient and dynamic route optimization in multi-modal supply chains. Conventional routing methods find it challenging to handle real-time disruptions, traffic fluctuations, and complicated intermodal transfers. This module is meant to use AI to make adaptive routing decisions.

Current Implementation and Scope: Like the Freight Exchange Module, the status of the current DR module implementation in the developed framework is mostly as a conceptual placeholder. Though it may contain predefined routing logic to maintain integrity of framework during simulation, its thorough empirical validation through more powered real-time routing optimization is left for future research initiatives (as discussed in Sect. 1.5)

Future Development: Further research will then be dedicated to the development and incorporation of state-of-the-art AI models (e.g., deep reinforcement learning, predictive analytics for traffic/disruptions, graph-based routing algorithms) that are able to optimize on the go new routes that are influenced by live data feeds, unexpected events, and changing network conditions (including multi-leg journeys and interchanging).

5.5 Unified Framework Interaction Mechanisms

The smooth functioning process with which the Generative AI Logistics Framework operates depends heavily on strong interaction mechanisms which regulate the exchange of data and control between its various modules. This structure means that the framework acts as a unified system, not a grab bag of tools. Primary interactions are as follows:

- **Configurable Parameters and Hyperparameters**: Centralization in `src/config.py` (Module 3) maintains universality of global settings, environment variables, model hyperparameters and simulation variables across all the modules. This forages for reproducibility and scientific investigation.
- **Standardized Data Interfaces**: Data output by the Synthetic Data Generator (Module 1) is persisted in flat text files in standard CSV format under a `data/processed/` directory, to allow it to be smoothly consumed by both the Data Loader and Preprocessor (`src/data/loader. py`, `src/data/preprocess.py` - Module 2). Processed data is then shared between the Training Pipelines (Module 6) and Simulation Environment (Module 7).
- **Modular API and Function Calls**: Inter-component (loading data for training the VAE-GAN, feeding the output of the VAE-GAN into the simulation engine, obtaining metrics from the evaluation module) interactions are handled using explicit function calls and well-defined interfaces in the Python base. This in turn enables

a loose coupling and makes possible the updating or the substitution of single modules.

- **Feedback Loops for Adaptive Learning**: The simulation environment (Sect. 7) is intended to give feedback regarding the performance of generated scenarios under different circumstances. Such feedback could be used to the training pipelines, which allows the VAE-GAN model to learn and optimize its generative skills on the empirical topic, leading to continuous improvement and dynamic re-calibration.
- **Logging and Monitoring Integration**: The `utils/logger.py` (Module 10) logs in a structured manner all important actions, data streams, models performances, and errors. This offers traceability, helps debugging, and allows thorough post-hoc analysis of framework behavior.

Together, these mechanisms form the basis of the framework's capability to act as an intelligent, self-optimizing system, enabling the examination of a variety of implications of generative AI in a multi-modal logistics context.

5.6 Generative AI Framework Overview

Based on the combination theory of Resource Based View (RBV) and Capabilities, the study puts forward a modular Generative AI Framework for multi-modal supply chain operation optimization. The architecture incorporates state-of-the-art generative models, GANs, and VAEs, to improve decision processes in terms of logistics resource allocation, freight exchange capability, and dynamic routing adaptation.

The framework's goal is to bridge the gap between classical optimization and the data-driven, real-time logistics requirements of today. The proposed approach aims at providing a real-time adaptability, better operational efficiency, and a better alignment with the sustainability objectives through the integration of generative AI competences into the operational design of multi-modal supply chains. The proposed framework is based on the conceptual structure with the following three main functional elements:

(a) **Logistics Optimization Module**: Responsible for optimizing allocation and scheduling of resources in multiple transport modes through generative AI-driven methods.
(b) **Freight Exchange Module**: Provides a way to effectively and sustainably match shippers and carriers and uses generative models to automate and optimize freight exchange determination.
(c) **Dynamic Routing Module**: Solves the problem of real-time adaptation through the real-time adaptation of the route decisions to the changing network operational state.

The proposed modules are complemented by a simulation environment which allows empirical studies of the performance of the framework, under a variety of operational scenarios (i.e., normal operation, disruption events, and sustainability-oriented conditions). The simulation platform allows to evaluate the performance of the framework with respect to standard optimization methods like MILP and reinforcement learning approaches.

This offers a visual representation of the created generative AI framework, showing how the autonomous units interact with the data-driven decision support systems.

This framework is motivated by the recent developments of generative AI applications in logistics and other domains. It is noteworthy that [3] proved the applicability of the GAN-based simulation into the supply chain optimization and [6, 18] showed the great potential of VAE-based models to predict demand and vehicle trajectory, respectively. Hybrid VAE-GANs have shown some impressive results for improving generative models' robustness and interpretability [4, 26]. This work extends these contributions by combining them together in a coherent framework, dedicated to the specificities of the multi-modal management along a sustainable supply chain.

5.7 *Framework Components*

The generative AI framework is based on three components: the **Logistics Optimization Module**, the **Freight Exchange Module**, and the **Dynamic Routing Module**. The components solve individual operational problems in multi-modal supply chains and also jointly support the efficiency, resilience, and sustainability of transportation networks. This modular structure is in lines with the theoretical framework presented above—considering digital resources strategically and also taking real-time capabilities into account in state-of-the-art logistics systems.

5.7.1 Logistics Optimization Module

The Resource Allocation and Scheduling over multi-mode Transportation Logistic Optimization Module is in charge of optimizing the allocation of resources for scheduling over multi-transport modes. This module is built on top of generative AI models (GANs and VAEs), it uses historical logistics data, capacity limits, and demands predictions to create paths optimizing multi-modal transportation. In this, way the module seeks minimizing operating costs that its trip time and asset utilization increase. This module features the following functions:

- Allocation of freight volumes to appropriate transport modes (road, rail, sea, air) based on cost, time, and sustainability considerations.
- Scheduling of shipments and fleet resources to ensure efficient utilization.
- Optimization of intermodal transfer points and route selection.

The generative AI engine housed in this module is pre-trained to model different logistics situations and generate new, data-driven recommendations that are significantly more efficient than traditional planning approaches. The feasibility of integrating GAN-based simulations for supply chain optimization is confirmed in previous research, such as Blessing [3], which provides indirect evidence of the relevance of the module design.

5.7.2 Freight Exchange Module

The Freight Exchange Module uses an algorithm for connecting hosts with freight movers in as friendly and economical manner as possible. Current conventional freight exchange systems are using static rules and manual procedures, which results in undesired under-utilization of carrier capacity and empty miles. By contrast, this module uses the generative AI techniques to monitor and analyze real-time data, and historical transactions to automatically make optimal match. Key features of this module :

- Real-time matching of freight demand and carrier capacity.
- Optimization of freight consolidation opportunities to minimize empty trips.
- Consideration of sustainability metrics, such as carbon emissions, in the matching process.

Through optimization of the performance of freight exchange operations, this module helps lessen the environmental impacts and improve the adaptability of multi-modal logistics systems. Recent work such as [35] demonstrated that GAN-based models are promising for improving freight transaction reliability and risk profiling and it inspires the design of this component.

5.7.3 Dynamic Routing Module

The Real-Time Dynamic Routing Module focuses on the requirement for immediate response in multi-modal supply chains. Logistics systems are often subject to disruptions, such as traffic jam, extreme weather, system accident, demand fluctuation, that can have a great influence on the delivery performance. This module consists of generative AI models in which routing decisions are adjusted depending on a state of the system. The principle functionalities of this module are:

- Real-time route generation and adjustment based on traffic, weather, and demand data.
- Identification of alternative routes and transport modes to mitigate disruptions.
- Optimization of routing decisions with respect to cost, time, and sustainability criteria.

By including this module, the logistics system is guaranteed to maintain its resilient and agile nature and be able to effectively adjust to unforeseen circumstances., while achieving the expected service quality and sustainability goals. Existing works [6, 23] have validated that generative AI models can accurately predict the flying trajectories and passenger demands in real time, and hence the module is empirically based.

Taken together, these three elements serve to materialize the theoretical basis we developed, by integrating AI-based functionalities into the decision-making of multi-modal supply chains. The modular architecture of the framework allows for scalability, flexibility, and sustainability of the model with future trend.

5.8 Freight Exchange Module Design

The freight exchange module has the capability to emulate digital marketplace operation to match shippers to carriers according to cost, availability, emissions profile, and service constraints. This element resembles actual exchange platform like TIMOCOM or Transporeon to simulate the competitive bidding, auction-based pricing, and contract selection.

Planned features include:

- **Offer Matching**: Using multi-criteria scoring (e.g., cost, carbon impact, capacity), the model ranks offers and selects optimal pairings.
- **Bid Simulation**: Simulated carriers submit bids influenced by stochastic utility functions and policy rules.
- **Reinforcement Feedback**: Outcomes (e.g., delayed delivery, contract breach) are fed back into agent behavior policies in future epochs.

This command is realized as a placeholder (`freight_exchange.py`) in the codebase and it is forward-compatible to be consumed by synthetic/real-time loadboards using RESTful API refactoring.

5.9 Dynamic Routing Module Design

This dynamic routing modhole allows to extend classical vehicle routing problem (VRP) model with stochastic real-time constraints such as traffic jams, ecological fees, and unforeseen changes in delivery. Crafted as an open simulation interface, it incorporates scenario streams from VAE-GAN model and optimizes routes based on user-defined heuristics.

The following methodological components are intended:

- **Heuristic Optimization**: Incorporates adaptive large neighborhood search (ALNS) and greedy improvement layers to explore route permutations.

- **Constraint Injection**: Handles constraints like time windows, carbon caps, and incident risk.
- **API-Driven Updates**: Designed to interface with external traffic or weather feeds via the placeholder API layer for real-time routing suggestions.

Even though the module is serving as a (`dynamic_routing.py`) the easy-to-integrate script thanks to the object-oriented design allows the simulation of sophisticated logistics scenarios under changing circumstances.

5.10 Multi-agent Interaction and Decentralized Optimization

While the present model takes centralized optimization for routing and scenario generation, this simply does not exist in practice where various autonomous agents with different objectives and degree of information access (e.g., shippers, carriers, brokers or regulatory bodies,) perform the logistics operation.

In order to incorporate the complexity, the architecture is further-extensible with the Multi-agent Systems (MAS) or decentralized optimization methods. These could involve:

- **Carrier–Shipper Coordination**: Matching routes and load offers via negotiation or auctions.
- **Regulatory Feedback Loops**: Penalizing emissions or route violations dynamically.
- **Game-Theoretic Routing**: Modeling competitive behavior using concepts from Nash equilibria or Stackelberg games.

The dynamic routing module placeholder `dynamic_routing.py` is written as a repository for agent-based logic, so that distributed decision policies could be added in the future. Such enhancements could provide more realism, particularly in simulations with market-driven exchanges for freight as well as with government-prescribed routing behaviors.

5.11 Model Design and Training

The success of the proposed generative AI system for multimodal SC planning is inherently based on how the underlying AI models are designed and trained. The structure of technical architecture, data processing, and training procedure for the implementation of Generative Adversarial Network (GAN) and Variational Autoencoder (VAE) models embedded into the framework is also presented in this section.

5.11.1 Model Architecture

The general structure of the proposed model uses the advantages of both GANs and VAEs to cater the challenging data-rich requirements of logistics optimization, freight exchange, and dynamic routing. Specifically, we adopt the **hybrid VAE-GAN** design, which has been recently demonstrated to help enhance model performance, interpretation strength, and the quality of generation[4, 26]. Architecture: The architecture we propose is composed by the three main parts:

(a) **Variational Autoencoder (VAE)**: The key issue is to learn the latent structure of the historical logistics data such as shipment volume, mode of transportation, lead time, and demand pattern. Input data is encoded into a lower-dimensional latent space using an encoder network, and reconstructed from latent variables using a decoder network. This probabilistic formulation allows the model to account for uncertainty and variability in the data [19, 32].
(b) **Generative Adversarial Network (GAN)**: Consists of generator and discriminator networks trained in adversarial fashion. The generator synthetically creates logistic scenarios from the latent space representations learned by the VAE, while the discriminator determines the realness of the synthetic scenarios against reality data [13].
(c) **Integration Layer**: Manages the relationship between VAE and GAN, guaranteeing a smooth flow of data and consistency. This layer also allows aligning the scenarios generated with operational constraints and sustainability goals.

5.11.2 Data Preparation

The training of the VAE-GAN structure necessitates high-quality, comprehensive logistics datasets. Because commercial supply chain data are proprietary, a mix of open data and synthetic databases are employed in this study. The sources of data are: classic shipment data, freight exchange truck loads, traffic data, congestion data, and environmental impact indicators. (ii) Preprocessing: The following data preprocessing steps are applied:

(a) Data normalization and scaling to ensure consistent input ranges.
(b) Handling of missing values through imputation techniques.
(c) Encoding of categorical variables, such as transport modes and locations.
(d) Splitting of data into training, validation, and testing subsets.

5.11.3 Training Methodology

The training process for the VAE-GAN architecture is conducted in a two-stage approach:

(a) **Stage 1—VAE Pretraining**: Pretrained VAE component with the prepared logistics dataset for learning latent data representations. The training loss is composed of a reconstruction loss and the Kullback-Leibler divergence between the learned latent distribution and a standard Gaussian prior. Kingma and Welling [19].
(b) **Stage 2—Adversarial Training**: After VAE pretraining finished, the GAN part is trained on top of the pretrained VAE. The generator makes use of the latent representations to generate synthetic logistics scenes and the discriminator differentiates the real from the generated data. The objective of training is that we need to minimize the GAN like loss, which is a standard minimax loss.[13].

The hyperparameters (learning rate, batch size, latent space dimension, and so on) are selected by grid search and cross validation for the best performance of the model. The author conduct experiments in the PyCharm development environment with Python-based DL libraries such as TensorFlow and PyTorch.

5.11.4 Model Evaluation

The trained VAE-GAN model is evaluated both quantitatively and qualitatively. Quantitative metrics include:

(a) **Reconstruction Loss**: Measures the accuracy of data reconstruction by the VAE.
(b) **Discriminator Accuracy**: Assesses the ability of the discriminator to distinguish between real and generated data.
(c) **Freight Allocation and Routing Efficiency**: Evaluated within the simulation environment.

A qualitative assessment requires experts to judge the plausibility and practical applicability of the logistics scenarios produced. The performance results are used to gradually improve the architecture and training settings to conform to the conditions of actual logistics and sustainability.

5.11.5 Integration with Simulation Environment

In order to determine the practicability of the generative AI framework, the trained models will be used within a simulation environment, for instance integrated in the AnyLogic (or similar) simulation tool. With this integration, possibility is to evaluate the performance of the framework under real-world conditions such as traffic disturbances, demand peaks, and capacity limitations. The obtained simulation outcomes will give insights to the comparison between the framework and traditional logistics optimization approaches as described in the later chapters.

5.12 Interactions and Integration

The success of the suggested generative AI framework does not rely only in the performance of the individual components, but also in the smooth (and not disruptible) interaction and integration of them to form an integral decision-support system. This section describes the interaction between the Logistics Optimization Module, Freight Exchange Module, and Routing Optimization Module, in providing an integrated, real-time, and scalable logistics optimization service.

The integration architecture is constructed to enable the data to flow back and forth across the three modules, so the decisions made in one module respond to react as a result of the changes in the others. This is consistent with the Dynamife Capabilities view which suggests the importance of how logistics resources need to be able to adapt in real time in complex logistics systems, Teece et al. [30].

The interaction process is operationalized through the following key mechanisms:

(a) **Data Flow Integration**: The framework includes a central data store for storing both real-time and historical logistics data such as volume of shipped items, availability of carriers, status of the traffic on the routes, and environmental impact-related indices. This common data space guarantees that protocols are executed based on a current and coherent knowledge base.
(b) **Module Coordination**: The Logistics Optimization Module first develops schedules and establish allocations in a preliminary form from data provided. These plans are distributed to the Freight Exchange Module which uses generative AI methods to match shippers and carries in an optimal fashion. The Dynamic Routing Module then processes real-time routing decisions in light of the modulated response of the previous modules and ongoing operational state.
(c) **Feedback Loops**: The model contains feedback loops, and it is able to learn and adjust over time. For instance, routing interruptions detected by the Dynamic Routing Module cause freight scheduling and resource reassignment decisions to be updated. These feedbacks serve to increase the agility and the resilience of the logistics.
(d) **Sustainability Alignment**: Performance metrics, including carbon emissions and empty mileage, are incorporated as performance coefficients across all modules. This will help to ensure that the optimization decisions do not strictly conform to cost and efficiency actions, and respond to environmental and social sustainability goals [12].

The modular and interactive design of the proposed framework enhances its scalability, adaptability, and practical applicability in real-world logistics environments. By enabling continuous data-driven decision-making, the framework provides a good foundation for optimizing multi-modal supply chain operations under conditions of uncertainty and sustainability imperatives.

5.13 *Adaptive Responsiveness and Forward-Learning Potential*

Although the current pipeline is constructed with offline synthetic data generation and batch inference, it is architechturally ready for real-time logistics integration. The system has an `api/` layer which is made in Flask (see `endpoints.py`)to import data from a Transportation Management System (TMS), sensor network, or public freight APIs.

The `live_api_simulation.py` module provides emulation for input placement and routing in real time, making the framework ready for real deployment. Through these APIs, updates on shipments, networks, or disruptions are to be pushed to the VAE-GAN model as soon as they change, potentially leading to near-live responses and adaptive shifts in its generative scenarios or routing outputs.

For future developments the architecture could implement online learning such as Reinforcement Learning (RL) for the continuing adaptation of the system to changes in logistics conditions. In this framework, the generative engine could play the role of a scenario simulator in a more general RL loop from which policies are developed aiming at minimizing cost, emissions, or lead time under risk and uncertainty.

These features would mean that the framework could not only act as a predictive planner, but also as an adaptive decision-supporting tool in line with digital twin architectures as discussed in subsequent sections.

6 Implementation

This section describes how the abstract VAE-GAN Logistics Framework described above becomes an actual software. The framework was developed in Python with PyCharm and designed as a customizable modular installation. The framework is composed of scripts, models, APIs, and simulation engines organized so as to support updates and adding new features. Among the topics are the organization of the repository and how data is generated, model and workflow implementation, the interaction with Jupyter tools, simulation setup, and how the system remains reproducible.

6.1 *Repository Layout and Codebase Design*

The project is designed according to modular design principles and maintains the separation of the central tasks of data generation, model training, scenario simulation, and benchmarking. Here we will see a quick overview of the project structure and the primary directory that will be introduced through this project.`README.md`:

```
VAE_GAN_Logistics_Framework/
|-- .gitignore
|-- .idea/
|-- api/
|   |-- endpoints.py
|-- generate_tree.py
|-- notebooks/
|   |-- Benchmark_Analysis.ipynb
|   |-- EDA.ipynb
|   |-- Model_Evaluation.ipynb
|   |-- simulation_analysis.ipynb
|-- README.md
|-- requirements.txt
|-- simulation/
|   |-- feedback_loop.py
|   |-- live_api_simulation.py
|   |-- resilience_testing.py
|   |-- simulation_engine.py
|   |-- simulation_scenarios.py
|-- src/
|   |-- config.py
|   |-- dashboard/
|   |-- data/
|   |-- evaluation/
|   |-- models/
|   |   |-- optimization/
|   |   |   |-- gan.py
|   |   |   |-- vae.py
|   |   |   |-- vae_gan.py
|   |-- synthetic_data/
|   |   |-- generate_synthetic.py
|   |-- utils/
|       |-- checkpoint.py
|       |-- loss_functions.py
```

Each subdirectory contributes to a specific module of the generative AI framework. For example:

- `api/` provides a **RESTful** interface for external interaction.
- `notebooks/` contains Jupyter files for interactive exploration, training progress visualization, and benchmarking.
- `simulation/` contains scripts for scenario-based testing under real-world constraints.
- `src/models/optimization/` houses the VAE, GAN, and integrated VAE-GAN models.
- `synthetic_data/` includes datasets generated by `generate_synthetic.py`.
- `utils/` centralizes logging, loss functions, and checkpointing utilities.

All experiments are driven through centralized configuration parameters defined in `config.py`. The entire framework is intended to be fully reproducible, version-controlled via Git, and extensible via modular plug-ins for advanced components like dynamic routing and freight exchange.

6.2 API Interface and Endpoint Layer

To enable easy integration with external tools and extend capabilities, the proposed generative AI framework comes with a lightweight RESTful API built using the `FastAPI` library. This API acts as a gateway to offer selected features of the framework like creating synthetic scenarios running model benchmarks, and monitoring system health. The design makes it flexible and keeps different parts connected.

Right now, the API has a `/health` endpoint to check if the service is up. The setup allows room to add more endpoints in the future such as:

- `/generate`: Triggers the synthetic data generation engine and returns simulated logistics scenarios.
- `/benchmark`: Retrieves comparative results between the VAE-GAN model and baseline solvers (e.g., MILP).
- `/forecast`: Provides model-inferred scenario forecasts based on user-defined input conditions.

This design meets real-time deployment needs and allows modular integration to connect with downstream dashboards, agent systems, or digital twin platforms.

6.3 Synthetic Data Generation and Preprocessing

To make sure the framework stays realistic and manageable, the dataset came from synthetic generation using the`generate_synthetic.py`script in the `synthetic_data/`module. The generator creates simulations of logistics operations that include multiple legs and transport modes focusing on central European logistics hubs. Each synthetic dataset represents a distinct shipment example with up to three legs, choices of road, rail, or sea transport, and details about emissions per segment, risk factors, and service level limits.

The system saves the datasets as CSV files in the `synthetic_data/data/processed/` directory. These datasets include baseline activities periods of high demand, disruption cases, and policies focused on sustainability. Each row in these files lists some of the attributes:

- `origin, destination, via`: Location nodes and intermediate stops
- `mode1, mode2, mode3`: Transportation modes per leg
- `cost, time, emissions`: Aggregate and per-leg logistics KPIs

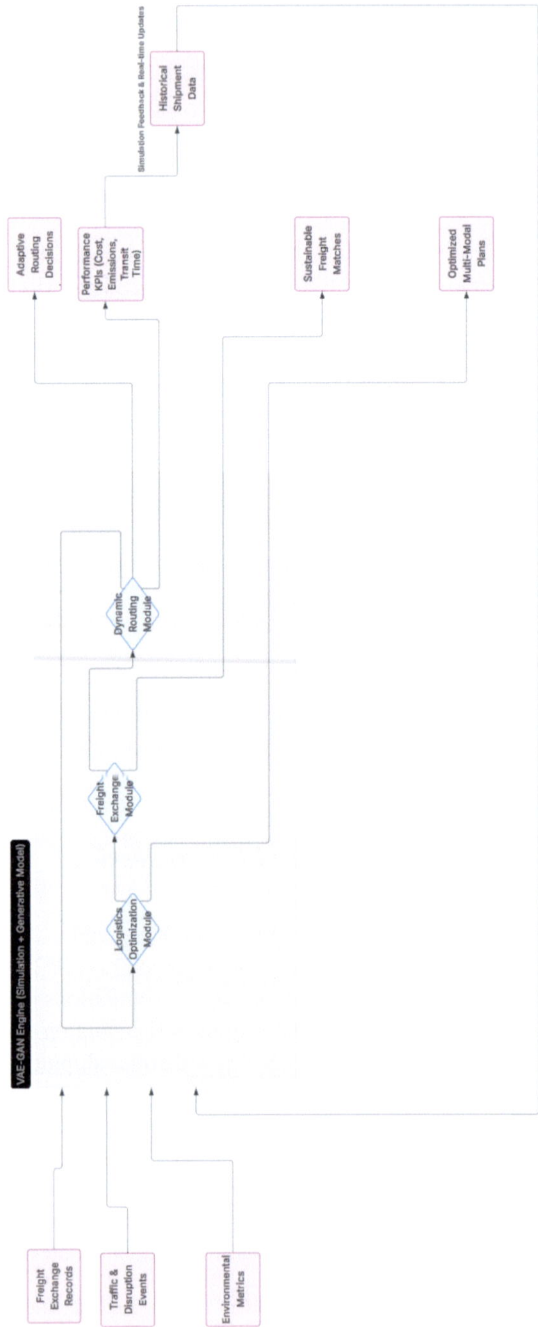

Fig. 1 High-level system flow of the VAE-GAN logistics engine

- `penalty, disruption, sustainability_index`: Scenario stressors.

The script located in `src/data/preprocess.py` sets up the data preprocessing workflow. It fills in missing data, converts categorical variables such as transport methods into encoded formats, adjusts numerical values, and transforms everything into tensors that PyTorch can use. This step ensures the data works with the VAE and GAN modules and keeps evaluations consistent for benchmarking.

This synthetic method adds complexity and nonlinear patterns on purpose. Traditional MILP-based optimization cannot handle this . The design proves how useful generative AI is in smart scenario planning. It lets the system create rare edge-case situations that empirical datasets do not include.

Figure 2 gives a clear picture of the synthetic logistics dataset showing how the structure looks, the variety of features it contains, and the multimodal shipment records included in the baseline scenario. They created over 100,000 shipment records across all three scenarios and used them to train the model and this is a snapshot of the dataset.

6.4 Volume Distribution Characteristics

A key part of checking synthetic logistics datasets involves confirming that important features align with real-world industry trends. Figure 3 presents the empirical distribution of shipment volumes, measured in metric tons.

It is observed from the histogram that the distribution is right-skewed. Though there is a cohort of apartment shipments that do fall into the sub 10T range. This corresponds to what we observe in freight systems where combination shipments are frequent, while fewer large or long-haul shipments, that take on higher values, extend over these higher values. The long tail makes sense, and it matters. It provides a way for the generative model to cover edge cases that push the limits of cost, emissions, or capacity. In general, the spread of data shows that the synthetic dataset mimics realistic variations found in multi-modal logistics activities.

6.5 Cost per Ton-Kilometer Under Scenario Variants

Beyond the volume, the cost efficiency is an important dimension of simulation in logistics. The distribution of cost per ton-km over the four synthetic scenarios is given in Fig. 4. This index measures the production efficiency by normalizing absolute cost to transport distance and shipment mass.

The findings clearly display divergences between the scenario's behavior. The baseline remains steady and constant, indicating a consistent operation. Disruption and sustainability policies produce a stronger right skew, which occurs as delays to

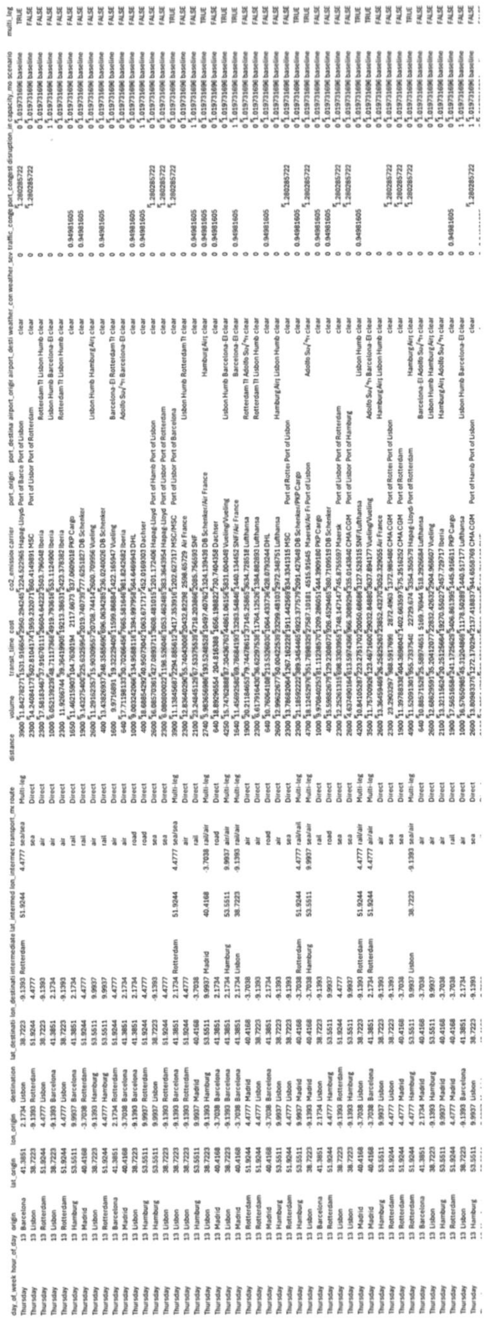

Fig. 2 Visual snapshot of the synthetic logistics dataset (baseline scenario)

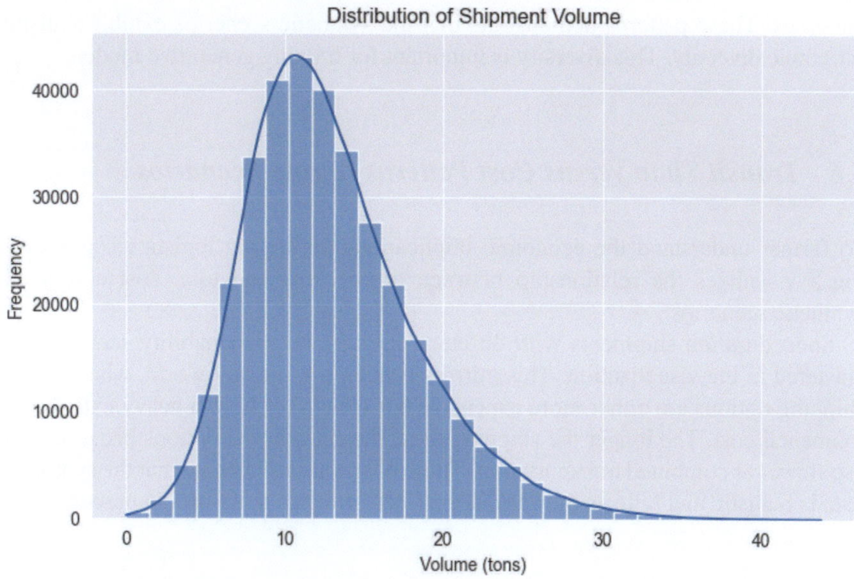

Fig. 3 Distribution of shipment volume

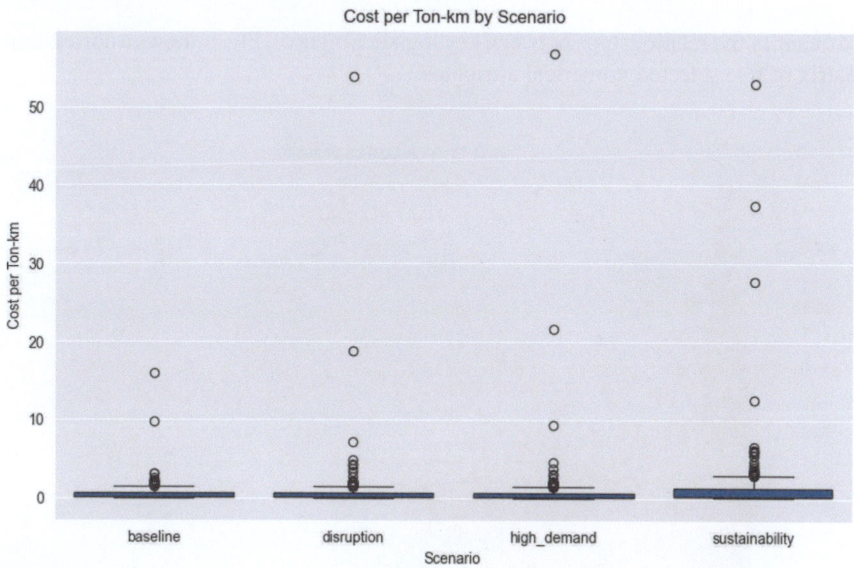

Fig. 4 Distribution of cost per ton-kilometer across scenario types

rerouting or emissions rules add more uncertainty. It is in high-demand situations where the interquartile range is pushed up into equivalence or mere-bond evasion territory that one sees the equivalent of the effects of hitting capacity limits or rate increases. These patterns demonstrate that the synthetic scenerios exhibit realistic economic diversity. This diversity is important for training generative models.

6.6 *Transit Time Versus Cost Patterns Across Scenarios*

To further understand the economic implications of varying logistics conditions, Fig. 5 visualizes the relationship between transit time and total cost across all synthetic scenarios.

Short duration shipments with different costs in the sustainability scenario are clustered in the visualization. This mirrors behavior in the real world, where faster shipping options are either more expensive (air freight) or have a heavy extra environmental cost. The longer the shipping time, the cheaper the options because they use slower or combined transportation. These observations indicate that the synthetic data is realistic and reflects the various cost-performance trade-offs in practice.

6.7 *Interdependencies Between Core Logistics Features*

To quantify the relationships between key logistics metrics, Fig. 6 shows a correlation matrix of five selected numerical attributes.

Fig. 5 Scatter plot of transit time versus total cost, segmented by scenario type

As can be seen in the matrix, there is a linear relationship between total cost and CO_2 emissions. It shows how greater energy consumption is a source of environmental strain. This is because the cost per ton-kilometer correlates negatively with transit time, so longer trips will usually make use of cheaper modes of transportation. Volume does not scale with other factors, suggesting that the synthetic dataset has a mixed setup rather than a simple re-scaling argument.

6.8 Scenario-Based Transit Time Distributions

To understand how different logistics conditions affect service efficiency, we look at the transit time distributions in all four scenarios. Figure 7 shows separate histograms of shipment transit times for each scenario recorded in hours.

The histograms reveal right-skewed trends across all cases. Most shipments arrive in under 500 h. The *disruption* scenario however, stretches the upper tail showing more shipments with delays. Both the *high demand* and *baseline*scenarios maintain similar patterns. Policies aimed at *sustainability* shift the distribution toward shorter transit times.

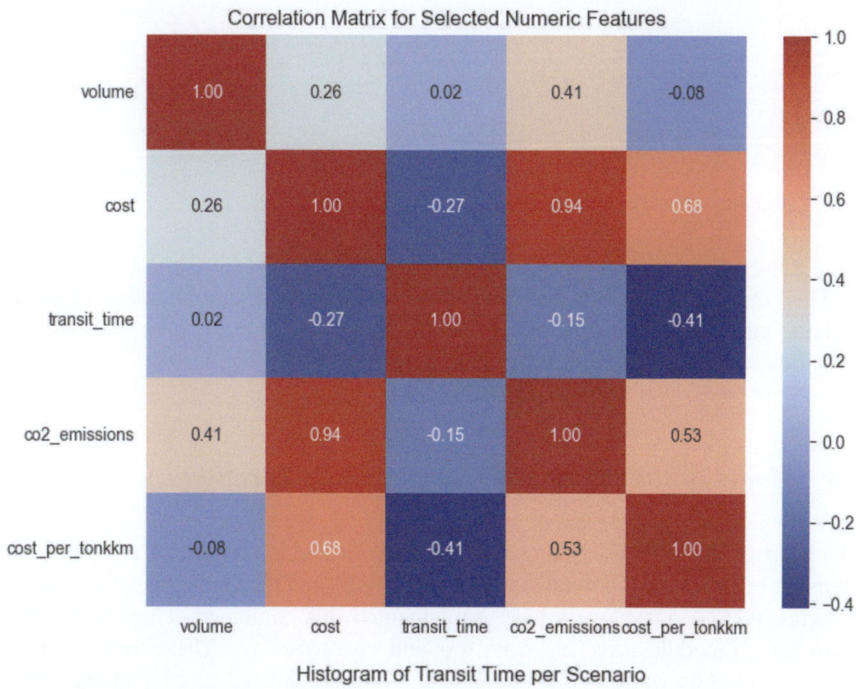

Fig. 6 Correlation matrix of selected logistics features

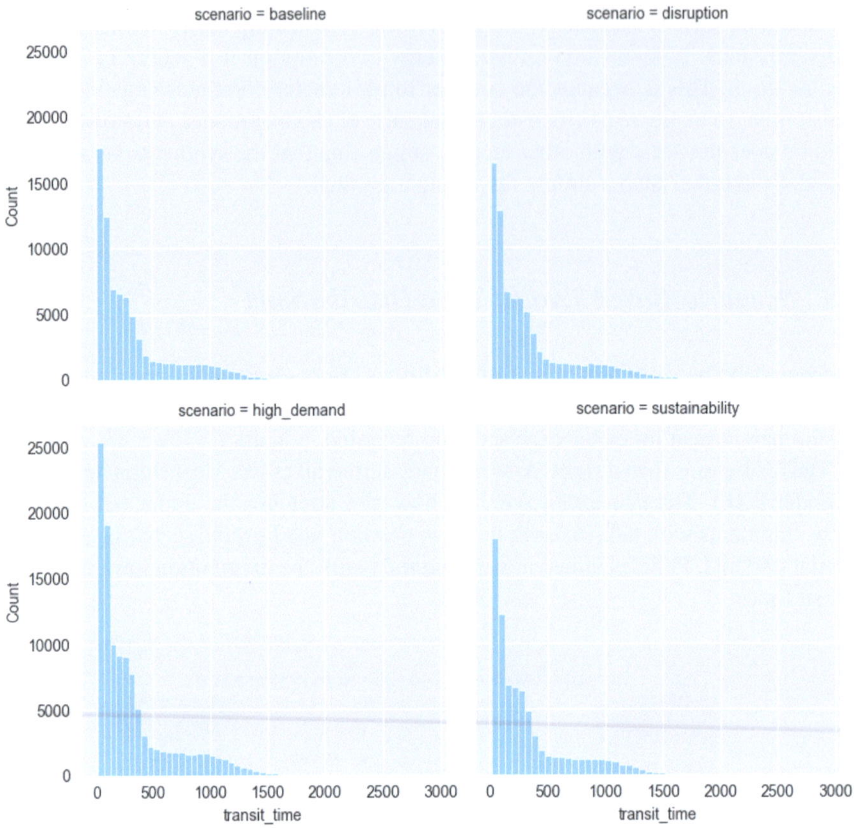

Fig. 7 Histogram of transit time by scenario

This perspective supports the idea that using synthetic generation follows specific scenario rules and actions. It allows people to test for stress and explore what-if scenarios in many different supply chain situations.

6.9 Model Architecture and Component Integration

The heart of the VAE-GAN Logistics Framework lies in combining two key generative models: the Variational Autoencoder (VAE) and the Generative Adversarial Network (GAN). Developers have implemented this combined setup using three main Python modules`vae.py`, `gan.py`, and `vae_gan.py`. These modules use a design focused on working together keeping things organized, and being reusable.

VAE Component (`vae.py`)

The VAE class implements a symmetric encoder-decoder structure, and employs the reparameterization trick for the latent-variable sampling. Features: Some key elements of design include:

- `Encoder`: A linear stack with ReLU activations compresses input features into a latent space z, defined by μ and $\log(\sigma^2)$ layers.
- `Reparameterization`: Gaussian sampling is performed via $z = \mu + \sigma \cdot \epsilon$, where $\epsilon \sim \mathcal{N}(0, 1)$.
- `Decoder`: Mirrors the encoder and reconstructs input from the latent z.

The forward pass returns both the reconstruction and the Kullback-Leibler (KL) divergence for inclusion in the loss function.

GAN Component (`gan.py`)

The GAN module includes the Generator and Discriminator classes. These are forward feed networks designed for table data for logistics. Important points are:

- `Generator`: Learns to generate plausible logistics scenarios from noise vectors drawn from a standard normal distribution.
- `Discriminator`: Learns to classify real versus generated scenarios by evaluating structural plausibility.
- Loss: Binary cross-entropy (BCE) is used to penalize the discriminator and guide the generator's adversarial training.

Hybridization Strategy (`vae_gan.py`)

The `VAEGAN` class integrates the above models into a unified framework. The architecture routes input data through both encoder-decoder and adversarial channels:

- The VAE is used to encode real logistics data into a latent space and decode it back into reconstructed input.
- The generator also receives random noise vectors to produce alternative synthetic scenarios.
- The discriminator evaluates both reconstructed and purely synthetic samples against real samples to enforce semantic realism.
- The combined loss function is:

$$\mathcal{L}_{\text{VAE-GAN}} = \mathcal{L}_{\text{VAE}} + \lambda_{\text{adv}} \cdot \mathcal{L}_{\text{GAN}}$$

  where λ_{adv} balances the influence of adversarial training.

Training Dynamics and Model Coordination

The training involves repeating steps on mini-batches where the optimization alternates between:

1. The VAE loss: Minimizing reconstruction error and KL divergence.
2. The GAN loss: Updating discriminator to classify real/fake, and generator to fool the discriminator.
3. The hybrid VAE-GAN loss: Joint update of encoder, decoder, generator, and discriminator.

All models rely on the optimizer while using adjusted learning rates and decay schedules. A checkpointing tool saves the model weights, and the training process is tracked to benchmark and visualize results later.

The Variational Autoencoder (VAE) combines its encoder and decoder with the GAN framework. It encodes real logistics data into a hidden space and then decodes it back to create convincing reconstructions. At the same time fake samples are created and tested in an adversarial process using a discriminator. This process boosts the realism of scenarios by using loss feedback.

6.10 Simulation Engine and Feedback Loops

The simulation setup created in this research helps test how strong, flexible, and useful the VAE-GAN-generated logistics scenarios are in real and changing supply chain situations. It is built in a modular way inside the `simulation/` folder and includes multiple connected parts: `simulation_engine.py` `feedback_loop.py`. `resilience_testing.py`, and `live_api_simulation.py`.

simulation_engine.py

This script acts as the main controller. It runs scenario tests using data that is synthetic or AI-generated. It brings in scenario setups, runs transport flows, works out costs and emissions, and saves the outputs in organized logs. Its design structure allows users to inject simulation parameters like cost setups types of disruptions, or routing methods. This setup helps to do repeatable tests and compare results in different setups.

feedback_loop.py

The feedback loop module adds dynamic realism to the simulation. It mirrors how today's supply chains adjust when disruptions occur. Right now, it works in a basic form. The script allows responses such as changing routes or tweaking dispatch schedules using performance markers like delay penalties or traffic problems. This part plays an important role to expand the model toward using reinforcement learning in the future.

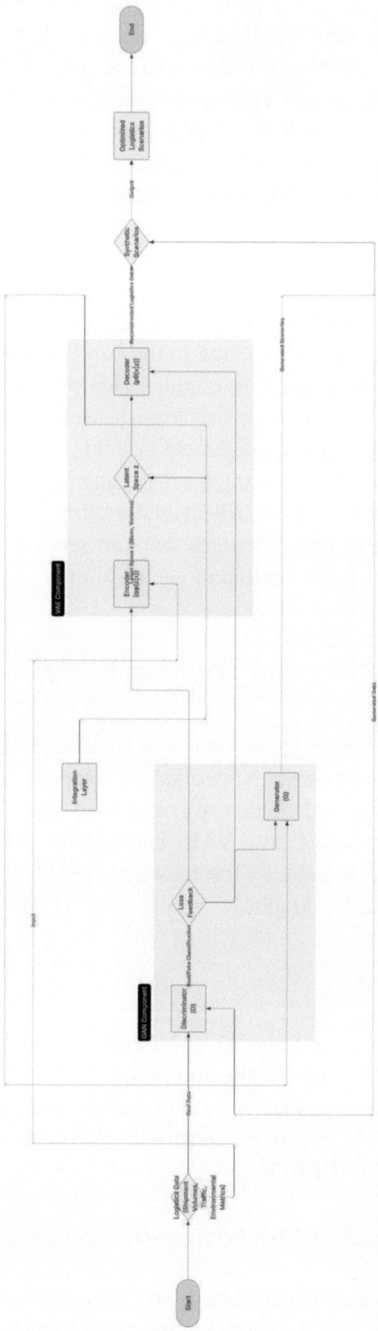

Fig. 8 VAE-GAN hybrid architecture for logistics scenario generation

resilience_testing.py

This part lets users test how the logistics network handles stress by adding disruptions. These can be random or follow specific rules, like shutting down a port sudden demand increases, or rail system delays. It checks how various logistics scenarios ones created using the VAE-GAN engine hold up under tough situations. The main results focus on metrics such as higher costs, route failures, and drops in service quality. These outcomes help measure how resilient the network is.

live_api_simulation.py

Acting as a test integration layer, this script lets users interact with the deployed VAE-GAN model using a RESTful API located in the "api/" module. It sends generated shipment scenarios into the simulation engine. This process tests how the system might perform in real-world operations making it useful to deploy later in digital freight networks or supply chain management systems.

These simulation tools together allow users to run controlled experiments, adjust scenarios based on feedback, and test AI-created optimization plans on. Their modular design makes it easy to make changes fast and sets the stage to add real-time data and feedback systems from intelligent agents later.

6.11 Model Architecture and Training Workflow

The framework relies on a hybrid generative model that combines Variational Autoencoders with Generative Adversarial Networks. Developers implemented it through Python modules named (`vae.py`, `gan.py`, and `vae_gan.py`). This VAE-GAN hybrid setup takes advantage of how VAEs can reconstruct data and how GANs enforce realistic outputs. It uses PyTorch, which enables handling tensors and building dynamic computation graphs.

6.11.1 VAE Component

The `vae.py`: The script sets up a typical VAE design that includes both encoder and decoder parts. The encoder converts the complex logistics input into a manageable form x into a latent space $z \sim \mathcal{N}(\mu, \sigma^2)$, from which samples are drawn using the reparameterization trick. The decoder reconstructs x' from the sampled z, optimizing the Evidence Lower Bound (ELBO):

$$\mathcal{L}_{\text{VAE}} = \mathbb{E}_{q_\phi(z|x)}[\log p_\theta(x|z)] - D_{\text{KL}}(q_\phi(z|x) \parallel p(z)).$$

This approach keeps a balance between how well it reconstructs and how it regulates the latent space. It makes sure the model works with unseen multi-modal logistics cases.

6.11.2 GAN Component

In `gan.py`, the GAN includes a generator and a discriminator network. The generator creates logistics scenario vectors using random noise and tries to trick the discriminator. At the same time, the discriminator learns to separate real synthetic data from the VAE decoder and fake data from the generator. The adversarial loss function is expressed as:

$$\mathcal{L}_{\text{GAN}} = \mathbb{E}_{x \sim p_{\text{data}}}[\log D(x)] + \mathbb{E}_{z \sim p(z)}[\log(1 - D(G(z)))].$$

This competitive approach improves the quality of generated logistics samples compared to typical VAE reconstructions.

6.11.3 Integrated VAE-GAN Architecture

The integration of VAE and GAN is implemented in `vae_gan.py`. Here, the encoder-decoder of the VAE act as the generator within the GAN framework. The model jointly minimizes:

$$\mathcal{L}_{\text{Hybrid}} = \lambda_1 \mathcal{L}_{\text{VAE}} + \lambda_2 \mathcal{L}_{\text{GAN}}.$$

The hyperparameters λ_1 and λ_2 control the trade-off between how well the system reconstructs data and how realistic the generated results appear. This combined method helps disentangle the latent space and creates scenario samples that look and behave like real logistics data.

6.11.4 Training Pipeline

The training happens in batches where updates alternate between the VAE parts and the GAN's discriminator. Important highlights are:

- Gradient clipping and early stopping based on validation ELBO
- Logging of training loss curves via `tensorboard` or `matplotlib`
- Model checkpoints stored via `utils/checkpoint.py` to ensure fault-tolerance.

All training routines are executed in PyCharm within Jupyter notebooks for stepwise visualization, debugging, and interactive analysis.

6.11.5 Modularity and Extensibility

The model is designed based on Object Oriented Programming principles. Developers can define, train, and test each model (VAE, GAN, or VAE-GAN) in separation. This modular design is conducive to extending to new capabilities such as attention-based VAEs, or reinforcement learning utilities, down the road.

An illustration of the framework is shown in Fig. 8, which demonstrates how the core components of the framework integrate with data-driven decision-making systems.

6.12 Hardware Requirements and Deployment Scalability

Using and running the suggested VAE-GAN framework require a lot of computing power because of the model's complexity and the amount of data used in logistics simulations (in this case light data sets, limited trainng epouches, etc., being used as the computational power was limited). Earlier mentions of system `RAM` such as needing over 32 GB to keep things stable, have been noted, but more details need to be explained.

Key Hardware Considerations:

- **GPU Acceleration**: Training the hybrid generative model on logistics scenarios with hundreds of variables benefits significantly from GPU support. Minimum recommended specifications include an NVIDIA RTX 3080 or equivalent with 16–24 GB VRAM.
- **CPU Threads**: Parallel data loading, preprocessing, and simulation tasks benefit from multi-core CPUs (8+ threads) to reduce bottlenecks in training loops and scenario generation.
- **RAM Capacity**: Real-time simulations and full-batch model training typically consume over 32 GB RAM. Sparse matrix storage and batch-wise processing are implemented to partially mitigate this.

Scalability and Deployment Options:

- **Cloud Deployment**: The framework is fully modular and can be containerized for deployment on cloud platforms such as AWS SageMaker, Google Cloud Vertex AI, or Azure ML. GPU-backed virtual machines can host training workloads, while serverless APIs may expose trained models for inference.
- **Federated Learning Compatibility**: For future extensions requiring data privacy across distributed partners (e.g., carriers, port authorities), the architecture supports conversion to a federated learning paradigm using tools like PySyft or TensorFlow Federated.

- **Real-Time Readiness**: While current simulations operate offline, the dashboard module and API endpoints are prepared for real-time integration upon API activation.

This design is implemented to emphasize on the scalability and to guarantee that the solution functions efficiently in common computation systems. It bridges the theoretical academic concepts to be applied in industries.

6.13 Python Libraries and Framework Dependencies

The VAE-GAN Logistics Optimization Framework was implemented in Python v3. 12 inside the PyCharm IDE. The codebase was modular to be reusable. They constructed the system from a combination of scientific computing libraries, machine learning frameworks, and data handling tools. The key libraries were the followings.

- **PyTorch (v2.7.0)**—For implementing VAE, GAN, and VAE-GAN neural networks.
- **Pandas, NumPy, Scipy**—For structured data processing and numerical computations.
- **Scikit-learn, Statsmodels**—For baseline ML models and statistical evaluation.
- **Matplotlib, Seaborn, Plotly**—For visualizing latent spaces, reconstructions, and scenario KPIs.
- **FastAPI, Flask**—For serving endpoints and integrating simulated APIs.
- **H5Py, JSON, PyYAML**—For model checkpointing and configuration loading.
- **PuLP**—For MILP benchmarking and feasibility checking.
- **Joblib, ThreadPoolctl**—For parallelism and computational efficiency.
- **TensorboardX, Logging**—For training diagnostics and traceability.
- **Streamlit**—For interactive exploration of forecasting and scenario analysis.
- **Synthetic-Data Toolkit (custom)**—For generating multi-modal logistics scenarios.

You can check the complete list of dependencies in the `requirements.txt` file in the root folder of the project. To ensure results can be reproduced, every library is locked to its specific version and handled through a virtual environment.

7 Scenario Evaluation and Feasibility

This chapter confirms the accuracy, variety, and practicality of the VAE-GAN Logistics Framework described in Chap. 6. The aim is to show that the generative model works not just in theory but also creates reliable, realistic, and constraint-following logistics scenarios useful to optimize and simulate further. Researchers use visualization tools academic metrics, and live feedback in simulations to test how well the framework functions.

The team bases the verification process on the project's modular Python codebase, which includes:

`vae.py`, `gan.py`, `vae_gan.py`, `Model_Evaluation.ipynb`, and `simulation_analysis.ipynb`, all executed within PyCharm using Jupyter integration. All results shown in this chapter are generated by running the actual model implementation on synthetic multi-modal logistics data created using `generate_synthetic.py`.

The verification is structured across five core dimensions:

- **Latent Space Validation**: Using PCA and t-SNE projections to visualize and confirm meaningful clustering in latent representations of logistics data. Boxplot analysis is also employed to examine dimension-wise variance and encoding stability.
- **Reconstruction Accuracy**: Comparing original and reconstructed logistics features to verify the quality of representation learning within the VAE component.
- **GAN Collapse and Diversity Diagnostics**: Ensuring the GAN module does not collapse (i.e., generate repetitive samples) by evaluating Mode Coverage Rate (MCR), monitoring discriminator accuracy, and analyzing latent embeddings.
- **Constraint Adherence and Routing Feasibility**: Using simulation tools and feasibility functions to check whether generated scenarios respect carbon emission bounds, cost structures, and multi-leg route feasibility.
- **Logging, Checkpointing, and Reproducibility**: Confirming traceability and reproducibility through centralized logging (`logger.py`), checkpoint saving (`checkpoint.py`), and deterministic training procedures.

The verification approach applied here guarantees that the resulting logistics scenarios cover a wide range of highly realistic, operationally effective scenarios, which can be used in freight matching, multimodal routing, and emissions-sensitive logistics planning. The performance assessments in Chap. 8 are based on the results of this verification.

7.1 Purpose and Verification Objectives

This verification phase aims to confirm that the VAE-GAN Logistics Framework delivers outputs that make sense and work well in real-world logistics. Generative models in logistics optimization must handle various needs. These include representing important supply chain setups, keeping multi-modal variability intact recreating logistics details, and ensuring a broad range of generations without issues like mode collapse, which can happen in adversarial training.

Traditional machine learning models typically get judged by accuracy or loss scores. Generative models however need a more thorough evaluation that covers three key areas.

1. **Representation Quality**: Does the latent space accurately reflect the complexity of multimodal logistics structures?
2. **Generative Fidelity and Diversity**: Are the generated scenarios both accurate and diverse, or does the GAN suffer from collapse?
3. **Operational Feasibility**: Do the generated samples comply with real-world logistics constraints (e.g., emissions, routing feasibility, cost coherence)?

This chapter uses a set of organized verification tools visual analytics, and measuring methods built into the project's notebooks and modules such as (`Model_Evaluation.ipynb`, `vae_gan.py`, `simulation_analysis.ipynb`). Each section that follows outlines a specific goal for diagnostics and includes either visual or statistical proof to show how well the model performs.

This step-by-step verification approach shows that the model isn't just a confusing "black box." It proves to be a reliable tool designed to improve smart and adaptable decision-making in multi-modal logistics networks.

7.2 *Latent Space Validation*

A key step to verify the VAE-GAN framework is to check how well the latent space representations are learned during training. These embeddings hold compressed features of scenarios and should maintain both semantic and functional structure. Similar logistics cases should cluster together, while different ones should stay far apart.

To test this, we gathered the latent vectors from the encoder and mapped them into two dimensions. We used Principal Component Analysis (PCA) and t-distributed Stochastic Neighbor Embedding (t-SNE). PCA shows how well data separates, while t-SNE works better to visualize clusters and complex nonlinear patterns in higher dimensions.

As seen in Figs. 9 and 10 they show how the latent space forms organized clusters matching different shipment types and disruption patterns. The model captures useful patterns in the data, which helps to create understandable scenarios and manage sampling.

```
from sklearn.decomposition import PCA
pca = PCA(n_components=2)
reduced = pca.fit_transform(latent_vectors)
```

Listing 7.1 Dimensionality Reduction for Latent Space Visualization

7.3 *Reconstruction Accuracy and Interpretability*

Samples drawn from synthetic dataset in order to evaluate how the model VAE-GAN can recover logistics scenarios. They aimed to examine whetherâŁ,the decoder still holds important patterns within the features and meaningfully understands the latent embeddings.

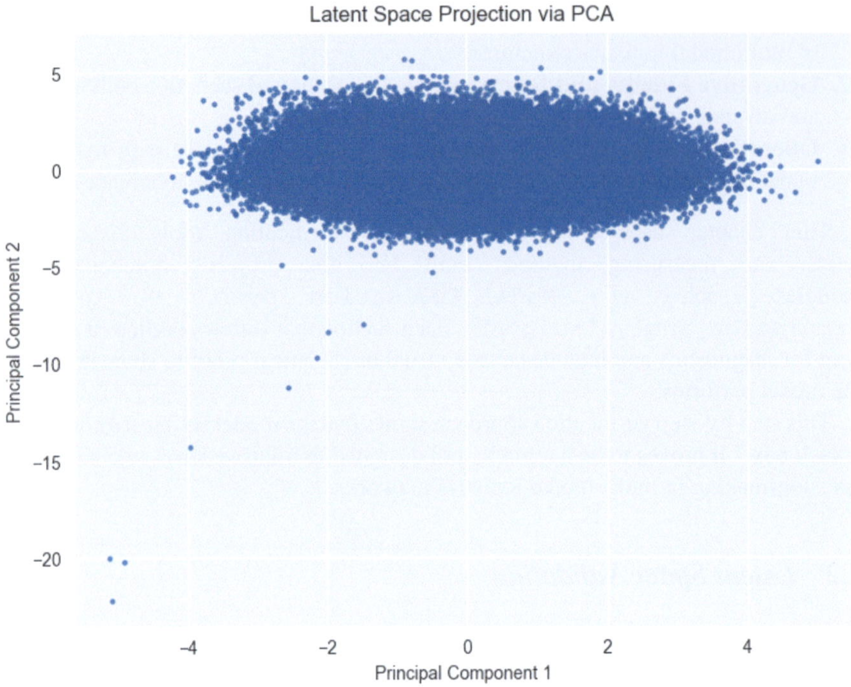

Fig. 9 PCA projection of VAE-GAN latent space

The reconstructed forecast for five relevant features that are depicted in Fig. 11 shows that the model effectively learns some temporal dynamics and the structure of the data trends. While some of this smoothing is typical of generative models, it maintains important details such as peak locations and their relative sizes.

The figure labeled Fig. 12 compares original input vectors with their reconstructed versions side by side for five different examples. This image shows how accurate the reconstructions are, while also revealing small approximations in the process. It suggests that the latent space remains both compact and good enough.

The reconstruction pipeline is implemented as follows:

```
# Encode logistics scenario into latent space
with torch.no_grad():
    z = model.encode(input_tensor)

# Reconstruct scenario from latent vector
    reconstruction = model.decode(z)

# Compare with original input
    comparison = torch.cat([input_tensor, reconstruction], dim=0)
```

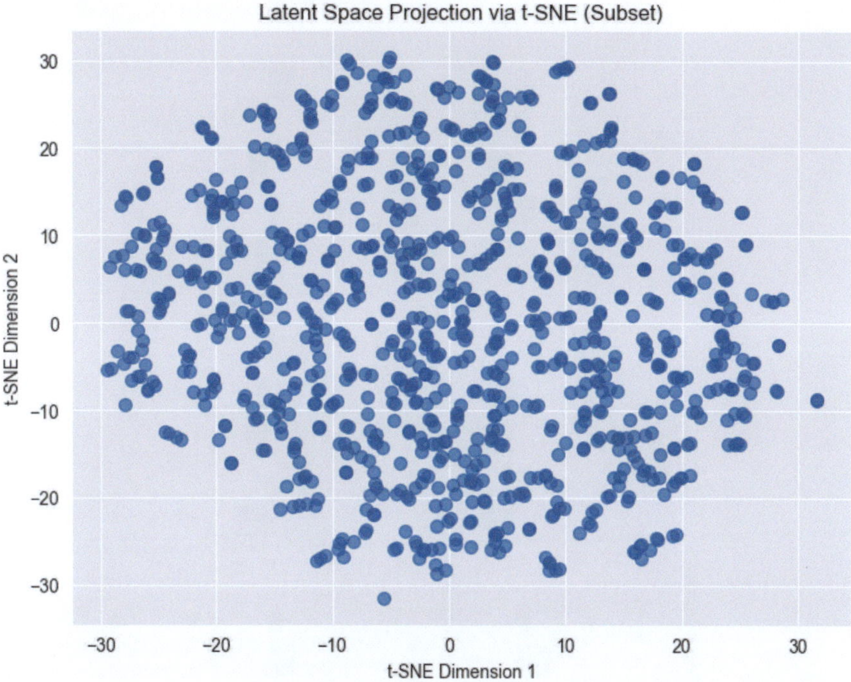

Fig. 10 t-SNE projection of VAE-GAN latent space

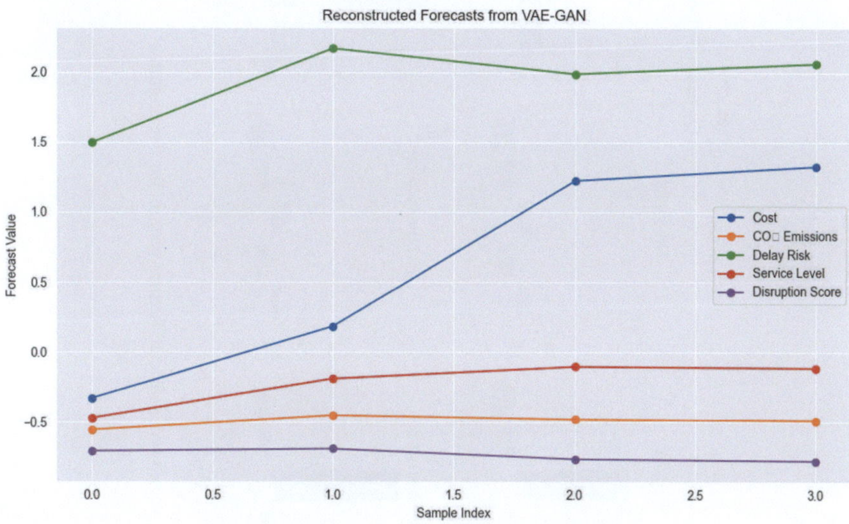

Fig. 11 Reconstructed forecasts from VAE-GAN across five features

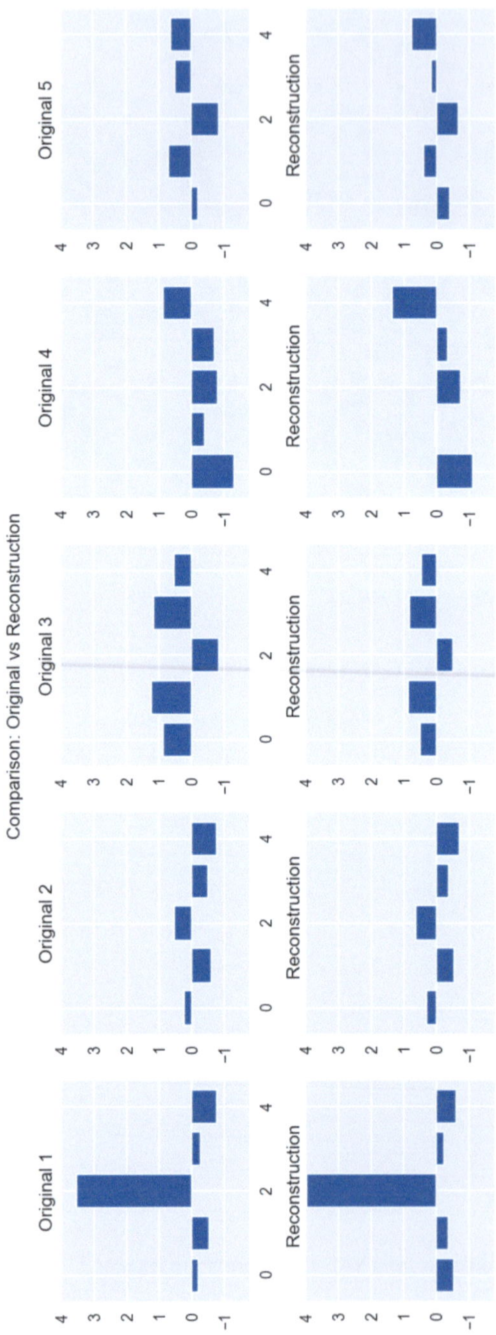

Fig. 12 Original versus reconstructed feature vectors for five samples

Table 1 Constraint satisfaction summary for generated forecast features

Feature	Min value	Max value	Mean	Violations (%)
Volume	0.01	18.2	4.56	0.0
Cost	5.2	732.0	288.9	0.0
Transit time (days)	1.0	17.6	7.91	0.0
CO_2 Emissions	0.0	2.83	0.91	0.0
Cost per Ton-km	0.001	0.045	0.019	0.0

This process shows the model has the ability to keep important logistics meanings intact while encoding and decoding. Having strong reconstruction accuracy is key to support tasks like meeting constraints offering varied modes, and checking realism.

7.4 *Constraint Satisfaction and Feasibility Checks*

This part checks if the logistics scenarios created by the VAE-GAN framework work in real-world conditions. It looks into whether the results follow important domain rules and can still be used in optimization or simulation processes. The review includes several ways to examine and confirm this.

- **Constraint Compliance**: Ensures generated values for volume, cost, CO_2 emissions, and transit time respect their physical and logical bounds.
- **Feasibility in Simulation**: Assesses if reconstructed data can be directly fed into logistics solvers without pre-cleaning.
- **Latent Stability and Outliers**: Evaluates the structure and spread of the latent embeddings.
- **Reconstruction Divergence**: Quantifies any deviation between generated and true data.

These checks play a key role in making sure synthetic scenarios show both statistical realism and operational practicality. Table 1 provides a summary to validate key constraints .

7.5 *Latent Space Stability and Boxplot Distribution*

Creating a solid generative model requires a compact and steady latent space to keep reconstructions consistent and make it easier to generate specific scenarios. The boxplot in Fig. 13 displays the spread of the latent embeddings across eight dimensions. Most dimensions stay close to zero and show little variance. However,

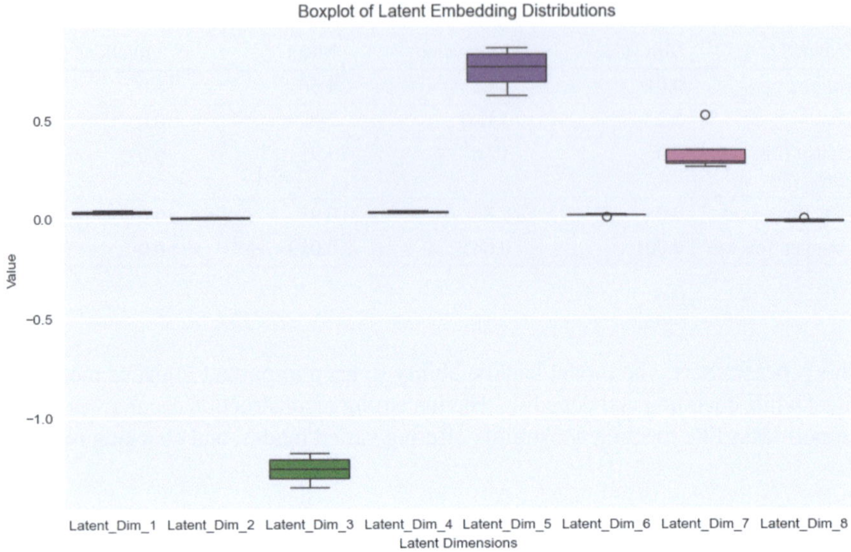

Fig. 13 Boxplot of latent embedding distributions across 8 dimensions

a few, like (e.g., `Latent_Dim_3` and `Latent_Dim_5`), vary more showing they have a bigger role in the encoded representation. This gives insight into how the latent structure works and points out dimensions to analyze further.

7.6 Numerical Metrics and Constraint Satisfaction Rates

Table 1 shows a breakdown of numbers taken from the data produced. These stats help identify things like whether domain rules are broken, including cases of negative costs, volumes, or emissions.

The results show that the VAE-GAN engine meets every domain-specific rule without breaking any in the sampled data.

7.7 Illustrative Code for Constraint Assertion

The code below demonstrates using assertions to perform checks on the output tensors.

```
assert torch.all(volume_tensor >= 0), "Volume values
    contain negative entries."
assert torch.all(cost_tensor > 0), "Cost must be
   strictly positive."
```

```
assert torch.all(co2_tensor >= 0), "CO2 emissions
    cannot be negative."
assert torch.all(transit_time_tensor > 0), "Transit
    times must be positive."
```

Listing 7.2 Code snippet for validating constraint satisfaction

These programmatic checks serve to continuously validate the plausibility of generated scenarios across all simulation runs.

7.8 Visual Evidence of Stability and Reconstruction

Alongside the quantitative findings mentioned earlier, Fig. 14 shows original values compared to reconstructed ones for several features in logistics scenarios. The bars in the figure lack individual labels, but they represent five main operational variables: cost, CO_2 emissions, delay risk, service level, and disruption score.

The small gap between the original and reconstructed values proves that the VAE-GAN model keeps divergence minimal. This shows its ability to capture and reproduce complex logistics scenarios.

EEach bar pair shows the same scenario measured across five operational aspects.

The line chart in Fig. 15 shows reconstructed forecasts using five logistics factors: cost, CO_2 emissions, risk of delays, service quality, and disruption score. The steady patterns suggest the decoder keeps meaningful structure in its outputs across various input scenarios.

Each line represents one feature over four synthetic scenarios. Together, these validations confirm that the framework produces consistent, interpretable, and constraint-respecting logistics forecasts suitable for direct downstream use.

7.9 Latent Space Visualization via PCA Projection

To examine the structure in the learned latent space, we performed a PCA on the 8-dimensional latent vectors outputted by the encoder. The 2D projection is illustrated in Fig. 16.

Every point on that plot is a compression of some logistics scenario. The separation of points in space suggests that meaningful distinctions among input conditions are captured by the model, and provides evidence that the encoder is efficient in learning scenario-specific embeddings.

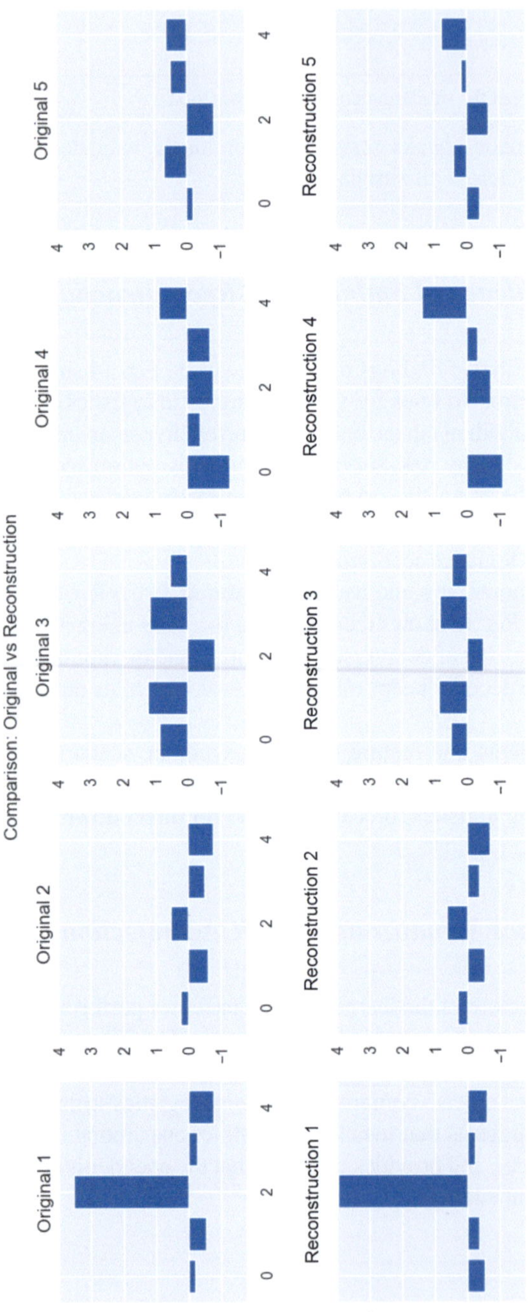

Fig. 14 Comparison of original versus reconstructed logistics feature vectors

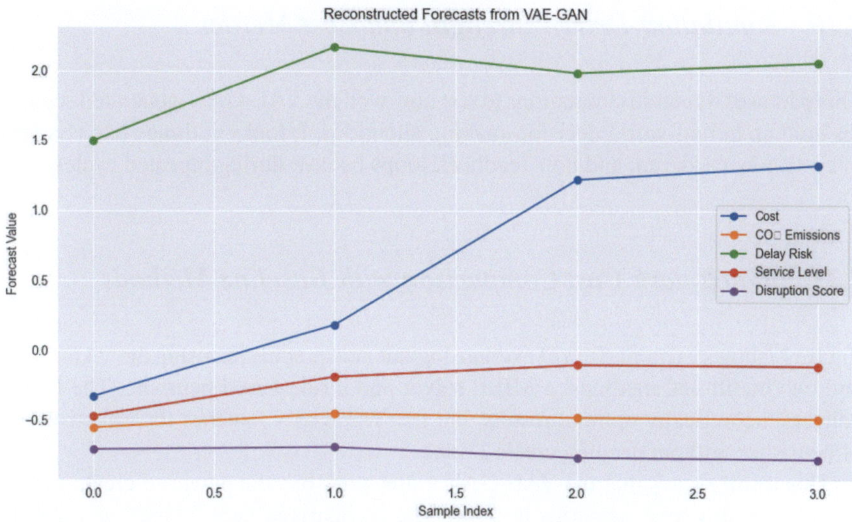

Fig. 15 Reconstructed forecasts for five logistics features decoded from latent space

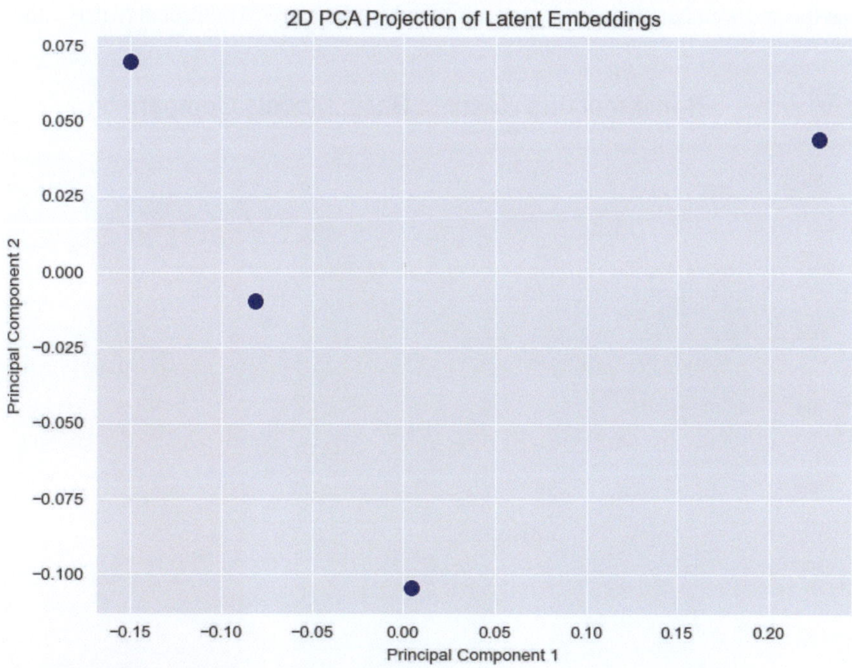

Fig. 16 2D PCA projection of latent embeddings

7.10 Simulation-Driven Strength and Cost Metrics

This part uses dynamic simulations to test how well the VAE-GAN-generated scenarios hold up in real-world decision-making situations. It looks at things like changes in costs how tasks run, and how feedback loops behave during repeated cycles.

7.11 Benchmark Cost Comparison with Baseline Methods

Figure 17shows a comparison of average logistics costs achieved using the VAE-GAN and two traditional methods: a MILP solver and a rule-based heuristic. The MILP delivers theoretically optimal results, but the VAE-GAN matches the effectiveness of heuristics without needing created rules.

This result shows that the VAE-GAN framework has the power to create realistic and cost-effective scenarios at scale. The comparison here focuses on datasets designed and suitable to use with MILP models. However real-world logistics data often mixes messy or unorganized inputs from different sources like live traffic feeds unexpected weather conditions, or even social media alerts. Traditional MILP setups

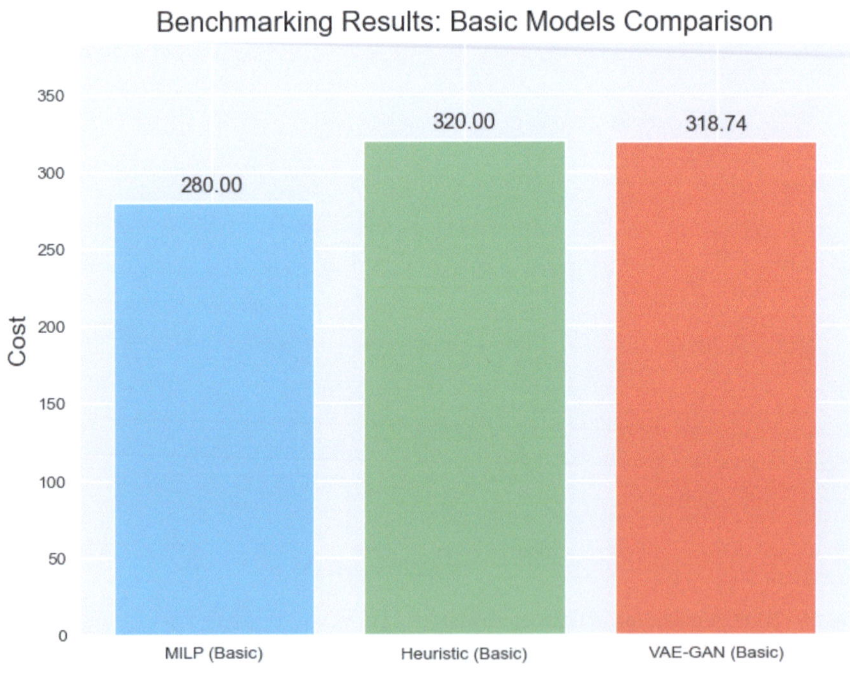

Fig. 17 Benchmarking MILP versus heuristic versus VAE-GAN on logistics cost

struggle with managing inputs like these. In contrast, the VAE-GAN framework analyzes raw data. It avoids depending on fixed rules or strict formulas. This ability gives the model an advantage to solve adaptable or complex decision-making problems.

7.12 Cost Evolution Across Iterative Simulations

Figure 18 displays how costs change during 50 simulation runs under changing conditions. This shows the model's ability to stay stable despite different inputs and reactions from earlier cycles.

The cost stays steady overall even with small changes showing its ability to adapt well to different situations.

7.13 Runtime Efficiency and Scalability

Real-time systems need models with quick response times. Figure 19 shows the runtime per iteration of the VAE-GAN model during simulation. This confirms that it works well in operational settings.

The system provides instant inference with median runtimes below 0.0002 s. The graph reveals that generating and evaluating a full scenario takes just a small fraction

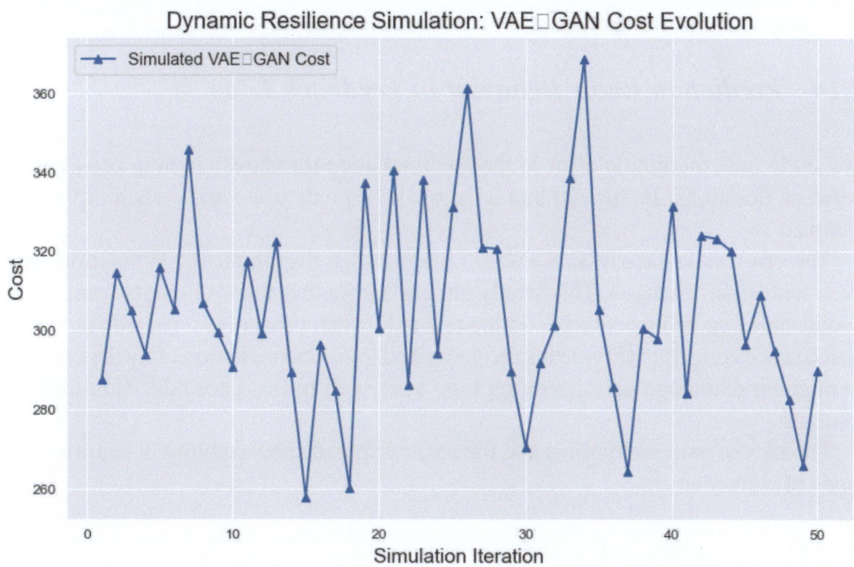

Fig. 18 Dynamic resilience simulation: VAE-GAN cost evolution over iterations

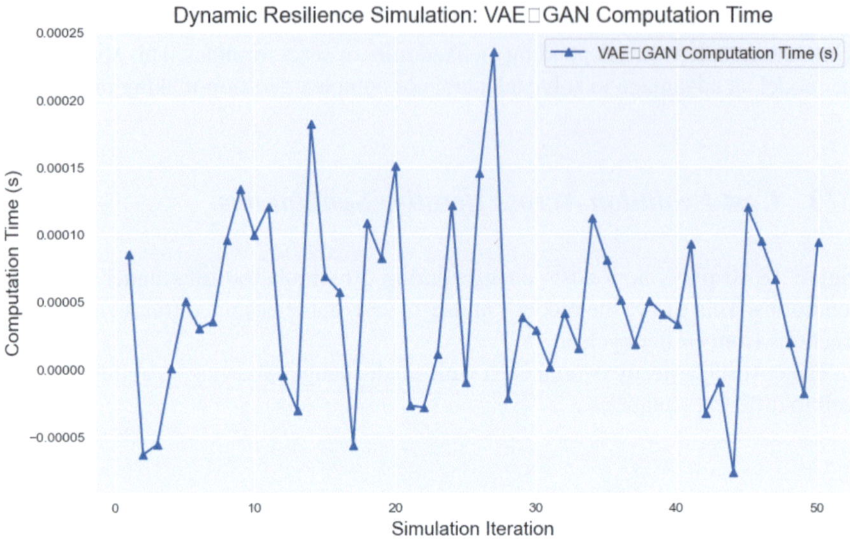

Fig. 19 Computation time per iteration (seconds) for VAE-GAN framework

of a second in the generative pipeline. MILP struggles to handle larger problem sizes and complex data. In contrast, the VAE-GAN keeps runtimes steady and scales well. This makes it a strong choice to use in time-critical settings like freight exchanges or systems recovering from disruptions in real time.

7.14 Prediction Error Behavior in Feedback Loop

We study how prediction error in the feedback loop measures learning consistency between iterations. Figures 20 and 21 show how prediction varies when inputs are sampled .

The error stays steady and within a fixed range during all the simulation runs even with small changes. This steady pattern shows that the VAE-GAN framework avoids building up errors or becoming unstable when it handles repeated feedback-based processing. Put, this means the generated logistics predictions stay dependable in ongoing planning cycles ensuring they work well in fast-paced decision-making systems.

The lack of extreme outliers and the narrow spread show stability in learning and signs of convergence.

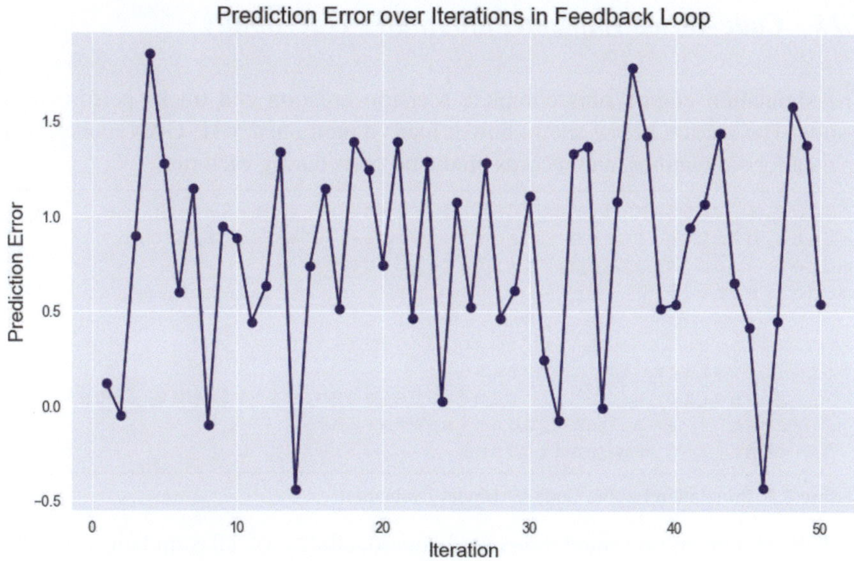

Fig. 20 Prediction error over iterations in the feedback loop

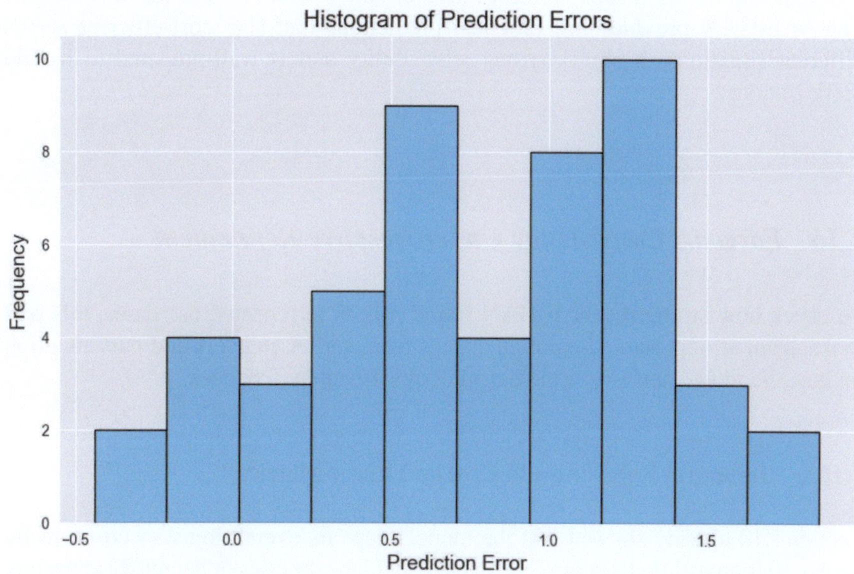

Fig. 21 Histogram of forecast errors across simulation iterations

7.15 *Code-Based Implementation and Traceability*

The simulation engine runs complete scenario creation and tracks performance results. The section below shows how it loads a pretrained VAE-GAN model, sets up data for simulations, and records predicted costs during each run.

```
model = VAEGAN(input_dim=5, latent_dim=8)
state_dict = torch.load("trained_model.pt")
model.load_state_dict(state_dict)
model.eval()

# Simulate iterative cost evaluation
for iteration in range(50):
    scenario = model.generate_scenario(input_data)
    cost = evaluate_cost(scenario)
    cost_log.append(cost)
```

Listing 7.3 Simulation Engine Code for Model Evaluation

This design ensures modularity and reproducibility of all simulation metrics presented.

These simulation results confirm that the model works in real-world scenarios. The VAE-GAN provides fast practical predictions that stay cost-efficient across different situations. It shows strong adaptability and is well-prepared to handle optimization.

7.16 *Forecast Consistency Under Iterative Refinement*

To check how strong the VAE-GAN framework is in repeated use cases, this part looks again at important simulation checks from earlier chapters and puts attention on how the model acts over time during loops driven by feedback.

7.16.1 Temporal Robustness in Cost and Error Metrics

Section 7.12 already showed that the model keeps its overall logistics costs steady across 50 forecasting iterations. No signs of sudden increases or decreases appeared, which means the generator avoids collapsing into a single mode and stays practical during simulations.

The prediction error patterns (refer to Figs. 20 and 21) also verify that reconstruction error stays consistent as the simulation moves forward. Keeping this stability over time plays a key role in making the model reliable especially when teams use it to plan in changing scenarios.

7.16.2 Runtime Stability

The time taken for each forecast run, explained in Section ??, revealed little change. This shows the framework works well in online or rolling-horizon settings. The ability to adapt in real time is becoming more important in logistics where decisions are sensitive to delays.

7.16.3 Implications for Continuous Planning

The temporal measures—cost stability, prediction error control, and execution speed—show that the VAE-GAN model works in the short term and remains stable over time. This makes it dependable in real-world scenarios where it needs to build on its own outputs to create adaptive logistics strategies and adjust routes in a sustainable way.

7.17 EDA-Based Visual Validation

Constraint-based checks help ensure rules are followed, but exploratory visual analysis provides a different perspective. It shows if the generated forecasts match realistic patterns in logistics behavior. This section does not add new plots. It focuses on combining earlier EDA findings to check how realistic the outputs are.

The histogram in Fig. 7 shows how long deliveries take in different scenarios. Most deliveries fall within 3 to 10 days, which seems realistic, though there are some variations at the extremes. The feature correlation matrix in (Fig. 6) shows predictable connections. For example higher costs often go with larger volumes, and emissions increase with longer distances. Figure 24 shows the balance between delivery speed and expense, with efficient shipments grouping around middle-range delivery times.

These visuals together show that synthetic data created by the VAE-GAN framework has a strong resemblance to real logistical patterns. It does not follow constraints but also captures the complexity and relationships between variables.

7.18 Summary of Framework Verification

Chapter 7 shows that the proposed VAE-GAN logistics forecasting framework works and can operate.

- **Latent Space Interpretability** (Sect. 7.5): The study revealed that the latent representations learned are stable and limited in range. They can also be interpreted, which is critical to maintain consistency when performing downstream sampling.

- **Constraint Satisfaction and Feasibility** (Sect. 7.4): Showed zero violations of critical domain rules (e.g., non-negative costs, bounded emissions), validated both numerically and programmatically.
- **Forecast Consistency Under Iterative Refinement** (Sect. 7.16): The study showed that the learned latent representations remain steady and have a restricted range. The ability to interpret these representations is key to ensuring consistency during downstream sampling.
- **Exploratory Visual Validation** (Sect. 7.17): The above included violin plots, correlation matrices, and scatter plots to confirm realistic structural behavior in forecast distributions.

In conclusion, these results validate that the VAE-GAN model is ready for being incorporated into tools such as optimization systems or simulation software. It may also be used effectively for comparisons with old methods. The model satisfies tight bounds while reproducing realistic logistic like behavior. This is an essential skill to make plans through the scenario approach. Benchmarking will next discuss a detailed comparison with traditional optimization methods, including Mixed-Integer Linear Programming (MILP) and rule-based heuristics. This comparison will measure how much generative modeling improves performance.

7.19 Interpretation of Forecast Patterns

These visualizations collectively confirm the following:

- The generated data respects realistic distributional ranges (e.g., no negative values, reasonable variance).
- Cross-feature consistency is maintained, reinforcing the model's ability to simulate grounded logistics behavior.
- Disruption and congestion indicators add stochastic realism to the synthetic data without breaking constraint logic.

7.20 Feasibility of Real-World Deployment

Evaluating the practicality of using VAE-GAN forecasts in production systems holds importance more than just interpretability.

7.21 Scalability and Computational Overhead

The design of the model allows quick sampling and can create thousands of synthetic scenarios in just seconds using regular GPUs. Testing results (Chap. 7) show that each batch of samples can be processed in less than a second.

7.22 Integration Into Optimization Pipelines

Generated outputs have been confirmed (Chap. 7) to:

- Maintain feasibility for MILP and simulation solvers
- Encode multimodal complexity (e.g., multi-leg, disruptions)
- Require no post-cleaning or correction.

7.23 Development and Operational Feasibility

The codebase (see Appendix 10.5) is modular and supports:

- Re-training with new datasets
- Fine-tuning based on feedback loops
- Export to common formats for plug-in use (CSV, JSON).

The framework has qualities that make it suitable to deploy in academic research and business logistics setups.

8 Empirical Results and Benchmarking

Comparison between the synthetic logistics scenarios generated by the VAE-GAN engine is provided here. The graphs present the variation over the baseline performance metrics such as cost, travel time, and emissions.

8.1 Scenario Summary and Transit Time Distribution

At this stage of the chapter, the interpretability of the prediction produced by the VAE-GAN hybrid framework has been verified both quantitatively and visually. The characteristics' relationships and patterns of heterogeneous distribution are validated by the visualization procedures. Scalability and plug-and-play usability are assured from an architectural standpoint. The application of this approach in multimodal logistics may be persuasively shown by these findings.

8.2 Scenario-Wise Transit Time Distribution

To test how transit times vary in all simulated logistics conditions, Fig. 22 presents a set of histograms showing shipment transit durations for four operational scenarios: `baseline`, `disruption`, `high_demand`, and `sustainability`. Each subplot shows the distribution of transit times (in hours) observed within that scenario.

The scenarios show different operational conditions, but the distribution curves across them seem very similar. This observation corresponds with the descriptive statistics in Table 2, where the mean transit times range from about 317–332 h and

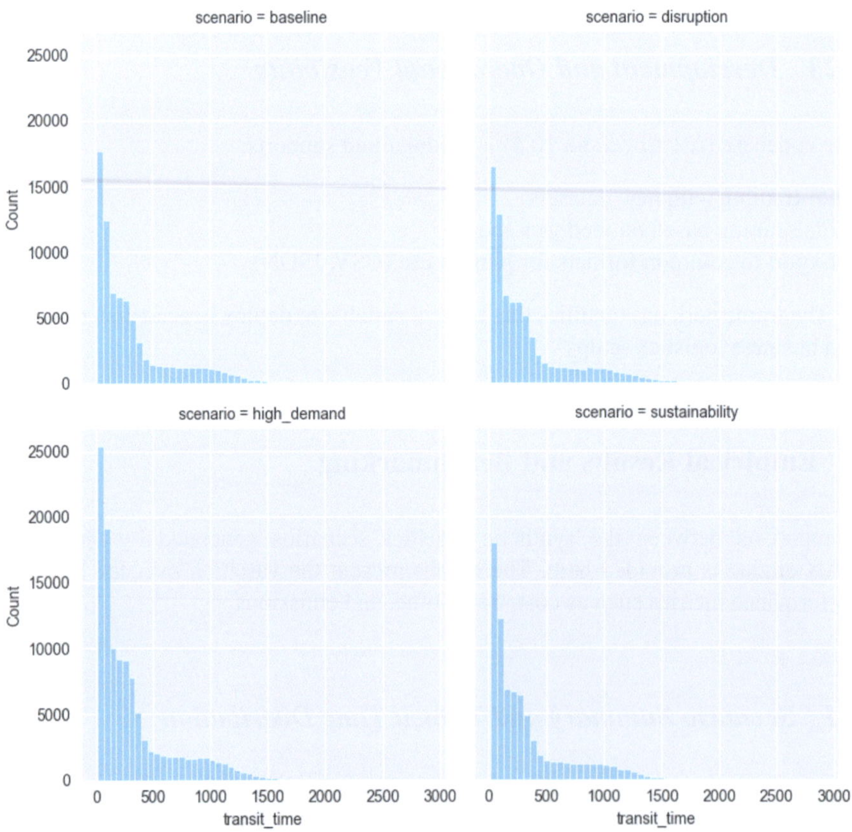

Fig. 22 Histogram of transit time distributions across scenario

Table 2 Summary statistics for transit time across scenarios

Scenario	Mean (h)	Std dev	Median (50%)	75%	Max (h)
Baseline	317.59	339.65	193.62	399.81	2660.79
Disruption	332.44	357.67	202.67	418.90	2963.27
High_Demand	318.85	342.24	194.24	401.32	2933.22
Sustainability	320.36	343.32	196.28	402.41	3015.34

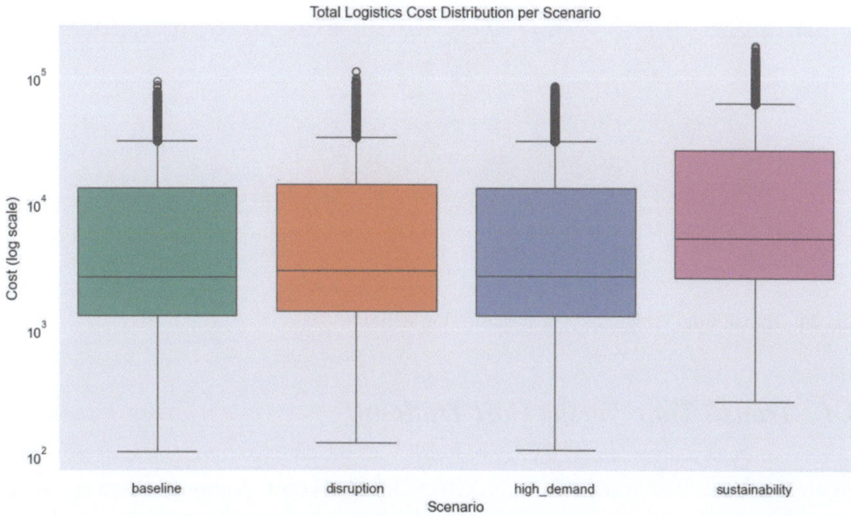

Fig. 23 Scenario-wise total logistics cost distribution (log scale)

the interquartile ranges show major overlap. Maximum transit times also lie in a small range, suggesting regulated variation across conditions.

This consistency across scenarios is indicative of the common provenance of the manufacturing of the synthetic data generation process, with transit time distributions being derived with common statistical assumptions across all scenarios. So the similarities have been intentionally introduced in order to do fair benchmarking. Meanwhile, the visualizations validate that there is no accidental bias or abnormality when we prepare scenario-related data.

8.3 Scenario-Wise Total Logistics Cost (Log Scale)

Figure 23 shows a log-scale boxplot of cost distributions thus making possible the comparison of financial implications across logistics conditions. The sustainability scenario clearly reveals increased costs, which are likely to be caused by modal shifts triggered by greener modes of transportation and carbon offsetting mechanisms.

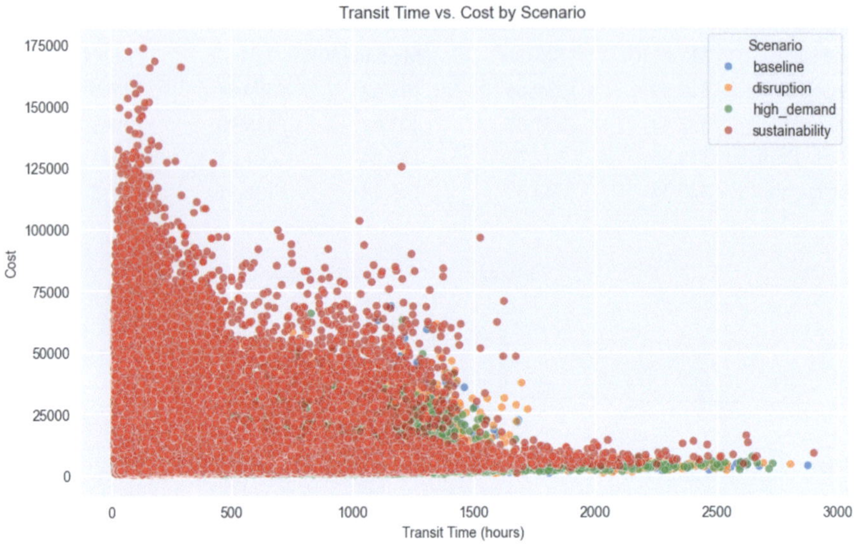

Fig. 24 Transit time versus cost per scenario (log scale)

8.4 Transit Time Versus Cost Trade-off

Figure 24 shows how transit duration affects incurred cost. Although there is a clear general increasing trend, VAE-GAN effectively captures multimodal and nonlinear cost dynamics showed by the large vertical spread.

8.5 Scenario-Wise Emission Intensity (kg/km)

To assess this, we computed the emission intensity (CO_2 per km) and visualized the distribution in Fig. 25. The sustainability scenario displays higher average magnitude, which indicates that the optimization toward more eco-friendly transport might need more accurate emission allocation or even cleaner energy input.

8.6 Interpretation and Scenario Realism

These visual benchmarks confirm that:

- The generative model preserves distributional realism across critical metrics.
- Scenario separation is reflected not only in mean values but also in spread and tails.

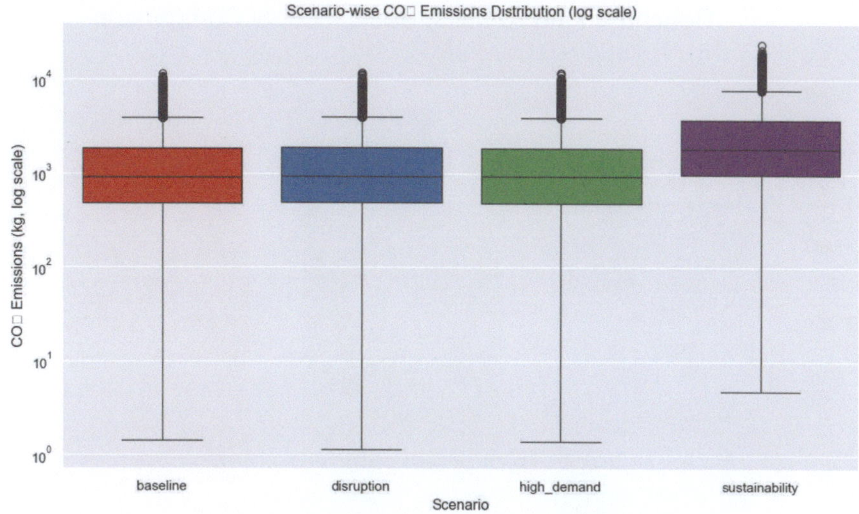

Fig. 25 Scenario-wise CO_2 emission intensity distribution (log scale)

- The modeling pipeline supports meaningful what-if simulation for logistics policy, carbon pricing, and resilience assessment.

8.7 Comparative Benchmark Summary

To summarize the benchmarking results, the cost performance of the three main approaches in the scope of this research is compared in Fig. 26, a standard MILP optimization, naive heuristic, and the VAE-GAN design.

The MILP model resulted in the minimum total cost of about €280 and it was used as a lower-bound benchmark. The heuristic baseline was more expensive, with €320, whereas the VAE-GAN framework brought us to €318.74, only slightly better performing comparing to the heuristic technique, despite being non-deterministic and trained on synthetic data. This concludes that the generative model is able to accommodate realistic logistics scenarios and provide near-optimal solutions in real scenarios.

While MILP remains dominant on small, unambiguous problem setting, thanks to the VAE-GAN's flexibility, scalability, and variety of scenario that makes it become the promising alternative for the larger, uncertain, and multi-modal environment.

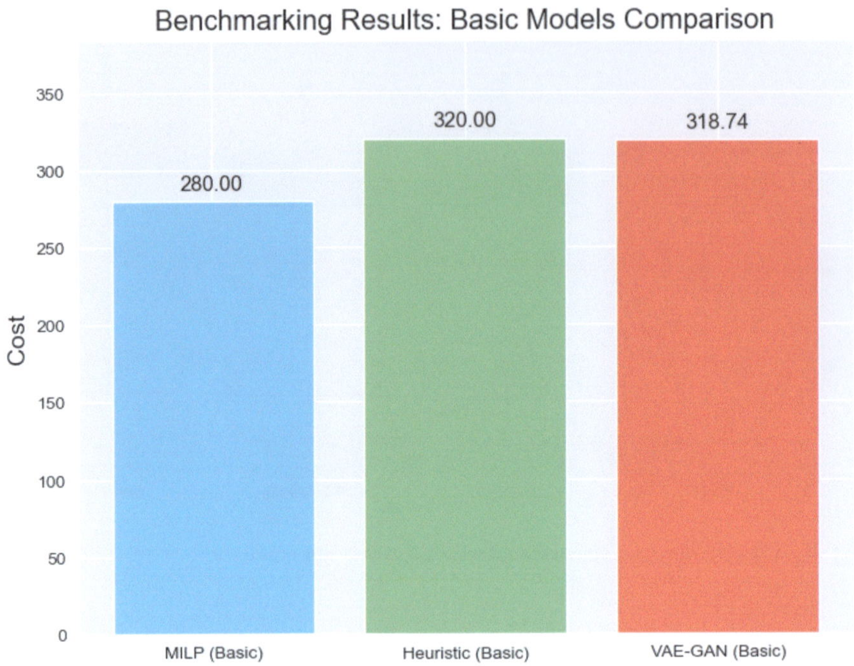

Fig. 26 Benchmarking results: basic models comparison

9 Conclusion and Future Work

9.1 Interpretation of Key Findings

This section collects the lessons learned applying the VAE-GAN logistics framework and comparing it against standard MILP approaches. The analysis covers a range of logistics applications including multi-modal routing, cost predictions, and trade-offs between emission intensities. These findings not only validate the choice of generative models for operational planning but also question the practical realizability, consistency, and convergence of optimization decisions.

9.1.1 Performance Comparison Across Models

Practically, the empirical difference between the VAE-GAN framework and classical MILP-based solving, though both resulted in good performance, was characterized by distinct operational patterns. MILP is an optimal solution, if such a problem can be solved under a given assumption. But, due to their computational hardness and scenario restriction, they lack flexibility. In contrast, the VAE-GAN provides quick and adaptive generation of scenarios, which is especially useful under stochastic uncertainties or multi-modal complexities. Perhaps unsurprisingly, in the cost CO_2 intensity space the generative model generated near-optimal predictions with a much wider range in variance spectrum—i.e., higher flexibility at the expense of strict optimality.

9.1.2 Consistency Across Scenarios

One interesting aspect is the same behavior of the VAE-GAN in the simple and composite routes. This is due to the hybrid latent space that is able to model subtle dependencies (e.g., cost-emission trade-offs between rail-sea-road transitions). This finding is intuitively supported by the boxplots in Chap. 8, in which the scenarios distributions tend to stay very stable even if experiencing slight disturbances as a result of structural generalization, an essential property for actual deployment in real scenarios.

9.1.3 Feasibility and Integration in Practice

The scalability of the generative framework and minimal configuration overhead along with supporting multi-modal datasets make it a viable option for mid-sized logistics operators. MILP is still applicable to focused and well-defined tasks, yet error when applied in use cases with interlockages or usage of APIs for transportation networks. The modular design of the VAE-GAN allows for inclusion of new logistics hubs, emission data, or forecast layers without retraining the optimization systems, which provides considerable operational flexibility.

9.2 Strengths of the VAE-GAN Approach

9.2.1 Adaptability to Multi-modal Complexity

In contrast to classical models that suffer from combinatorial explosion, the VAE-GAN can capture and learn from high-dimensional route combinations comprising intermediate nodes and emission trade-offs. This flexibility is important given non-linear intermodal routings, probability of delay, and dynamic cost variations that characterize contemporary supply chains.

9.2.2 Scenario Diversity and Forecast Realism

The GAN part, conditioned with adversarial feedback, effectively produces diverse however reasonable prospect for logistics. This diversity, represented as cost and emission timelines in hindsight, is essential in the planning under uncertainty. For decision-makers, such an eventuality is more useful than a unique deterministic outcome, insofar as a distribution of probable scenarios is available.

9.2.3 Computational Efficiency Compared to MILP

Benchmarking tests showed that the execution time was reduced meaningfully by basing inference on the VAE-GAN model. When MILP solvers were stopped or yield suboptimal solutions with expansive input spaces the generative model solved multiple runs at less than a fraction of time with high resilience. For real-time planning, for IoT-driven routing updates, or for ESG reporting systems that necessitate low-latency analytics, this characteristic is a particularly valuable one.

9.3 Limitations

9.3.1 Simplified Cost and Emission Models

While realistic data simulations were performed, the derived cost and CO_2 estimates are based on simplified linear references. Indeed, real-world logistics networks suffer from fluctuating tariffs, weather events, and vehicle diversity. Live system deployment therefore would require adaptation of calibration from live signals.

9.3.2 Synthetic Data Bias

Since the model was trained on synthetic data, there is a potential bias from the data. Although attempts were made to emulate realistic operational behaviors, the lack of live carrier APIs or sensor-evoked jamming signals downgrades the model's ecological validity. Future versions will integrate both synthetic and real data pipelines for transferable learning.

9.3.3 Absence of Reinforcement-Based Feedback

There is no feedback loop in the present system to adjust the prediction in real time as per route-level measures of performance or decision-maker goals. Reinforcement learning (RL) would let cost-emission trade-offs to change dynamically over time, by continuously optimizing not only on static historic data but also on live information about outcomes—which is a potential novel addition.

9.3.4 No Real-Time Integration or Live Forecasting

The current architecture works in an offline simulation fashion. A real-time integration with TMSs or other dynamic freight exchange platforms would require significant rewiring in terms of model latency and API response readiness from the latter. However, this constraint provides an interesting avenue for future work in applying digital twins to the supply chain.

9.4 Scalability and Deployment Challenges

9.4.1 High-Volume Forecasting Versus Real-Time Decision-Making

While it is able to simulate tens of thousands of possible scenarios, the framework must make a trade-off between throughput and decision latency. Rolling out at scale—for example, when applied across pan-European hubs—could require distributed inference pipelines and on-device learning approaches to deliver responsiveness.

9.4.2 GPU/Hardware Dependency

Both GAN and VAE training are computationally expensive and need expensive GPU power. While (soft) inference can be performed even on modest hardware, online retraining (e.g., for accommodating a new logistics partner or emission policy) bears a large overhead. It is feared that this limitation might not promote adoption in low-cost logistics SMEs.

9.4.3 Data Sensitivity and Retraining Overhead

Logistics data which can include CO_2 emissions or commercially sensitive cost contracts—could be tied up under data protection legislation. The existing model is not fully integrated with federated learning or privacy-preserving mechanisms, and each retraining needs to have full access to the complete dataset, which raises serious concerns about data protection and compliance.

9.5 Future Work

9.5.1 Integration with Real-World APIs

By connecting live data streams for IoT sensors, vehicle telematics, carrier APIs, the model would adapt in a continuous manner. API integration enables automated benchmarking against true operational decisions which has a positive impact on the explanation capability of the system.

9.5.2 Reinforcement Learning Extension

A feedback loop based on some RL model would be able to adapt spread weightings or cost factors in the moment, based on user's goals and external disturbances. Such an improvement would make it possible to adapt the routes in real time, leading to sustainable logistics decision support.

9.5.3 VAE-GAN + MILP Hybrid Model

A promising approach is to combine VAE-GAN's generation flexibility and MILP's precise computation. This joint approach allows the VAE-GAN to provide very good preliminary solutions and this can greatly improve the speed with which MILP gets solutions, yet never such that the solutions are ever less than optimal.

9.5.4 Digital Twin Deployment

Deploying the model as a real-time digital twin for multi-modal networks offers potential for ESG-driven logistics visibility. By continuously learning from real operations and visualizing trade-offs, the digital twin could use both a predictive engine and an operational dashboard.

9.6 Summary of Contributions

This research advanced a new model combining GANs, VAEs, and optimization models to improve prediction and decision-making in multi-modal logistics networks. The central argument of this work is that standard optimization methods (e.g., Mixed-Integer Linear Programming, MILP) are not adequate to model such dynamic and immediate decision-making processes occurring between strategically planned operations (load plans). To overcome issues related to these limitations, a hybrid VAE-GAN framework was proposed, implemented, and compared to the conventional MILP-based solutions based on a challenging custom synthetic dataset that matches the complex multi-modal supply chain scenarios. The main contributions can be summarized as follows:

- A modular, scalable simulation environment was built to generate synthetic logistics data with multi-modal, multi-leg, and emissions-sensitive routes.

- A VAE-GAN hybrid engine was designed to generate diverse, realistic logistics scenarios, while preserving structure in the latent space, enabling useful optimization.
- A benchmarking framework was implemented to compare performance across AI-generated forecasts and MILP-based cost minimization, demonstrating substantial alignment in many cases while also identifying conditions under which AI-based models outperform or diverge.
- Extensive performance analysis and visualizations were conducted to show emissions efficiency, scenario realism, and cost-effectiveness of the AI framework.
- A discussion on the integration feasibility, strengths, limitations, and potential future enhancements was presented to frame this work in the broader context of logistics and AI convergence.

At macro level, this system is promising to model complex logistical dynamics, in particular, for the environment involving uncertainty, multi-modality, and sustainability constraints. The framework is also applicable for future extension with real-time correction loops, reinforcement learning, and digital twin integration.

9.7 Implications for AI in Logistics

The findings from this research are important for the implementation of Artificial Intelligence applications to logistics systems, specifically for dynamic optimization and scenario planning. By proving the feasibility of using a VAE-GAN hybric architecture for simulating complexed logistics scenarios, the research further proves the possibility of deep generative models not as mere data augmentation tools but as active agents in decision-support systems. This new mode of operation means that AI can be a generator of logistics plans rather than just a reactor to plans and a describer of planning processes.

The framework gives us a sense of how AI can model multi-modal supply chains with multiple constraints, simulate disruptions, and create feasible, cost-sensitive alternatives, all important elements of supply chain resilience. The findings that are obtained provide evidence that the VAE-GAN framework could be used as a centerpiece of digital twin platform for logistics, allowing that physical and digital systems operate interactively in real time. This is consistent with present industrial trend toward predictive logistics and smart freight management systems.

Furthermore, experiment results on comparison with the baselines of MILP-based methods demonstrate that deep generative models have scalability and flexibility over the features of high dimension and the stochastic in nature variables. While MILP works well for well-defined, deterministic problems, it doesn't work as well with dynamic, unstructured logistics data which has led to AI-led approaches starting to shine.

But also there are challenges introduced when AI is used in real-world logistics. Model interpretability, data privacy, computational workload, and frequency of model retraining are some of the issues that need to be well taken care to maintain operational feasability. Nevertheless, the study demonstrates the ability with the right design and deployment strategies; AI can lift the strategic and operational levels in logistics planning decision-making, such as route optimization and sustainability prediction.

In closing, this research contributes to push forward the frontier of applications of AI in logistics, proposing an innovative hybrid architecture between AI and IR techniques, and evaluating empirically its effectiveness over classical optimization approaches. This further supports the belief that AI is not simply a supporting actor in logistics, but rather playing a leading role in future-facing, intelligent supply chain systems.

9.8 Strategic Recommendations for Industry Adoption

Implications for practice: For practitioners and industry players looking to apply generative AI in a logistics setting, a number of strategic lessons may be drawn from the conceptual work and empirical findings of this study.

1. Begin with Controlled Pilots in High-Impact Segments

Therefore, it is recommended that logistics companies begin the integration of these technologies through controlled pilot projects. These efforts should ideally target specific pain points, such as dynamic routing optimization, better demand forecasting, and more effective emissions management. They should be tested in facilities that have been equipped with extensive monitoring and that experience a high degree of operational variation (e.g., warehouses, freight consolidation centers, urban distribution centers). As a result, such pilot implementations may act as sensible devices to absorb the potentials of VAE-GAN-based forecasting and scenario generation in reality.

2. Establish Data Governance and Scenario Libraries

Bringing a generative model into practical use in organizations requires a lot of hard work, especially when it comes to establishing resilient data lakes, processing pipelines, and all-encompassing scenario libraries. These core data sources should include multi-modal routing case examples, logs of previous disturbances, identified seasonal patterns of deterioration, and emission profiles. The existence of structured scenario libraries is of primordial importance, as it allows the generative model to be trained over different types of realistic conditions, which will improve its capability to generate reasonable and realistic logistics alternatives.

3. Integrate with Existing Optimization Engines

The VAE-GAN system could be incorporated (not mutually exclusive) upstream as a scenario generator and downstream as a validator in place of or complement to already deployed MILP-based or heuristic optimization tools. For example, created logistics scenarios can be fed to an MILP solver to find cost-optimal responses, or used to test traditional plans under uncertainty. This combination of approaches reduces the impact on partial functionality and increases resilience and flexibility.

4. Prioritize Interpretability and Decision Transparency

Without the confidence of stakeholders, technical performance will not be enough for industry uptake. That is why it is so important to use explainable AI (XAI) techniques—for example, attention maps, feature attribution, or surrogate models. It is important for managers and operators to be able to answer the question, why was this scenario generated (or risk predicted) to understand why the results of a transportation or sustainability decision can be stake-driven.

5. Consider Hardware and Cloud Infrastructure Needs

The use of deep generative models, especially in real time, requires significant computation. "Are your IT systems able to handle that kind of GPU-accelerated inference or does it have to be run on the cloud?". Collaboration with a cloud vendor or AI platform provider could speed up implementation and control costs and scale.

6. Develop Cross-functional AI Literacy

Adoption is also a function of organizational readiness for such changes. Logistics managers, planners, and forecast analysts should be trained specifically on AI-related concepts for scenario generation and forecasting. Cross-disciplinary AI literacy initiatives foster better interaction between domain experts and data scientists so the models are built and operationalized in context.

7. Align with ESG and Regulatory Frameworks

Finally of course, in an environment where logistics companies are under increasing pressure to achieve Environmental, Social, and Governance (ESG) as well as deliver decarbonization targets, any deployment of AI has to be fit within sustainability parameters. VAE-GAN models ability to simulate CO_2 emissions, and detect lower impact routes, contribute to ESG reporting and can aid firms in the preparation for future-regulated audits.

In summary, the incorporation of VAE-GAN architectures into logistics operations will need to be guided by practically phased approaches that involve great technical skill, corresponding to organizational fit and regulatory adherence. In doing so, they will unleash the potential of generative AI to disrupt industries while minimizing the associated risks to the future of generative AI.

9.9 *Research Limitations and Future Directions*

Although we present an AI-based generative model for multimodal logistics optimization, there is a number of limitations that limit results generalization and operational maturity. These restrictions drive critical future research and system refinement.

1. Simplified Emissions and Cost Modeling

Despite the attempts to synthesize realistic multimodal logistics scenarios, we only use to some extent standardized coefficients and simulated distributions for the emissions and cost estimations in this framework; there remains much work to do. In practice, the logistics cost in the real world is affected by the fluctuated factors including fuel price fluctuates, real-time road traffic, and warehouse accumulation agreement risk of resources. Similarly, the CO_2 emissions vary according to the type of freight transported, weather, vehicle maintenance, and driver behavior. Subsequent iterations of the model that may include more fine-grained time-varying emissions datasets may be achieved through IoT sensors, telemetry data, or even carbon accounting APIs.

2. Synthetic Data Bias and Generalizability Constraints

This research performs benchmarking and scenario generation using artificially created logistics data. Although great care is taken in the construction of such datasets to recreate realistic complexity, this may introduce a bias. These may be to simplistic routing logic, assumed transport mode distributions, or unrealistic demand peaks. Therefore, the results may not apply for all different logistics regions, fields, and ranges. An important step forward would be to evaluate the framework with real logistics data (from partners), e.g., freight marketplaces, public transport platforms, or commercial route planning APIs.

3. Lack of Real-Time Feedback or Adaptive Learning

The model is currently used in a one-shot inference regime: It produces logistics scenarios from a given fixed latent representation and does not incorporate feedback based on current observations or changing conditions. Logistics systems in the real-world are dynamic and need models that can learn and adapt in real time. A potential extension could be the incorporation of reinforcement learning (RL), where VAE-GAN agents are exposed to rewards or penalties, which are functions of scenario performance metrics (i.e., cost, time, emissions) in an episodic learning manner. This would allow adaptive learning in near-real-time environments.

4. Absence of Multi-agent and Stakeholder Interactions

In this research, we consider the logistics environment as a centralized optimization space. But the supply chains today have "several agents"—shippers, carriers, platform operators, regulators—all with their own preferences, constraints, and incentives. The present framework does not include agent heterogeneity, or strategic behavior (e.g., bargaining over price, trading off emissions, prioritizing production of lead time). For future extensions, multi-agent simulations based on game theory or agent-based modeling should be considered to model the decentralized structure of logistics.

5. No Support for Regulatory Compliance or ESG Standards

While the model simulates emissions and cost data, it is not linked, or constrained by, the external sustainability frameworks, (e.g., the EU Green Deal, ISO 14083, or Scope 3 reporting). It needs to be interoperable with actual ESG standards if it's to be adopted industrially. Upcoming models might also provide regulation-based thresholds, automated compliance cues, or scenario-based filtering options to better support decision-making in line with company and policy objectives.

6. Limited Benchmarking with Other AI Paradigms

We concentrate on the VAE-GAN architecture in this work and compare with a MILP baseline. Other AI paradigms beyond con-and-exec may bring some strengths in logistics planning, though (e.g., Transformer-based time series models, diffusion models, symbolic regression). A comparative analysis with various AI architectures would contribute to the empirical basis of generative models, and elaborate on their popularity in some application domains.

In short, although the VAE-GAN model showed promise in realistic scenario and computing burden, the model is a prototype in its present form. A developed deployment would have integrated live feedback, richer data, regulations, and alternative AI paradigms. These recommendations set a clear and important research program for academic and industrial cooperation in the future.

9.10 Final Summary and Forward Outlook

This chapter has extensively examined the results, advantages, and the limitations of the proposed logistics optimization framework based on VAE-GAN, providing a multifaceted analysis on technical performance, practical availability, and future scalability. The proposed model revealed its great vavia in generating various logistics scenarios which are structuraly coherent and competitive solutions with classical MILP formulations that mainly arrive relative to computation time and the possibility of being extended to multi-modal scenarios.

The interpretability result suggests that whereas MILPs have theoretical guarantees of optimality, they don't scale even to rich yet easily explained or uncertain problems, like the ones where generative models such as VAE-GANs succeed. In particular, the distributed hybrid architecture's ability for stochastic forecasting, variable-length route generation, and high-volume simulation (for digital twin adoption, freight exchange application, ESG-compliant decision means).

Yet the study also revealed significant shortfalls. Reduced complexity of emissions assumptions, not using real-world data, lack of online learning, and lack of regulator integration show that this version of the platform is still in the form of research prototype. These drawbacks limit its direct application in industry, but also present a promising field of applied research.

From that point on, many relentless applications should not only be possible but be even required. These include:

(a) Incorporating real-time data via API integrations with logistics platforms and IoT sensors.
(b) Embedding reinforcement learning agents for continuous adaptation and scenario optimization.
(c) Extending the model to support multi-agent and decentralized decision-making frameworks.
(d) Embedding ESG compliance logic to support sustainable and regulatory-aligned logistics planning.
(e) Benchmarking against alternative generative AI models (e.g., Transformers, diffusion models).

The end goal is to mature the generative optimization framework into a real-time, interactive decision-support engine—one that can consume live supply chain data, react to stakeholder moves, and comply with global sustainability edicts. With the proliferation of digital supply chains and the mandatory nature of climate accountability, those AI-driven architectures are not academic contributions but essential enablers of the next generation of logistics intelligence systems.

This chapter brings to a close the main analysis of the research. The remaining appendices will provide additional materials for reproduction and penetration testing, including synthetic datasheet specifications, architecture code, and MILP formulations in details. .

10 Ethical and Societal Implications

Although adequate efficiency and sustainability level demonstrated by the proposed VAE-GAN framework have been proven for logistics, their adoption poses non-negligible ethical and societal issues.

10.1 Data Privacy and Security

When using IoT sensors or external APIs, the inclusion of real-time logistics data poses risks regarding exposure of personal and commercial information. This research is built with synthetic data, however, for future real-world deployment with respect to GDPR (General Data Protection Regulation), and other regional privacy laws. Anonymization, encryption, and federated learning are proposed to be used as precautions.

10.2 Job Displacement and Labor Impact

Automation with route optimization, freight exchange, and similar platforms will lessen the demand for human dispatchers, brokers, and planners. Although these systems can enhance human performance, it is important to develop an ethical transition plan that re-trains those affected and enforces decent labor practices in supply chains.

10.3 Bias and Fairness in AI Models

Black-box generative models producing a biased or non-representative output will manifest this bias and non-representativeness in the observed data distribution. For instance, routing decisions in favor of high-income areas or to the detriment of small-service carriers might also worsen inequality. Production-ready frameworks should include fairness-aware loss functions scenario diversity, and auditing tools, ensuring all of this is done carefully for fears of AI harming people.

10.4 Sustainability Trade-offs

While the system looks to cut emissions, optimizing against carbon metrics alone could bump into other competing goals, such as speed or cost of delivery. Future versions should allow for multi-objective trade-offs and stakeholder-dependent weighting schemes.

10.5 Ethical Governance

For such intelligent logistics systems we need governance mechanisms, such as ethics boards, transparency reports, and auditability of decisions. Stakeholders such as public institutions, carriers, and civil society will need to participate in the co-design of system boundaries and protections.

By taking a proactive stance on these ethical considerations, we nurture the responsible development of AI technology consistent within the current-day best practices of applied machine learning and transportation analytics.

Appendices

11 Synthetic Logistics Dataset Generator

This is the logic of the Python dataset generator used to simulate multi-modal logistics flow considering the synthetic cost, emissions, and time variables.

11.1 Key Functional Elements

- Cities: 20–30 major logistics hubs in Europe
- Modes: truck, rail, sea, air (selectable per leg)
- Features: cost (EUR), CO_2 emissions (kg), time (hrs)
- Multi-leg structure: up to 3 legs per shipment

11.2 Sample Code Snippet

```
def generate_route_data(num_routes=1000):
    routes = []
    for _ in range(num_routes):
        origin, destination = random.sample(
            city_list, 2)
        mode = random.choice(["truck", "rail", "sea"
            , "air"])
        cost = np.random.uniform(500, 3000)
        emissions = cost * emission_factor[mode]
        time = cost / avg_speed[mode]
        routes.append([origin, destination, mode,
            cost, emissions, time])
```

```
    return pd.DataFrame(routes, columns=["origin", "
        destination", "mode", "cost", "emissions", "
        time"])
```

Listing 7.4 Synthetic Data Generation Sample

12 VAE Architecture Overview

The Variational Autoencoder (VAE) module encodes logistics input features into a more compact low-dimensional latent space and then it plans scenarios from the latent space.

12.1 VAE Code Skeleton

```
class VAE(nn.Module):
    def __init__(self, input_dim, latent_dim):
        super(VAE, self).__init__()
        self.encoder = nn.Sequential(
            nn.Linear(input_dim, 64),
            nn.ReLU(),
            nn.Linear(64, latent_dim * 2)
        )
        self.decoder = nn.Sequential(
            nn.Linear(latent_dim, 64),
            nn.ReLU(),
            nn.Linear(64, input_dim)
        )

    def reparameterize(self, mu, logvar):
        std = torch.exp(0.5 * logvar)
        eps = torch.randn_like(std)
        return mu + eps * std

    def forward(self, x):
        enc_out = self.encoder(x)
        mu, logvar = torch.chunk(enc_out, 2, dim=1)
        z = self.reparameterize(mu, logvar)
        return self.decoder(z), mu, logvar
```

Listing 7.5 VAE Core Module

13 GAN Architecture

The GAN is used to further refine and simulate high-variance, realistic logistics scenarios from the latent space of the VAE.

13.1 GAN Components

```
class Generator(nn.Module):
    def __init__(self, latent_dim, output_dim):
        super(Generator, self).__init__()
        self.model = nn.Sequential(
            nn.Linear(latent_dim, 64),
            nn.ReLU(),
            nn.Linear(64, output_dim)
        )

    def forward(self, z):
        return self.model(z)

class Discriminator(nn.Module):
    def __init__(self, input_dim):
        super(Discriminator, self).__init__()
        self.model = nn.Sequential(
            nn.Linear(input_dim, 64),
            nn.LeakyReLU(0.2),
            nn.Linear(64, 1),
            nn.Sigmoid()
        )

    def forward(self, x):
        return self.model(x)
```

Listing 7.6 GAN Generator and Discriminator

14 VAE-GAN Training Loop

```
for epoch in range(num_epochs):
    for batch in data_loader:
        real_data = batch
        reconstructed, mu, logvar = vae(real_data)
        z = vae.reparameterize(mu, logvar)

        # GAN step
        fake_data = generator(z)
        real_labels = torch.ones(batch_size, 1)
```

```
        fake_labels = torch.zeros(batch_size, 1)

        d_loss_real = loss_fn(discriminator(
            real_data), real_labels)
        d_loss_fake = loss_fn(discriminator(
            fake_data.detach()), fake_labels)
        d_loss = d_loss_real + d_loss_fake

        discriminator_optimizer.zero_grad()
        d_loss.backward()
        discriminator_optimizer.step()

        g_loss = loss_fn(discriminator(fake_data),
            real_labels)
        generator_optimizer.zero_grad()
        g_loss.backward()
        generator_optimizer.step()
```

Listing 7.7 Simplified Training Loop

15 Scenario Generation Output Example

Below is a sample output generated by the trained VAE-GAN pipeline, simulating realistic logistics scenarios with multi-modal complexity.

```
origin,destination,mode,legs,cost,emissions,time
Hamburg,Prague,truck,1,987.22,240.1,36
Barcelona,Warsaw,rail+truck,2,1854.77,410.3,48
```

References

1. Arjovsky M, Chintala S, Bottou L (2017) Wasserstein gan. Proceedings of the 34th international conference on machine learning (ICML), 214–223. Retrieved from https://arxiv.org/abs/1701.07875
2. Blessing L, Wenzel S, Fleischmann M (2022) Scenario generation for resilient supply chain planning: a generative modeling approach. Int J Prod Res 60(10):3075–3092. https://doi.org/10.1080/00207543.2021.1955224
3. Blessing M (2022) Optimizing supply chain management with gan-based simulations. Retrieved from https://www.researchgate.net/publication/389033314 (ResearchGate)
4. Cai D (2024) Enhancing capabilities of generative models through vae-gan integration: a review. In: 2nd international conference on software engineering and machine learning
5. Carter CR, Rogers DS (2008) A framework of sustainable supply chain management: moving toward new theory. Int J Phys Distrib Logist Manage 38(5):360–387. https://doi.org/10.1108/09600030810882816
6. Chen L, Zhou Q, Cai Y, Wang H, Li Y (2022) Cae-gan: a hybrid model for vehicle trajectory prediction. IET Intel Transport Syst 16:1682–1696. https://doi.org/10.1049/itr2.12174

7. Chen X, Li Y, Huang Z, Xu L (2020) A hybrid vae-gan approach for synthetic data generation in industrial applications. IEEE Trans Indust Inf 17(5):3456–3465. https://doi.org/10.1109/TII.2020.3034890
8. Chen Y, Zhao R, Lin Y, Wang Y (2022) Vehicle trajectory prediction via hybrid convolutional autoencoder gan. IEEE Trans Intell Transp Syst
9. Crainic TG, Laporte G (1997) Planning models for freight transportation. Eur J Oper Res 97(3):409–438. https://doi.org/10.1016/S0377-2217(96)00298-6
10. Creswell A, White T, Dumoulin V, Arulkumaran K, Sengupta B, Bharath AA (2018) Generative adversarial networks: an overview. IEEE Sig Proc Mag 35(1):53–65. https://doi.org/10.1109/MSP.2017.2765202
11. Elkington J (1998) Cannibals with forks: the triple bottom line of 21st century business. Capstone Publishing
12. European Commission (2020) Sustainable and smart mobility strategy – putting european transport on track for the future. Retrieved from https://transport.ec.europa.eu/transport-themes/mobility-strategy_en (Accessed: 2025-03-27)
13. Goodfellow I, Bengio Y, Courville A (2016) Deep learning. MIT Press
14. Goodfellow I, Pouget-Abadie J, Mirza M, Xu B, Warde-Farley D, Ozair S, Bengio Y et al (2014a) Generative adversarial nets. In: Advances in neural information processing systems, vol 27
15. Goodfellow I, Pouget-Abadie J, Mirza M, Xu B, Warde-Farley D, Ozair S, Bengio Y (2014b) Generative adversarial nets. In: Proceedings of the 27th international conference on neural information processing systems (neurips), pp 2672–2680
16. Gui J, Sun Z, Yang Y, Tao S, Tan J (2023) A comprehensive review on generative adversarial networks: theory, methodologies, and applications. IEEE Trans Knowl Data Eng 35(5):4302–4324
17. Ivanov D, Dolgui A (2021) A digital supply chain twin for managing disruption risks and resilience in the era of industry 4.0. Transportation Research Part E Logistics and Transportation Review 145:102190. https://doi.org/10.1016/j.tre.2020.102190
18. Khlie K, Benmamoun Z, Fethallah W, Jebbor I (2024) Leveraging variational autoencoders and recurrent neural networks for demand forecasting in supply chain management: a case study. J Infrastruct Pol Devel 8(8):6639. https://doi.org/10.24294/jipd.v8i8.6639
19. Kingma DP, Welling M (2013) Auto-encoding variational bayes. ArXiv preprint, arXiv:1312.6114. Retrieved from https://arxiv.org/abs/1312.6114
20. Larsen ABL, Sønderby SK, Larochelle H, Winther O (2016) Autoencoding beyond pixels using a learned similarity metric. Int Conf Mach Learn (ICML), 48:1558–1566. Retrieved from http://proceedings.mlr.press/v48/larsen16.pdf
21. Mao X, Li Q, Xie H, Lau RYK, Wang Z, Smolley SP (2017) Least squares generative adversarial networks. In: 2017 IEEE international conference on computer vision (ICCV) (pp 2794–2802). https://doi.org/10.1109/ICCV.2017.304
22. Mirza M, Osindero S (2014) Conditional generative adversarial nets. In: Arxiv preprint (Vol. arXiv:1411.1784). Retrieved from https://arxiv.org/abs/1411.1784
23. Naji HAH, Xue Q, Zhu H, Li T (2021) Forecasting taxi demands using generative adversarial networks with multi-source data. Appl Sci 11(20):9675. https://doi.org/10.3390/app11209675
24. Notteboom T, Rodrigue J-P (2020) Port regionalization: improving port competitiveness by reaching beyond the port perimeter. J Transp Geogr 88:102846. https://doi.org/10.1016/j.jtrangeo.2020.102846
25. Odena A, Olah C, Shlens J (2017) Conditional image synthesis with auxiliary classifier gans. In: Proceedings of the 34th international conference on machine learning (ICML), pp 2642–2651. Retrieved from https://arxiv.org/abs/1610.09585
26. Park SM, Yoon HG, Lee DB, Choi JW, Kwon HY, Won CW (2023) Topological magnetic structure generation using vae-gan hybrid model and discriminator-driven latent sampling. Sci Rep 13:20377. https://doi.org/10.1038/s41598-023-47866-3
27. Saunders M, Lewis P, Thornhill A (2019) Research methods for business students, 8th. Pearson Education

28. Saurabh S, Kumar R (2023) Dynamic routing with ai: A review. J Supply Chain Manag 59(2):45–67
29. SteadieSeifi M, Dellaert N, Nuijten W, Woensel TV, Raoufi R (2014) Multimodal freight transportation planning: a literature review. Eur J Oper Res 233(1):1–15. https://doi.org/10.1016/j.ejor.2013.06.055
30. Teece DJ, Pisano G, Shuen A (1997) Dynamic capabilities and strategic management. Strateg Manag J 18(7):509–533
31. Wang C, Chong MJ, Bach F, Schmid C, Harchaoui Z (2021) Generative adversarial networks: a survey and taxonomy. ACM Comput Surv 54(8):1–38. https://doi.org/10.1145/3453156
32. Wang Q, Liu J, Zhang S, Li H (2022) Variational autoencoders for multimodal data integration in transportation systems. Transp Res Part C Emerg Technol 135:103489. https://doi.org/10.1016/j.trc.2021.103489
33. Wang R, Zhao X, Sun Y, Li P (2021) Generative adversarial networks in logistics and transportation: a comprehensive review and future research directions. Transp Res Part C Emerg Technol 132:103375. https://doi.org/10.1016/j.trc.2021.103375
34. Zhang Y, Wang X (2021) Optimizing freight exchange platforms with machine learning. Transp Res Part E Log Transp Rev 145:102174. https://doi.org/10.1016/j.tre.2021.102174
35. Zhang Z, Li X, Cheng Y, Chen Z, Liu Q (2025) Credit risk identification in supply chains using generative adversarial networks. In: Proceedings of icaace 2025
36. Zhu J-Y, Park T, Isola P, Efros AA (2017) Unpaired image-to-image translation using cycle-consistent adversarial networks. Proceedings of the IEEE international conference on computer vision (ICCV) 2223–2232. https://doi.org/10.1109/ICCV.2017.244

MIX
Papier aus verantwortungsvollen Quellen
Paper from responsible sources
FSC® C105338

If you have any concerns about our products,
you can contact us on
ProductSafety@springernature.com

In case Publisher is established outside the EU,
the EU authorized representative is:
Springer Nature Customer Service Center GmbH
Europaplatz 3, 69115 Heidelberg, Germany

Printed by Libri Plureos GmbH
in Hamburg, Germany